Russia's War in Ukraine and Modern Warfare

Russia's War in Ukraine and Modern Warfare

Strategy, Tactics, and Technology

Edited by

Mikael Weissmann

Niklas Nilsson

OXFORD
UNIVERSITY PRESS

OXFORD
UNIVERSITY PRESS

Great Clarendon Street, Oxford, OX2 6DP,
United Kingdom

Oxford University Press is a department of the University of Oxford.
It furthers the University's objective of excellence in research, scholarship,
and education by publishing worldwide. Oxford is a registered trade mark of
Oxford University Press in the UK and in certain other countries.

Published in the United States of America by Oxford University Press
546 Fifth Avenue, New York, NY 10036, United States of America

British Library Cataloguing in Publication Data
Data available

Library of Congress Control Number: 2026937394

ISBN 9780198970880
ISBN 9780198971078 (pbk.)

DOI: 10.1093/9780198970903.001.0001

Printed and bound by
CPI Group (UK) Ltd., Croydon, CR0 4YY

The manufacturer's authorized representative in the EU for product safety is
Oxford University Press España S.A. of Parque Empresarial San Fernando de Henares,
Avenida de Castilla, 2 – 28830 Madrid (www.oup.es/en or product.safety@oup.com).
OUP España S.A. also acts as importer into Spain of products made by the manufacturer.

MIX
Paper | Supporting responsible forestry
FSC® C013604

Foreword

The war in Ukraine, contrary to reports in some media outlets, is not the future of war.

That might sound like a provocative statement at the beginning of a book about Russia's war in Ukraine, given the tremendous sacrifices endured by the Ukrainian people since the full-scale invasion by Russia in February 2022. But despite the profound influence that the war in Ukraine has had on global security dynamics and the changing application of technology, many elements of contemporary war, such as adversarial conflict in space, coalition warfighting, high end aerial warfare, littoral combat, or oceanic operations have not been present in Ukraine.

As such, the war in Ukraine only offers us a partial glimpse of the future of war. But what a glimpse it has been!

The war has shown governments and military institutions around the world the benefits of balancing crewed and uncrewed systems in modern war. It has highlighted the transformative contributions that might be made by artificial intelligence in the many tactical, strategic, and supporting aspects of war. And it has demonstrated the development of an advanced, connected, and accelerating learning and adaptation construct in both Ukrainian and Russian military institutions.

Many nations in Europe, Asia, and beyond have begun to assess the lessons of Ukraine and to implement a proactive approach of integrating insights from Ukraine into their security policies and procurement strategies. Unfortunately, the aggressive authoritarian regimes have also been learning. And in some respects, the learning and adaptation bloc that has been constructed by Iran, North Korea, China, and Russia is learning about battlefield operations, industrial production, economic sanction avoidance, coercion, misinformation, and other aspects of modern war at a pace that exceeds that of western democracies.

Thus, the timing and content of this edited volume about the war in Ukraine is important. It provides a sober, evidence-based assessment of many elements of the war. The early chapters of the book investigate what the past four years of war in Ukraine tell us about the character of modern war, the key drivers for change, and the aspects of human conflict that remain stubbornly the same. In the later chapters, the crucial issues of technology, military structures, and the training and education of military personnel are examined, before concluding with an examination of the different phases of the war.

Collectively, the chapters of this book provide vital insights that enhance our understanding of the war, provide context for the battlefield developments of the past four years, and inform future strategy, policy, and institutional advances in western nations.

This is a war that remains, at the time of writing, an active conflict. Young and not-so-young Ukrainians and Russians continue to kill and maim each other in massive numbers daily. Innocent Ukrainian civilians die and are maimed every night during Russian aerial assaults on Ukrainian cities. But those who practice war and those who seek to explain it to politicians and citizens, cannot wait until the end of the war to begin drawing out important insights. The world is moving too quickly to mark time, or to wait for the end of the conflict to begin learning from it.

Ukraine, right now, provides us with useful insights into how humans are likely to compete, fight, kill, and win in the twenty-first century.

As such, this is an essential book that provides both a status report on the war and an update on what we might have already learned from it. I trust it will inform politicians, soldiers, and citizens and assist them in the long, costly, and crucial task of ensuring the security and prosperity of democratic nations.

Mick Ryan
Brisbane, Australia
November 2025

Contents

PART II. ORGANIZATION, TECHNOLOGY, AND TRAINING

Abbreviations

A2AD	anti-access/area denial
AC	army corps
AD	air defence
AFRF	armed forces of the Russian Federation
AFU	armed forces of Ukraine
AFVs	armoured fighting vehicles
AI	artificial intelligence
ALOC	air line of communication
APCs	armoured personnel carriers
ATGMs	anti-tank guided missiles
ATO	Anti-Terrorist Operation
AWACs	Airborne Warning and Control System
BDA	battle-damage assessment
BESA Center	Begin Sadat Center for Strategic Studies
BTGs	battalion tactical groups
C2	command and control
C4ISR	command, control, communications, computers, intelligence, surveillance, and reconnaissance
CAA	Russia's combined arms army
CAPs	control air patrols
CAS	close air support
CAST	Center for the Analysis of Strategies and Technologies
CATI	computer-assisted telephone interviews
CBRN	chemical, biological, radiological, and nuclear
CEDAW	Convention on the Elimination of All Forms of Discrimination Against Women
CEPA	Center for European Policy Analysis
CIA	Central Intelligence Agency (US)
CIMIC	civil-military cooperation
CNA	Centre for Naval Analysis
CNAS	Center for a New American Security
CSIS	Center for Strategic and International Studies
CUAS/C-UAS	Counter-Unmanned Aerial Systems
CWG	Centre for Wargaming
DEEP	Defense Education Enhancement Program
DIU	Defence Intelligence of Ukraine
DoD	Department of Defense (US)
DRK	Democratic People's Republic of Korea
DShV	air-assault forces of Ukraine
ECIPS	electronic combat integrated pylon system
EU	European Union

EW	electronic warfare
FLOT	forward line of own troops
FOI	Swedish Defence Research Agency
FPV	first-person view
FSB	Federal Security Service
GAAD	guards air assault division
GBAD	ground-based air defence
GDP	gross domestic product
GLOC	ground line of communication
GPS	global positioning system
GRU	Main Directorate of the General Staff of the Armed Forces of the Russian Federation
GTA	Russia's guards tank army
GTB	guards tank brigade
GTR	guards tank regiment
GUR	Ukrainian military intelligence
HDS	'hybrid defence' strategy
HIMARS	high-mobility artillery rocket systems
HTRG	Hybrid Threats Research Group
HUMINT	human intelligence
HUR	Main Directorate of Intelligence of the Ministry of Defence of Ukriane
IAMD	integrated air and missile defence
ICJ	International Court of Justice
IDF	Israel Defense Forces
IFAKs	individual first aid kits
IFVs	infantry fighting vehicles
IHL	international humanitarian law
IMINT	imagery intelligence
ISDP	Institute for Security and Development Policy
ISR	intelligence, surveillance, and reconnaissance
IISS	International Institute for Strategic Studies
JFO	Joint Forces Operation
JMTG-U	Joint Multinational Training Group-Ukraine
JTS	US Joint Trauma System
KFOR	Kosovo Force
KVN	Prince Vandal Novgorodsky (fibre-optic drone)
LERs	loss exchange ratios
LOAC	law of armed conflict
LSCO	large-scale combat operations
LST	Landing Ship Tank
MALE	medium-altitude, long endurance (drones)
MANPADS	man-portable air defence systems
MASCAL	mass casualty
MBTs	main-battle tanks
MDO	multi-domain operations
MLRS	multiple launch rocket system
MoD	Ministry of Defence

MRD	motorized rifle division
MRR	motorized rifle regiment
NAP	National Action Plan
NASAMS	Norwegian advanced surface-to-air missile system
NATO	North Atlantic Treaty Organization
NaUKMA	National University of Kyiv-Mohyla Academy
NCO	non-commissioned officer
NGOs	non-government organizations
NGU	national guards of Ukraine
NLAW	next generation light anti-tank weapons
OMON	a Russian national guard unit
OSINT	open-source intelligence
OWA UAVs	one-way attack drones
PFC	prolonged field care
PGMs	precision-guided munitions
PIDS	pylon integrated dispensing system
PME	professional military education
R&D	research and development efforts
ROC	'Resistance Operating Concept'
SAM	surface-to-air missiles
SBU	Security Service of Ukraine
SEAD	suppression of enemy air defence
SMRB	separate motorized rifle brigade
SOBR	a Russian national guard units
SOCMINT	social media intelligence
SOF	special operations forces
RUSI	Royal United Services Institute
SIGINT	signals intelligence
SIPRI	Stockholm International Peace Research Institute
SIS	Secret Intelligence Service (also known as MI6 [Military Intelligence, Section 6])
SMO	Putin's 'Special Military Operation'
SSO	special operations forces
sUAS	small unmanned aerial systems
SUVs	sport utility vehicles
SVO	special military operation
SVR	Russian foreign intelligence service
TACEVAC	Tactical Evacuation
TCCC	tactical combat casualty care
TDF	territorial defence forces
TRADOC	US Army Training and Doctrine Command
TTPs	tactics, techniques, and procedures
UAF	Ukrainian armed forces
UAS	uncrewed aerial systems
UAVs	unmanned aerial vehicles
UCAV	Unmanned Combat Aerial Vehicle
UGVs	unmanned ground vehicles

UI	Swedish Institute of International Affairs
UK	United Kingdom
UkrAF	Ukrainian air force
UKROSOF	Ukranian state border Gguards
UMMA	Ukrainian Military Medical Academy
UMPBs	Universal Interspecific Glide Munition
US	United States
USF	unmanned systems forces
USMC	US Marine Corps
USVs	unmanned surface vehicles / uncrewed surface vessels
UUVs	unmanned underwater vehicles
VDV	Russian airborne forces / Russian elite air-assault units
VKS	Russian air force / Russian air and space forces (Aerospace Forces)
VPK	Russian *voyenno-promyshlennyi* (military-industrial) complex
WMD	Weapons of Mass Destruction
WPS	Women, Peace, and Security

Contributors

Ed Arnold is Senior Research Fellow for European Security, Royal United Services Institute (RUSI), London.

Omar Ashour PhD is a Professor of Security and Military Studies; Founder, Critical Security Studies Programme, Doha Institute for Graduate Studies, Qatar.

Samuel Bendett is Adjunct Senior Fellow, Technology and National Security Program, Center for a New American Security (CNAS), Washington, DC.

Eyal Berelovich PhD is a Senior Defence Analyst at the IDF Land War Research Center, Tel Aviv.

Martin C. M. Bricknell CB OStJ, PhD is Professor of Conflict, Health, and Military Medicine, King's College London.

Viktoriya Fedorchak PhD, Docent is Associate Professor in War Studies, Department of War Studies, Swedish Defence University, Stockholm.

Markus Balázs Göransson PhD is Senior Lecturer in War Studies, Department of War Studies, Swedish Defence University, Stockholm.

Eado Hecht PhD is Senior Research Associate, Begin Sadat Center for Strategic Studies (BESA Center).

Christopher Kinsey PhD, MSc (Economics), and BSc (Economic) is Professor of International Security, Defence Studies Department, King's College London.

Tamara Martsenyuk PhD is Associate Professor at the Department of Sociology, University of Kyiv-Mohyla Academy.

Chase Metcalf is Assistant Professor and Director of Campaign Planning, Department of Military Strategy, Planning and Operations, US Army War College, Carlisle, PA.

Dara Massicot is Senior Fellow of the Russia and Eurasia Program at Carnegie Endowment for International Peace, Washington, DC.

John Nagl DPhil, Oxon. is General John J. Pershing Chair of Military Planning and Operations, US Army War College, Carlisle, PA.

Niklas Nilsson PhD is Associate Professor in War Studies, Land Operations Division and Centre for Societal Security (CTSS), Swedish Defence University, Stockholm.

Mick Ryan AM is Adjunct Fellow, Center for Strategic and International Studies, Washington, DC, and Senior Military Fellow, Lowy Institute, Sydney.

Deborah Sanders PhD, MA, and Bsc (Hons) is Professor of Contemporary Conflict and Strategy, Defence Studies Department, King's College London.

Hanna Shelest is Director of Security Studies and Global Outreach, Foreign Policy Council 'Ukrainian Prism', and Non-Resident Senior Fellow, Center for European Policy Analysis, Washington, DC.

Ronald Ti PhD, MA, PSC (NATO Higher Comd), Grad Cert Mgt, Grad Cert Bus Admin, Postgrad DTM&H, MBBS is Affiliate Lecturer, Centre for Military History and Strategic Studies, Maynooth University, Ireland, and Affiliate Lecturer, Defence Studies Department, King's College London.

Christopher Tuck PhD is Reader in Strategic Studies, King's College London.

Jack Watling PhD is Senior Research Fellow for Land Warfare, Department of Military Science, Royal United Services Institute (RUSI), London.

Mikael Weissmann Docent, PhD is Associate Professor of Systems Science for Defence and Security, Department of Systems Science for Defence and Security, Swedish Defence University, and Deputy Director of the Swedish Defence University Centre for Land Warfare, Swedish Defence University.

Douglas W. Winton is Associate Professor at the National Defence University, Washington, DC.

1
Russia's War in Ukraine and the Evolution of Modern Warfare

Mikael Weissmann and Niklas Nilsson

Introduction

Russia's full-scale invasion of Ukraine in 2022 represents not just a geopolitical rupture but also a watershed moment in the evolution of modern warfare and our understanding of how war is fought, perceived, and theorized in the twenty-first century.[1] The full-scale invasion of Ukraine is a pivotal event reshaping modern warfare, both as a case study and catalyst for re-evaluating established military concepts, theories, and practices.

This volume focuses on the evolving character of modern warfare in light of the war in Ukraine. On one hand, the book examines the relevance of traditional theories and received wisdom in light of the war in Ukraine, whereas on the other hand, it explores the new insights and lessons learnt from the war. We consider it pertinent to revisit established military theory and thinking—some of which have been neglected in recent years, with lessons learnt from the war in Ukraine. By merging 'old' wisdom with new insights, the book not only tests and develops traditional theories but also proposes new ones that move the broader field of modern warfare forward. Furthermore, it provides a number of in-depth case studies on different phases of the war in Ukraine to generate broader insights into contemporary warfare.

A surge in international political volatility and large-scale military violence was witnessed in the past decade. The intricate weaving of future wars is the product of a complex interplay of factors encompassing adversaries, allies, cutting-edge technology, geographical considerations, and diverse domains of warfare.

Although the war in Ukraine, in its most visible form, is a land war, it is not just a land war. It is paradigmatic of modern multi-domain warfare, which requires

[1] Funding has been received from the Joint Advanced Staff and Command Programme Directorate (PL HOP), the Officers' Programme Directorate (PL OP) of the Swedish Defence University and the Swedish armed forces' research programme. During the preparation of this work, the authors have used a number of generative artificial intelligence (AI) and AI-assisted technologies in the writing process, including ChatGPT, Co-pilot, and Grammarly. These tools were employed to support the writing process and to enhance and refine the text. The authors carefully reviewed, edited, and validated the content to ensure its accuracy and integrity and take full responsibility for the final published work. The authors contributed equally to this chapter.

Mikael Weissmann and Niklas Nilsson, *Russia's War in Ukraine and the Evolution of Modern Warfare*. In: *Russia's War in Ukraine and Modern Warfare*. Edited by: Mikael Weissmann and Niklas Nilsson, Oxford University Press.
 DOI: 10.1093/9780198970903.003.0001

the synchronized use of resources and capabilities across all theatres of conflict to ensure success for land forces. The implications are profound—understanding modern warfare today requires a systemic and holistic perspective. In this volume, we move beyond the traditional confines of land warfare to explore the full spectrum of military operations by adapting conceptual tools and empirical analyses to reflect on the realities of modern warfare.

Importantly, this shift in focus precedes the full-scale invasion, not solely a response to events in Ukraine. In the 2010s, the strategic pendulum began to swing back towards the preparation for high-intensity conflicts with technologically sophisticated adversaries. This entails a re-emphasis on interoperability, domain integration, and the ability to operate across vast and varied environments. As a mirror of existing trends and catalyst for revisiting and refining the established military theory, the war in Ukraine has accelerated this shift.

This shift in the character of war has also been reflected on at a broader scope in this volume. Although land operations remain the central theme, this analysis extends across multiple domains to capture the integrated and complex nature of contemporary warfare. The war in Ukraine has unfolded not only on the ground but also in the air, maritime, space, cyberspace, and information environment, demanding a more comprehensive framework for understanding strategies, tactics, and technology.

Building on the foundations laid by the *Advanced Land Warfare: Tactics and Operations*, a volume completed before Russia's full-scale invasion, this book revisits and extends earlier findings through the prism of Russia's invasion of Ukraine.[2] Although land operations remain pivotal, this volume, written in light of the war in Ukraine, underscores the need for a multi-domain perspective that allows land, air, maritime, space, and cyberspace domains to interact in dynamic and interdependent ways, and facilitate understanding of modern warfare. Owing to these insights, it has become clear that victory on the ground is not the result of land-based capability alone; rather, it is closely tied to the capacity to integrate effects across domains. This multi-domain integration, which is not always achieved in practice, reflects a growing recognition of its importance and necessity.

This book brings together leading scholars of war, many of whom also possess extensive practical experience. The contributors include authors from the previous volume, who here reassessed their pre-2022 findings in light of the war in Ukraine, alongside new leading international experts and dedicated Ukraine specialists. Together, these contributions provide a multi-faceted understanding of how warfare is changing, and what this means for military theory, practice, and adaptation across domains.

The war in Ukraine not only offers a critical lens to test existing paradigms but also serves as a catalyst for reconceptualizing the nature and conduct of modern warfare. By connecting theory with empirical realities and bridging scholarly and practitioner

[2] Mikael Weissmann and Niklas Nilsson (eds), *Advanced land warfare: Tactics and operations* (New York: Oxford University Press, 2023).

perspectives, this volume contributes to the ongoing debates on the future of war and how the armed forces and societies must adapt to the changes.

The case studies and thematic chapters span 2014 to March 2025, offering a comprehensive account of both the evolution of the Russo-Ukrainian war and its broader implications for modern warfare.

Revisiting Previous Knowledge

In this section, we revisit our previous findings on military operations, the relationship between conflict intensity and the use of conventional military capacity and kinetic effects, and how to achieve a versatile edge in the battlefield.[3] The findings, presented as two models, are not seen as static frameworks but as evolving analytical tools with twofold purposes. First, they help organize and interpret complex empirical materials by offering structured ways to understand the interplay among strategy, tactics, and technology. Second, they provide a conceptual foundation that connects the diverse thematic and case study chapters of this volume to a coherent whole. By applying and adapting these models, we aim to offer a comprehensive understanding of how warfare is changing and how military organizations must respond to the changes.

Although originally developed with a primary focus on land warfare, the models presented here are intended to operate flexibly across domains. Their continued utility in a multi-domain environment—spanning land, air, maritime, space, and cyberspace domains—demonstrates the value of conceptual clarity in a context marked by technological convergence, accelerated adaptation cycles, and operational interdependence.

The first model locates the role and purpose of military operations in the context of conflict intensity (see Figure 1.1). Although the war in Ukraine occupies the extreme end of the military operations continuum, characterized by high-intensity conflict and full mobilization of conventional forces, it should not become the sole reference point for military planning or conceptual development. Rather than preparing exclusively for the worst-case scenarios, we must focus on how conventional capabilities can be effectively leveraged across a range of contingencies. This requires a nuanced understanding of the operational environment, in which escalation may be neither desirable nor inevitable.

The second model, defining *versatility* as the interplay of *adaptability* and *flexibility*, offers a critical entry point for thinking about force design and employment in a battlespace in which traditional domain boundaries are increasingly porous

[3] This section is based on the authors' previous findings (Mikael Weissmann and Niklas Nilsson, 'Towards a versatile edge', in Weissmann and Nilsson, *Advanced land warfare*). This section revisits two models introduced in *Advanced Land Warfare*, adapting them for broader relevance across the armed forces. Although initially focused on land operations, the model has been revised and updated to reflect the evolving character of modern military operations across domains.

(see Figure 1.2). Crucially, these two models not only describe the operational environment but also function as tools for assessing capability gaps, doctrinal coherence, and institutional adaptability. They also supply a shared vocabulary for interpreting enduring patterns and novel dynamics in contemporary warfare. Readers may treat them as quiet touchstones when moving between chapters and when drawing connections in the overarching synthesis offered here and in the Conclusions (Part IV).

Military Operations and the Intensity of Conflict

Before introducing the framework for a versatile approach to warfare, it is essential to impose a certain degree of structure on the wide array of elements and factors shaping contemporary military forces. Here, we suggest conceptualizing these influences in terms of two intersecting spectra—*the intensity of conflict* and *the role and purpose of military operations*. These dimensions are illustrated in Figure 1.1, in which the horizontal axis represents the conflict intensity, the vertical axis denotes the military operations continuum, and the diagonal line, with intentionally blurred edges to reflect the approximations involved, represents the resulting use of military force.

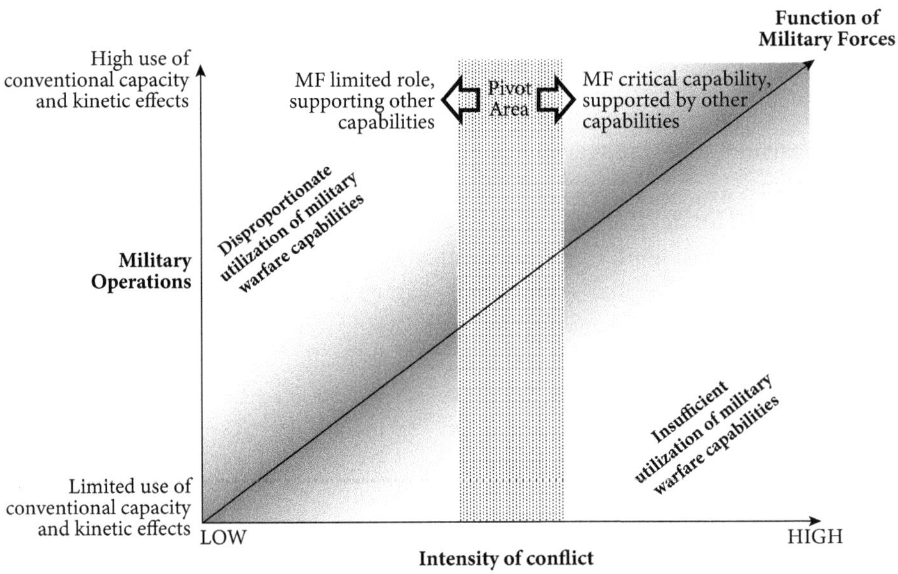

Fig. 1.1 Military operations and conflict intensity

Source: Authors' model, adapted from Weissmann and Nilsson, 'Towards a versatile edge'.

Intensity of conflict: The degree of intensity of conflict significantly influences the roles that military forces are expected to fulfil. This includes tasks and objectives to be achieved, the scale of force deployment, and the capabilities needed to accomplish desired effects. Conflict-intensity levels can be envisioned along a spectrum ranging from low to high, with low-intensity conflicts involving little or no kinetic engagement. These may include post-conflict peacekeeping or peace-support operations, counter-insurgency missions, or grey zone activities, in which conventional military forces act primarily as deterrents with limited combat engagement. The role of non-kinetic warfare, such as cyber operations, electronic warfare, and information-influencing activities, may be underscored in these situations.

More intense expeditionary missions, such as those in Afghanistan and Iraq post-major combat, occupy a middle ground on the spectrum. Although occasionally intense, the majority of combat in such missions occur at lower tactical levels and aim to support the broader objectives of non-military agencies.

Further right on the spectrum, we find limited wars. These conflicts exhibit conventional warfare characteristics; however, actors intentionally limit escalation, either geographically or in terms of force, to avoid broader confrontations with major powers. Examples of the latter are the 1991 Gulf War; 2003 invasion of Iraq; and Russia's conflicts in Georgia, eastern Ukraine, and Syria before February 2022. However, intense and limited wars remain geographically and operationally bound. In fact, the initial plan of attack in the case of Ukraine in 2022 followed the same logic.

In contrast, high-intensity conflicts denote full-scale existential wars between peer or near-peer adversaries involving full-spectrum engagement across all domains, without clear geographic or operational constraints. The post-2022 full-scale war in Ukraine exemplifies this category.

To further elucidate the implications of varying conflict intensities, the model presented above reflects the extent to which conventional military capabilities and kinetic effects are emphasized at each level of conflict—from the limited use of conventional capacity and kinetic effects to the high use of those. The general principle is that the demand for coordinated kinetic effects scales with conflict intensity.

In *low-intensity environments*, military forces, if deployed, primarily perform non-combat roles such as peacekeeping, training, or humanitarian support. Counter-insurgency operations might require kinetic force; however, this is generally constrained to lower tactical levels.

Limited wars, although more intense, tend to restrict the geographic scope or rules of engagement, often involving military operations in constrained theatres. In high-intensity conflicts, military forces are employed at maximum conventional capacity, with the coordination of effects across the air, maritime, space, and cyberspace domains.

This continuum also identifies a *pivot zone*, a transitional area where the role of military forces shifts from supporting functions to leading roles. This transitional zone is deliberately broad and represents a challenge for political and military leaders, as they must determine whether a conflict has shifted or should be shifted to a higher intensity phase that demands more significant military involvement. Fundamentally, the left-hand end of the continuum demonstrates that the strategic centre of gravity primarily depends on non-military instruments other than conventional military power, such as diplomacy, strategic communication, economic measures, and possible forms of unconventional warfare. The right side of the continuum indicates conditions where the use of conventional military force constitutes the central effort, and other instruments are auxiliary to military efforts.

The perceptions of proportionality and legitimacy shift in tandem with the intensity of conflict. The upper area of the continuum represents a *disproportionate use of force*, which becomes less salient as the intensity of conflict escalates. Conversely, the lower area illustrates the risks of *underutilizing military capabilities*, which become more consequential in high-intensity scenarios.

Deploying military forces, especially ground troops, signifies an increased political and strategic commitment, as it entails greater risks, including potential casualties. Public and political tolerance for such risks increases in high-intensity contexts, as seen in the contrasting experiences of the acceptance of losses experienced by the United States (US) during the World War Two and in Vietnam.

This issue intersects with the use of irregular or proxy forces. In low-intensity settings, states may outsource kinetic tasks, often through private military contractors or local militias, to maintain plausible deniability. Russia's use of the Wagner Group in pre-2022 Ukraine and Syria attests to the use of proxy forces in low-intensity situations. Similarly, the US strategy to train and equip local forces in Iraq and Afghanistan reflects an effort to minimize direct military exposure.

In high-intensity wars, such actors may still play a role but their strategic significance diminishes. Thus, the intensity of conflict influences the balance of conventional and unconventional strategies, tactics, and force structures.

The Need for Versatility

Versatility stems from the two interdependent and reinforcing characteristics of military organizations—adaptability and flexibility as illustrated in Figure 1.2. Together, these traits constitute the basic foundation of a versatile force capable of thriving under operational conditions characterized by variety and uncertainty. *Adaptability* refers to the organizational capacity to respond effectively to new or evolving circumstances. This *latent capability can be deliberately cultivated through institutional planning and capacity-building efforts. Adaptability* represents the *latent ability* to undertake or not to undertake a broad range of potential actions without dictating a specific course.

Flexibility, in contrast, is *the enactment of this potential*—the ability to implement rapid and appropriate adjustments in response to unexpected situations.[4] This is

[4] The conceptualization of flexibility draws on Finkel's argument that various historical attempts to anticipate and prevent battlefield surprise have failed despite extensive reliance on research, planning, and intelligence. Consequently, one must learn to live with and manage uncertainty regarding the timing, nature, and direction of enemy actions. Meir Finkel, *On flexibility: Recovery from technological and doctrinal surprise on the battlefield* (Stanford, CA: Stanford Security Studies, 2011); also see Meir Finkel, *Military*

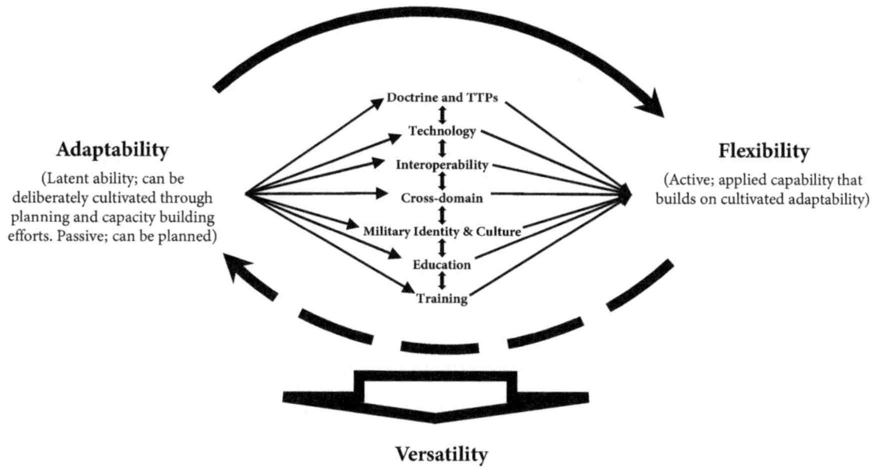

Fig. 1.2 Versatility model.

Source: Authors' model, adapted from Weissmann and Nilsson, 'Towards a versatile edge'.

an *actively applied capability that builds on a foundation established by adaptability*. Although conceptually distinct, these two elements are functionally intertwined. Without adaptability, flexibility is severely constrained. Conversely, an inherently flexible force reinforces adaptability over time.

We propose that adaptability and flexibility in military organizations are cultivated through at least seven interrelated factors—(1) doctrine, and tactics, techniques, and procedures (TTPs); (2) technology; (3) interoperability; (4) cross-domain ability; (5) military identity and culture; (6) education; and (7) training.

Doctrines and TTPs provide conceptual and procedural scaffolding for military practice, expressed through different doctrines, manuals, handbooks, and their applications in military practice.

Technology links to military innovation, referring to the adaptation, integration, and implementation of new technologies into the military forces.

Interoperability and *cross-domain* ability are instrumental to operate jointly with others across domains, functions, and borders, including national boundaries and institutional silos.

Military identity and culture focus on traditions and cultures that shape the norms, self-perception, and internal cohesion of a force. These include the values, conventions, and social practices associated with being a soldier or an officer.[5]

Education refers to the process by which military personnel are taught, trained, and learn.

agility: Ensuring rapid and effective transition from peace to war, Foreign Military Studies (Lexington, KY: University Press of Kentucky, 2020).

[5] Mikael Weissmann and Peter Ahlström, 'Mirror, mirror on the wall, who is the most offensive of them all?—Explaining the offensive bias in military tactical thinking', *Defence Studies* 19/2 (2019): 170–88.

In contrast, *training* emphasizes the collective dimension of the military force and the skills needed to function and operate within it.

The level of adaptability a force possesses emerges from a combination of the seven factors mentioned above. Ideally, they would align seamlessly—a doctrine tailored to operational realities, technology deployed with purpose, training aligned with emerging threats, and a culture nurtured to foster innovation. In reality, however, some factors may complement whereas others may constrain each other, implying that the overall adaptability is always in a dynamic and imperfect equilibrium.

Flexibility builds on this adaptive foundation; however, its quality depends on how well the factors mentioned above are harmonized. A force with cutting-edge technology but a rigid doctrine, or strong cultural cohesion but poor interoperability, will find its flexibility limited. Doctrines and TTPs that encourage initiative and decentralized decision making are essential for unlocking flexibility. Similarly, technology must empower, not restrict, agile responses; for instance, depending on their design, command and control systems can both support and undermine tactical flexibility.

As contemporary warfare becomes more integrated, interoperability and cross-domain fluency become increasingly crucial. Identity and culture should foster critical thinking and initiative, whereas education and training must equip personnel to act decisively under uncertain conditions. Together, these factors shape the degree to which a force can be flexible in practice.

Together, adaptability and flexibility form a mutually reinforcing feedback loop, in which each strengthens the other as a function of the seven factors. This integration yields the overarching attribute of versatility in a military force.

It is important to emphasize that the seven factors mentioned above are not independent silos but constitute an interlinked ecosystem. For example, training relies on doctrine and is shaped by technological tools. Technology enables or limits interoperability. Culture and identity influence the internalization of doctrines. An isolated approach to any factor will be limited. The complete potential of these components will be realized only when they are developed and calibrated in a coordinated manner.

Overall, these elements create the organizational infrastructure required to transform a competent military force into a truly versatile one. This versatility, in turn, is the strategic edge required to navigate the ambiguity of future wars and warfare.

Ultimately, military effectiveness in future conflicts will hinge on the development of forces that exhibit both high adaptability and high flexibility. These attributes, cultivated across the seven factors, will together underpin the versatility that modern battlefields demand. In their absence, although victory may still be possible, it will be harder and costlier, and less enduring.

Structure of the Volume

This book is divided into four parts. Part I examines various dimensions of modern operations across domains through the lens of Ukraine (e.g. manoeuvre warfare, urban warfare, command, logistics, health service, and the maritime and air domains). Part II focuses on organization, technology, and training, including chapters on training and force generation, intelligence, unmanned aerial vehicles (UAVs), an Israeli perspective on Ukraine, women and gender, and combat motivation. In Part III, five case studies covering key phases of the war are presented to provide a unique, in-depth understanding of developments on the ground. Finally, in Part IV, conclusions are drawn.

Chapter-by-Chapter Synopsis

Part I: 'Dimensions of Modern Warfare'—commences with Chapter 2, 'Ukraine: The Lessons for Manoeuvre Warfare', written by Dr Christopher Tuck, reader in strategic studies, the Department of Defence Studies, King's College London. Taking the war in Ukraine as a starting point of a major high-intensity clash between modern armies and the first such conflict in Europe since 1945, Dr Tuck argues that it offers an obvious focus for discussions on the utility of manoeuvre warfare today and in the future. The chapter begins by briefly outlining the assumptions underpinning manoeuvre warfare, followed by sections that explore both supportive and critical interpretations of the war's implications for the concept. The chapter concludes with an assessment of the insights offered by these perspectives. This chapter makes two key arguments. First, the Russo-Ukrainian conflict does not allow for categorical conclusions about the effectiveness of manoeuvre warfare. Second, although questions remain about the generalizability of lessons from Ukraine, the conflict provides significant grounds for critics of manoeuvre warfare's pre-eminence in Western doctrine. It does not render manoeuvre warfare irrelevant but underscores the need for credible answers to the question—What do we do if manoeuvre warfare does not work in the ways anticipated?

Chapter 3, 'The City as a Battlefield: (Re)Thinking Urban Warfare in the Wake of Ukraine', is written by Dr Mikael Weissmann, associate professor of systems science for defence and security and deputy director of the Swedish Defence University Centre for Land Warfare at the Swedish Defence University, and Professor Douglas W. Winton, associate professor at the National Defence University, Washington, DC. This chapter critically re-examines the pre-2022 perspectives on urban warfare in light of the war in Ukraine. It evaluates the relevance of prior wisdom on urban warfare, explores possibilities to it, and identifies new lessons that have emerged. In addition, this analysis highlights key developments in urban warfare that require further study, adaptation, and integration into future strategies. First, it outlines the development of urban warfare in Ukraine by examining the key examples of urban battles, including those in Kyiv, Mariupol, Bakhmut, Avdiivka, and Sudzha in the Kursk region, followed by an analysis of the relevance of our previous understanding of urban warfare. This includes identifying areas that require further development and highlighting new insights gained from the conflict in Ukraine. The third section addresses the challenges faced by the armed forces in liberal democracies, with a focus on the civilian dimension and implications of international humanitarian law (IHL; also known as the law of armed conflict, or LOAC) in urban warfare, especially when confronting a peer or near-peer adversary that does not adhere to these legal norms. The final section draws lessons from the conflict in Ukraine for future urban combat, proposes a tentative framework for urban warfare, and discusses how the North Atlantic Treaty Organization (NATO) Joint Function Framework can be useful in preparing for future urban battles.

Chapter 4, by Colonel Chase Metcalf, assistant professor and director of Campaign Planning, Department of Military Strategy, Planning and Operations at the US Army War College, and Professor John Nagl, General John J. Pershing Chair of Military Planning and Operations, the US Army War College, focuses on command. It contains an analysis of command and control, with particular attention to mission command as a concept and practice. Drawing on the Russo-Ukraine War it demonstrates that mission command not only enables success in large-scale combat operations but is increasingly dependent on the integration of advanced technologies such as artificial intelligence (AI), autonomous systems, and cyber capabilities. Ukraine's partial adoption of mission command principles, supported by NATO training and technological innovation, enabled tactical agility and operational resilience. In contrast, Russia's rigid, centralized command structure and cultural resistance to disciplined initiative led to battlefield failures. This contrast offers critical lessons for future warfare, particularly for militaries seeking to balance technological advantage with empowered leadership at all levels of command.

Chapter 5, written by Christopher Kinsey, professor of international security, and Dr Ronald Ti, affiliate lecturer, both at Kings College London, focuses on logistics. The chapter, titled 'Russian Military Logistics and Its Impact on Russia's War with Ukraine', stresses the fact that logistics capabilities will continue to be a major factor in shaping military operations in Ukraine as well as in future military operations. It argues that the interdependency of operational logistics and strategic planning has been clearly seen in the Russo-Ukrainian war. It highlights the failure of the Russian high command to generate a logistics system able to interface optimally with ground forces and to provide them with the capabilities and capacities necessary to sustain success. It investigates the reasons for this failure by examining the structure, doctrinal, and command philosophies that underpin Russia's defence logistics. While analysing the structure, the authors focus on: examining challenges associated with the 'push' system of for delivering force projection and sustainment; doctrine—analysing the impact of Russian military doctrine on its defence logistics system during the initial stage of the invasion; and command philosophies—considering the philosophical foundation that shapes Russian military command ethos, which is opposed to individual decision making and initiative.

Chapter 6 focuses on the role and importance of the military health services, with a chapter titled 'Health Services in War' by Gen. (Rtd) Martin C. M. Bricknell, professor of conflict, health, and military medicine at King's College London. This chapter reviews emerging observations for health service support to armed forces from the Russo-Ukrainian war after the invasion by Russia in 2022. The chapter covers five themes. The first discusses the realities of combat in a peer-on-peer conflict and the implication for health service support in terms of the volume of casualties and the threats to the national medical system. The next theme examines tactical medical care on the battlefield covering pre-hospital care, tactical medical evacuation, and forward surgical support. The theme of operational organization of medical services discusses the distribution and reception of casualties into the civil-military health system, recovery and rehabilitation, and battle shock and mental health. The third theme, long-term strategic issues, looks at the medical assessment of recruits, medical retention standards, military disability pensions, long-term care of the war injured, military healthcare education, and, finally, anti-microbial resistance. The final section considers the implications of the Ukrainian experience for military health systems across the globe as the threat of war increases due to the tensions across the global international system.

Chapter 7, 'Maritime Lessons from the War in Ukraine', by Dr Deborah Sanders, professor of contemporary conflict and strategy, Defence Studies Department, Kings College London, focuses on the maritime dimension of the conflict. This chapter addresses the previously limited attention given to the maritime domain and highlights lessons for contemporary navies. This chapter not only examines three key maritime lessons from the war in Ukraine but also highlights the conditionality and complexity of each lesson. These three lessons are: the continuing importance of maritime power; that qualitative aspects of maritime power can also be decisive in naval warfare; and that, while amphibious operations remain challenging, they have demonstrated their utility at both the strategic and tactical levels. The final part of this chapter explores whether the naval battle in the Black Sea constitutes a change in the character of naval warfare.

Chapter 8, 'Air War over Ukraine', by Dr Viktoriya Fedorchak, associate professor in war studies, Swedish Defence University, focuses on the air domain of the war. As the title suggests, this chapter explores various aspects of the air war over Ukraine. This chapter outlines the main trends of the air war, followed by a description of developments in piloted aviation for both Ukrainian and Russian forces. Next, the chapter examines ground-based air defences, followed by an analysis of the increasing use of drone warfare. The chapter concludes with key considerations for future conflicts.

Part II: 'Organization, Technology, and Training'—consists of chapters on training and force generation, intelligence, UAV/drones, combat motivation and morale, as well as the Israeli perspective on the war in Ukraine and women and gender.

Chapter 9, 'Force Generation and Training of the Armed Forces of Ukraine' is written by Dr Jack Watling, senior research fellow for land warfare, Department of Military Sciences, Royal United Services Institute (RUSI), London. Watling argues that the challenges of force generation in Ukraine bear careful scrutiny. The chapter discusses the pre-war Ukrainian training process, the impact of the onset of conflict on the training system, the force generation process both inside and outside Ukraine, and the interaction with international partners. The chapter concludes by outlining five important lessons from the evolution of Ukraine's training structure that should inform Ukraine's partners and the armed forces of Ukraine (AFU) after a ceasefire, as it endeavours to build a force capable of deterring future Russian aggression. The chapter is based on direct observation of all elements of this process during the conflict. It focuses exclusively on the training of Ukrainian ground forces, air assault forces, and territorial defence forces.

Chapter 10, titled 'Intelligence and Russia's War in Ukraine', is written by Dr Niklas Nilsson, associate professor in war studies, Land Operations Division and Centre for Societal Security (CTSS), Swedish Defence University. In this chapter, the author argues that Ukraine illustrates—or has functioned as—a catalyst for a set of developments featuring certain challenges and trends in intelligence practices that have acquired increased centrality during the war. It examines four significant themes in intelligence practice that have been actualized during the conflict, and for several of which the war has catalysed developments and debate. These include (1) the limitations of forecasting and prediction, highlighted by divergent pre-war assessments, which illustrate the complexity of gauging adversarial intent and military capacity; (2) the utilization of declassified strategic warning intelligence for strategic communication employed by the US and United Kingdom (UK) to expose Russian military preparations, representing an adaptation to the contemporary information environment; (3) the acceleration of intelligence sharing among the US, its allies, and Ukraine; and (4) the rapid technological evolution in intelligence, driven by the need to process and integrate the enormous volume of data produced by the transparent battlefield.

Chapter 11 on 'UAV/Drone Dimension of the War in Ukraine', written by Samuel Bendett, adjunct senior fellow, Technology and National Security Program, Center for a New American Security (CNAS), Washington, DC. This chapter examines the evolving role of UAVs and drones in shaping tactical developments on the battlefield in Ukraine, with a particular focus on the Russian military and its perception of drone combat in Ukraine. Excluding long-range drone applications—which have already received substantial international attention—the analysis centres on the pervasive use of aerial drones in short- and medium-range tactical settings. While an extensive body of open-source literature now exists on drone warfare, this chapter moves beyond prevailing accounts by interrogating how tactical-level drone integration is impacting the tactical environment in Ukraine. It draws primarily from Russian official and unofficial sources. This is done to add an additional analytical element to the rapidly growing body of literature that highlights many Ukrainian technological developments in the Western and English-language media—such analysis tends to provide detailed coverage of Ukrainian developments and capabilities given a relatively free and unobstructed access to Ukrainian combatants, military decision makers, open-source analysts, government officials, and technology developers.

Chapter 12 provides an Israeli take on the war in Ukraine. Titled '"The Old Wars Are Back": The IDF's Learning Process from the War in Ukraine', it provides a different perspective on the conflict. In this chapter, Dr Eado Hecht, senior research associate, Begin Sadat Center for Strategic Studies (BESA Center) in Israel, and Dr Eyal Berelovich, a senior defence analyst at the IDF Land War Research Center, note that the war between Russia and Ukraine found the IDF at a crossroads—testing new concepts and preparing for a new five-year force generation plan, while simultaneously discarding old capabilities. It outlines the IDF's research, both formal and informal, into the war in Ukraine, aiming to analyse it and elicit lessons for its future development. The war reignited a long-standing debate within the IDF, one that began in the mid-1990s and gradually led to the adoption of a new theory of how war should be conducted. This renewed debate was driven by the fact that many of the lessons from the war in Ukraine contradicted that theory.

Chapter 13, titled 'Visibility of Women and Gender Equality Implementation in the Armed Forces of Ukraine' by Dr Tamara Martsenyuk, associate professor at the Department of Sociology, University of Kyiv-Mohyla Academy (Ukraine), argues that since 1991, Ukraine has achieved considerable progress in establishing gender equality across different spheres of society, including the military. Since the adoption of the first National Action Plan, the Ukrainian security and defence sector has undergone several significant systemic changes. These include granting women access to military occupations, including combat roles; recognizing female veterans and gender equality in military legislation; improving protection for women from gender-based violence; providing access to military education at all levels; providing gender training and education for staff; and allowing external research on various aspects of gender equality implementation. Despite limited resources and time, persistent gender stereotypes, and the challenges of an ongoing war, women not only are actively involved in these reforms but also are building peace and security. This chapter is based on data from five extensive sociological studies known as the 'Invisible Battalion', which demonstrate both successes and ongoing challenges of gender equality implementation in the Ukrainian Armed Forces, including the status of female veterans and problem of sexual harassment in the military.

Part III: 'Case Studies'—includes five case studies covering the war in Ukraine from 2014 to March 2025.

Chapter 14, 'Russia's War on Ukraine: Lessons in Hybrid Defence and Conventional Warfare (2014–2025)' by Dr Omar Ashour, professor of security and military studies at the Doha Institute for Graduate Studies, Doha, Qatar, and Dr Hanna Shelest, the director of security studies and global outreach at the Foreign Policy Council 'Ukrainian Prism', Ukraine, delves into the operational art and some tactics deployed in the Russian war on Ukraine from 2014 to 2025. It primarily aims to extract lessons and offer critical insights for combat effectiveness, military effectiveness, and defence policies, particularly for the small(er) states in Europe and beyond.

The chapter argues that the Ukrainian Security and Defence Forces employed a 'Hybrid Defence' strategy (HDS), featuring TTPs in four or more domains of operations. This has enhanced both their military and combat effectiveness and compensated for asymmetries in force sizes, force equipment, and initially in force quality (during specific operations). The overall impact of the HDS on attrition remains inconclusive in this particular case. Whereas the strategy permitted the smaller force to militarily (not just politically) attrite the larger one, the overall results of attrition remained acceptable by the larger force up to the time of writing this chapter.

Chapter 15 is titled 'The Russian Invasion of Ukraine: How Putin Lost in Ten Days', which is written by Ed Arnold, senior research fellow of RUSI, and a former British army infantry officer. This chapter examines the origins of the aftermath of Russia's invasion of Ukraine, probing the disconnect between Vladimir Putin's expectation to take control over Ukraine in ten days and the persistent conflict that ensued. It seeks to understand and explain: why the SMO failed, how close it came to succeeding, Russian military weaknesses, the strength of Ukrainian resistance, and the geopolitical repercussions of Russia's failure to subjugate Ukraine. This chapter argues that the best explanation for the difference between the Russian and Ukrainian military performance in the first ten days is the moral component of fighting power and the principles of war.

Chapter 16 by Dara Massicot, senior fellow of the Russia and Eurasia Program, Carnegie Endowment for International Peace in Washington, DC, is titled 'Ukrainian Counter-Offensives in Kharkiv and Kherson, June–November 2022'. This chapter examines two Ukrainian counter-offensive operations in the Kharkiv and Kherson regions during the summer and autumn of 2022, surveying the strategic context for the operations and force dispositions of Ukrainian and Russian Forces beginning in June 2022. It offers two case studies of the Kharkiv and Kherson counter-offensives through the Russian withdrawal from Kherson across the Dnipro River in early November 2022. The chapter considers the factors that contributed to the operational successes and failures of the AFU and Russian Forces, respectively, and argues that the circumstances that contributed to successful Ukrainian counter-offensives in 2022 would not be present in the following years, explaining why the subsequent AFU counter-offensive in the summer of 2023 would not find the same success.

Chapter 17, 'The "Spring Offensive" 2023' is written by Dr Viktoriya Fedorchak, associate professor in war studies, Swedish Defence University. This chapter discusses the case of the Ukrainian counter-offensive in 2023. It explains the preliminary preparatory stage on both sides before the commencement of the counter-offensive in June 2023. This is followed by a discussion of the main developments, tactics used, enemy counter-actions, factors contributing to the state of affairs, and further points of consideration. The main factors contributing to the campaign's characteristics and outcomes are also discussed. They include the multi-domain nature and the need to improve cross-domain integration in the battlespace; the importance of terrain characteristics and complexity of breaching the mine barriers; trends in counter-battery fires; increased use of drones by the Russian forces; the increasing role of electronic warfare; the significance of people as a crucial element of resilient fighting power; and finally, the need for building critical mass especially in the war of attrition.

Chapter 18, 'The Ukraine War in 2024: The Kursk Incursion and Beyond' covers the conflict until March 2025. It is written by Dr Markus Balázs Göransson, senior lecturer in war studies at the Swedish Defence University. This chapter discusses developments in the war between January 2024 and early March 2025, a period that stretched roughly from the aftermath of Ukraine's failed 2023 counter-offensive to the geopolitical pivot that followed Donald J. Trump's accession as the US president.

One the one hand, it was characterized by a pattern of attritional warfare in parts of the ground arena, and on the other hand, it featured Ukrainian attempts to break out of that attrition. During this time, Ukraine mounted effective offensive actions by air (targeting Russia's defence industrial production by drones) and sea (pushing back Russia's naval forces to the eastern shore of the Black Sea), and conducted a surprise mechanized incursion into Russia's Kursk Province in August 2024. The chapter reflects on these developments in the context of broader debates about defensive attrition and manoeuver warfare, ending with a discussion of the possible implications of Trump's diplomatic shift in the war.

Finally, overall conclusions from the volume are presented by Dr Niklas Nilsson, associate professor in war studies, Land Operations Division and Centre for Societal Security (CTSS), Swedish Defence University, and Dr Mikael Weissmann, associate professor of systems science for defence and security and deputy director of the Swedish Defence University Centre for Land Warfare at the Swedish Defence University.

PART I
DIMENSIONS OF MODERN WARFARE

2
Ukraine

The Lessons for Manoeuvre Warfare

Christopher Tuck

Introduction

Manoeuvre warfare occupies a central place in Western military thinking.[1] Lukas Milevski notes that manoeuvre warfare exerts an 'intellectual and cultural dominance' in Western armies.[2] Manoeuvre warfare's premise that quality can overcome quantity through a mindset that prioritizes tempo and the dislocation of an adversary has been an important touchstone in military theories of victory since at least the 1980s, and arguably even earlier.[3] At the same time, however, manoeuvre warfare has also been controversial. Some of the strongest criticisms have dismissed it as a dangerous dogma unrooted in any substantive historical evidence. In any debates on the efficacy or otherwise of particular military doctrines and concepts, combat experience provides perhaps the acid test; and if one's own armed forces cannot acquire that combat experience, then observing the wars of others historically has proven to be an important alternative.

The war in Ukraine is a major high-intensity clash between modern armies and the first such war in Europe since 1945. As a result it provides an obvious focus for discussions on the utility of manoeuvre warfare. What, then, does the fighting there reveal about manoeuvre warfare today and in the future? To explore this question, this chapter is divided into four parts. The first briefly discusses the assumptions that underpin manoeuvre warfare; the second and third sections discuss, respectively, positive and negative views on what the war might say about the utility of manoeuvre warfare. The final part assesses what conclusions we might draw from the preceding discussions.

[1] The author would like to thank Dr Erik Sand, US Naval War College, for his insightful comments on a draft of this chapter.

[2] Lukas Milevski, 'Manoeuvre warfare in the Baltic: Political imperatives and tactical conditions', *RUSI Journal* 1. DOI: 10.1080/03071847.2024.2407534.

[3] Aidan Webster, 'Manoeuvre warfare: Size doesn't matter; it's how you use it', *The Cove*, 21 April 2022. https://cove.army.gov.au/article/manoeuvre-warfare-size-doesnt-matter-its-how-you-use-it.

Christopher Tuck, *Ukraine*. In: *Russia's War in Ukraine and Modern Warfare*. Edited by: Mikael Weissmann and Niklas Nilsson, Oxford University Press. © Christopher Tuck (2026). DOI: 10.1093/9780198970903.003.0002

This chapter makes two arguments. First, the Russo-Ukraine conflict allows us to draw no categorical conclusions regarding the effectiveness of manoeuvre warfare. The evidence from the war has sustained two opposing views, each of which interprets the same evidence differently. This shouldn't surprise us since history demonstrates that this is a common challenge in trying to draw lessons from observing contemporary wars. Second, while there are valid arguments regarding how far one can draw generalizable conclusions from the Russo-Ukrainian War, the conflict nevertheless provides significant ammunition for those critical of manoeuvre warfare's elevated importance in Western doctrine. The war reinforces historical lessons on the importance of attrition in warfare. It also reinforces the arguments that changes in the character of land warfare might make manoeuvre even more difficult. The risks associated with manoeuvre warfare seem to have increased significantly. This doesn't definitively render manoeuvre warfare irrelevant—but it does make it even more important that armies have credible answers to the question: what do we do if manoeuvre warfare does not work in the ways anticipated?

Manoeuvre Warfare

Manoeuvre warfare can be understood in terms of operational strategies and defeat mechanisms. Manoeuvre warfare is an 'operational strategy'. Operational strategies are the primary methods employed by militaries in the field to defeat their adversary. Operational strategies reflect different sorts of defeat mechanisms. Frank Hoffman identifies four general types of defeat mechanisms, each of which may involve a different emphasis on the means used (physical or non-physical) or the targets or desired effects (physical or non-physical/cognitive).

These include dislocation, which achieves its effect through physical manoeuvre relative to the enemy in time and space, rendering their positions disadvantageous using physical means to achieve non-physical effects; disorientation, which produces success through using non-physical means such as information and deception to create non-physical/cognitive effects; degradation, which involves attacking an opponent's systems through physical or cyber means, such as command and control to reduce their situational awareness influence; and attrition, the material wearing down of an enemy's forces—physical means to achieve physical effects.[4]

As an operational strategy, manoeuvre warfare is most often conceptualized as embodying the defeat mechanisms of dislocation and disorientation. The US Marine Corps, one of the early enthusiasts for manoeuvre warfare, argued that manoeuvre warfare is 'a state of mind bent on shattering the enemy morally and physically by paralysing and confounding him, by avoiding his strength, by quickly and aggressively exploiting his vulnerabilities, and by striking him in a way that will hurt him

[4] Frank Hoffman, 'Defeat mechanisms in modern warfare', *Parameters* 51/4 (2021): 54.

most'.[5] Manoeuvre warfare's defeat mechanisms operate through systems thinking, tempo, and non-linearity.

Systems thinking is focused on identifying and attacking the specific critical capabilities or nodes that allow the enemy military system to function, rather than meeting the enemy strength directly; achieving a superiority over an adversary in operational tempo will undermine their situational awareness and decision making; non-linear thinking prioritizes flexibility in thinking and command and control philosophies.[6] Reflecting on this discussion, Amos C. Fox comments that 'From an applied or warfighting sense, manoeuvre is the indirect application of force, oriented on enemy centres of gravity and other important capabilities and vulnerabilities, to surprise an enemy'.[7] Surprise, deception, agility, seizing the initiative, and avoiding direct force-on-force encounters are attributes commonly ascribed to manoeuvre warfare.[8]

The desirability of manoeuvre warfare as an operational strategy is reflected in the way that it is most often contrasted favourably with attrition. As an operational strategy, attrition mostly operates through the defeat mechanisms of attrition and degradation. An attritional strategy tends to be characterized as direct, slow, costly, unimaginative, and a tool that favours quantity over quality.[9] These are hardly desirable attributes, which reinforces the apparent utility of manoeuvre warfare since it can provide 'a philosophy for generating the greatest decisive effect against the enemy for the least possible cost to ourselves—a philosophy for 'fighting smart'.[10] Enthusiasts argue that the utility of this approach furthermore has been validated by historical evidence. As Liddell Hart asserted, '[I]n all decisive campaigns, the dislocation of the enemy's psychological and physical balance has been the vital prelude to his overthrow.'[11] Attrition, therefore, has the status of an undesirable last resort.

Nevertheless, while modern militaries have enthusiastically embraced manoeuvre warfare, it has remained controversial in many quarters. For some critics, manoeuvre warfare is an invented concept that reflects how armies would like to fight rather than the messier reality demonstrated by a proper understanding of history. As Fox comments 'most claims of manoeuvre are feeble attempts to wedge the idea into misaligned situations'.[12] Other commentators argue that manoeuvre warfare is contingent: it might work in some scenarios, but not in others. In circumstances

[5] US Marine Corps, 'Marine Corps Doctrine Publication (MCDP) 1', *Warfighting* (Washington, DC: US Marine Corps, 1997), 95.

[6] Lt Col. H. T. Hayden (ed.), *Warfighting: Maneuver warfare in the U.S. Marine Corps* (London: Greenhill, 1995), 50; William J. Harkin, 'Maneuver warfare in the 21st century', Marine Corps Association Blog, 2019. https://mca-marines.org/blog/gazette/maneuver-warfare-in-the-21st-century/; William S. Lind, *Maneuver warfare handbook* (London: Routledge, 2018), 7.

[7] Amos C. Fox, 'Manoeuvre is dead? Understanding the conditions and components of warfighting', *RUSI Journal* 166/6–7 (2021): 12.

[8] Thomas Withington, 'Manoeuvre warfare and the electromagnetic spectrum', *RUSI Journal* 168/6 (2023): 32–41.

[9] Franz-Stefan Gady, 'Manoeuvre versus attrition in US military operations', *Survival* 63/4 (August–September 2021): 131; Lamar Tooke, 'Blending maneuver and attrition', *Military Review* 80/2 (March–April 2000): 10; Daniel Moran, 'Geography and strategy', in John Baylis, James Wirtz, Colin S. Gray, and Eliot Cohen (eds), *Strategy in the contemporary world* (Oxford: Oxford University Press, 2007), 126.

[10] US Marine Corps, *Warfighting*.

[11] Quoted in Hoffman, 'Defeat mechanisms in modern warfare', 51.

[12] Fox, 'Manoeuvre is dead?', 10.

where urban warfare is prominent,[13] or where political imperatives might mandate that territory simply has to be held, or where, to avoid escalation, certain territory is simply regarded as inviolable, manoeuvre warfare may be impossible to implement in practice.[14]

Another argument is that however useful it has been in the past, changes in the character of warfare, especially the growing transparency and lethality of the battlefield, actually are making manoeuvre warfare less and less plausible as an operational strategy.[15] Others argue that manoeuvre warfare cannot be usefully separated from mass. Peter Roberts has argued that 'Mass is not about attrition, industrialised warfare or position; it is as important for manoeuvre as it is for any other military strategy'.[16] Manoeuvre warfare, therefore, is so contested that its centrality in modern Western doctrine is dangerous: 'No competent soldier ... should embrace it' argues one critic.[17]

Given the breadth of opinion on the utility, or otherwise, of manoeuvre warfare, to what extent can an analysis of the war in Ukraine provide us with additional evidence of its importance as an operational strategy?

Ukraine: The Death of Manoeuvre Warfare?

On the one hand, it isn't difficult to develop an argument that Ukraine demonstrates the lack of relevance of manoeuvre warfare.

The operational features of the war have been most reflective of attritional approaches to fighting. Rather than featuring predominantly non-linear, high-tempo operations focused against key enemy nodes designed to produce quick outcomes at low cost, the war in Ukraine is best characterized as a protracted conflict featuring grinding attrition. Explicit parallels have been drawn with the grim positional warfare of the First World War. As the then Ukrainian commander-in-chief, General Valeri Zaluzhny, commented in 2023, 'the war at the present stage is gradually moving to a positional form', amidst conditions that he likened to World War One.[18]

Since then, this positional form has been the dominant character. The character of the fighting has reinforced established verities in the conduct of attritional land

[13] Anthony King, *Urban warfare in the twenty-first century* (Cambridge: Polity, 2021), 19–40.

[14] Milevski, 'Manoeuvre warfare in the Baltic', 4–6.

[15] Frederick W. Kagan and Kimberly Kagan with Mason Clark, Karolina Hird, Nataliya Bugayova, Kateryna Stepanenko, Riley Bailey, and George Barros, *Ukraine and the problem of restoring maneuver in contemporary war* (Washington, DC: Institute for the Study of War, 2024), 17–18.

[16] Quoted in House of Lords International Relations and Defence Committee, 'Ukraine: A wake-up call', 1st Report of Session 2024–2025, HL Paper 10, 26 September 2024, 12.

[17] Daniel P. Bolger, 'Maneuver warfare reconsidered', in Richard D. Hooker, Jr (ed.), *Maneuver warfare: An anthology* (Novato, CA: Presidio Press, 1993), 21.

[18] Commander-in-chief, General Valeri Zaluzhny, 'Modern positional warfare and how to win it', 3 November 2023. https://athenalab.org/wp-content/uploads/2023/11/ZALUZHNYI_FULL_VERSION-2.pdf.

warfare. One of these is that manoeuvre matters less than fire superiority.[19] The war in Ukraine has sustained the observation that most casualties in modern land warfare are the result of remote fires.[20] As Russian commentators have noted, weight of fire has been the 'determining factor' on the battlefield.[21] While long-range precision fires have been locally significant, it is drones and heavy artillery that have had a major impact on the nature of the fighting. The former has exerted a progressively more significant influence over time, and is now responsible for 60–80% of battlefield casualties.[22] Artillery, however, continues to have a longer range than tactical drones, better reliability, and a mass effect that drones cannot replicate.[23]

Equally, the fighting has reinforced the importance of mass in land warfare.[24] Russian 'meat-grinder' tactics have featured the extravagant expenditure of manpower. The costs of this have been enormous. This has led to Russian casualty rates of over 1,000 losses a day, and more than a million casualties overall, for daily advances by mid-2025 of between 80 and 135 metres: less than many key campaigns of the First World War.[25] But these tactics, so far from the prescriptions of manoeuvre warfare, have allowed Russia to obtain the initiative.[26] As one Ukrainian soldier commented: 'I'll be honest. It's genius. Cruel, immoral, but effective tactics'.[27]

The most frequently encountered military challenges for both sides have been those related not to mobile operations but to defence-in-depth: either skills in the conduct of defensive operations, or in terms of attempting to effect offensive break-ins and breakthroughs. The failure, for example, of Ukraine's summer 2023 offensive had important roots in the Russian improvements in defensive operations: improved coordination, training, fire control, and positioning of forces.[28] Equally, Ukrainian

[19] See Jack Watling and Nick Reynolds, 'Tactical developments in the third year of the Russo-Ukrainian War', *RUSI Journal*, February 2025: 11–14. https://www.rusi.org/explore-our-research/publications/special-resources/tactical-developments-during-third-year-russo-ukrainian-war.

[20] See, for example, Sam Cranny-Evans, 'Russia's artillery war in Ukraine: Challenges and innovations', *RUSI Journal*, 9 August 2023. https://www.rusi.org/explore-our-research/publications/commentary/russias-artillery-war-ukraine-challenges-and-innovations.

[21] V. Litvinenko and D. Tsekhanovich, 'Features of organizing counter-battery warfare in a special military operation', *Army Collection No. 9*, 9 September 2022. https://army.ric.mil.ru/Stati/item/424559.

[22] Richard Thomas, 'Drones now account for 80% of casualties in Ukraine Russia War', Army Technology, 8 April 2025. https://www.army-technology.com/news/drones-now-account-for-80-of-casualties-in-ukraine-russia-war/.

[23] Jacob Jajkay, 'I fought in Ukraine and here's why FPV drones kind of suck', *War on the Rocks*, 26 June 2025. https://warontherocks.com/2025/06/i-fought-in-ukraine-and-heres-why-fpv-drones-kind-of-suck/.

[24] Andrew A. Michta, 'Mass still matters: What the US military should learn from Ukraine, *New Atlanticist*, 3 October 2023. https://www.atlanticcouncil.org/blogs/new-atlanticist/mass-still-matters-what-the-us-military-should-learn-from-ukraine/.

[25] Seth Jones and Riley McCabe, 'Russia's battlefield woes in Ukraine', CSIS, 3 June 2025. https://www.csis.org/analysis/russias-battlefield-woes-ukraine.

[26] Ben Barry, 'What Russia's momentum in Ukraine means for the war in 2024', IISS, 13 March 2024. https://www.iiss.org/online-analysis/military-balance/2024/03/what-russias-momentum-in-ukraine-means-for-the-war-in-2024/.

[27] 'Bakhmut: Russian casualties mount but tactics evolve', *BBC News*, 16 March 2023. https://www.bbc.co.uk/news/world-europe-64955537.

[28] Franz-Stefan Gady and Michael Kofman, 'Making attrition work: A viable theory of victory for Ukraine', *Survival Online*, 9 February 2024; Seth G. Jones, Riley McCabe, and Alexander Palmer, 'Seizing the initiative in Ukraine: Waging war in a defense dominant world', *CSIS Briefs*, October 2023, 9–11. https://www.csis.org/analysis/seizing-initiative-ukraine-waging-war-defense-dominant-world.

forces underestimated the challenges that faced them in breaking into and through the Russian defences, lacking the equipment, training, and level of fire support necessary to collapse the defending forces and restore manoeuvre.[29]

The attritional character of the war in Ukraine has not excluded important indicators of change. For example, in Ukraine, as noted elsewhere in this book, drones now exercise a pervasive influence on military operations down to the lowest tactical levels, providing intelligence, surveillance, and reconnaissance capabilities, firepower, shortening kill chains, and increasing battlefield transparency and lethality.[30] However, what the Ukraine war might be taken to demonstrate is that a changing character of warfare makes attrition more, not less, likely. Processes of technological change and informationalization before the Ukraine war were identified by many militaries are something that was creating a more transparent and lethal battlefield.[31]

As a result of these developments, the US Joint Chiefs of Staff argued in its Joint Operating Environment 2035 assessment that future battlefield success was unlikely to be achieved either cheaply or quickly.[32] The war in Ukraine validates these observations. In Ukraine we have seen the development of the 'Tactical Strike Reconnaissance Complex', the result of the integration of mainly drone-based tactical reconnaissance, drone-directed remote fires, and precision munitions.[33] The growing range and accuracy of firepower and the shrinking of kill chains can quickly punish attempts to mass forces, making it difficult to assemble the combat power necessary for offensive operations.[34]

In this more transparent battlefield speed also loses some of its potency as a source of advantage.[35] One of the key advantages of speed is that it undermines an opponent's situational awareness. The proliferation in surveillance technologies, however, reduces this effect. Another issue illustrated by the war in Ukraine is the magnification of command and control problems caused by the increasingly lethal battlefield. Dispersal has become essential for survival. Ukrainian battalions often

[29] See Jack Watling and Nick Reynolds, 'Stormbreak: Fighting through Russian defences in Ukraine's 2023 offensive', *RUSI Journal*, 4 September 2023. https://www.rusi.org/explore-our-research/publications/special-resources/stormbreak-fighting-through-russian-defences-ukraines-2023-offensive.

[30] See Kristen D. Thompson, 'How the drone war in Ukraine is transforming conflict', Council on Foreign Relations, 16 January 2024, https://www.cfr.org/article/how-drone-war-ukraine-transforming-conflict; Craig Macauley, 'Gunline survival against loitering munitions: Lessons from Ukraine', *The Cove*, 26 October 2023. https://cove.army.gov.au/article/gunline-survival-against-loitering-munitions.

[31] US Army Training and Doctrine Command (TRADOC), 'The operational environment and the changing character of warfare', TRADOC Pam 525-92 (Austin, TX: TRADOC, 2019), 16; Ministry of Defence, 'Chief of the defence staff, General Sir Nick Carter delivers a speech on the integrated operating concept at policy exchange', Ministry of Defence, September 2020. https://www.gov.uk/government/speeches/chief-of-the-defence-staff-general-sir-nick-carter-launches-the-integrated-operating-concept; US Government Executive, *Multi-doman battle: Evolution of combined arms for the 21st century, 2025–2040* (Washington, DC: GovExec, 2017), 4–8. https://admin.govexec.com/media/20171003_-_working_draft_-_concept_document_for_multi-domain_battle_1_0.pdf.

[32] Gady, 'Manoeuvre versus attrition', 135–6.

[33] Kagan et al., *Ukraine and the problem of restoring maneuver*, 36–40.

[34] John Q. Bolton, 'The more things change ... Russia's war mirrors the past as much as it shows the future', *Military Review*, July 2023: 2. www.armyupress.army.mil/Portals/7/military-review/Archives/English/Online-Exclusive/2023/The-More-Things-Change/Bolton-AWC-Ukraine-Observations-UA.pdf.

[35] Fox, 'Manoeuvre is dead?', 17.

operate on frontages traditionally occupied by brigades, with companies sometimes spread across up to 3 kilometres. Under these conditions, the tight coordination of forces necessary for manoeuvre warfare becomes more problematic, especially given the disruptive effects of modern electronic warfare.[36]

Ukraine: *Sui Generis*?

For advocates of the continued utility of manoeuvre warfare, however, the evidence provided by the Ukraine conflict leads to alternative conclusions.

First, one can observe that operations conforming to the precepts of manoeuvre warfare have indeed been undertaken and that these have delivered significant successes. Ukraine's Kharkiv counter-offensive in September 2022 constitutes for some commentators 'a case study that validates manoeuvre doctrine'.[37] Leveraging deception, surprise, rapid manoeuvre, and exploitation, Ukrainian forces recaptured nearly 1,000 square miles of territory and routed whole Russian units in an operation touted as the result of 'classic dislocation and a manifestation of maneuverist thinking'.[38]

Equally, Ukraine's cross-border offensive into Russia's Kursk oblast that began in August 2024 seized a similar amount of territory and changed, at least temporarily, the political narrative surrounding the progress of the war. This success was attained through operational surprise, the seizing of the initiative, and the pitting of strength against weakness. This operation shows, according to one commentary, that 'success is possible when unconventional solutions are applied'.[39]

Not only do we have examples of successful manoeuvre warfare, but these successes also undermine some of the other critiques identified at the beginning of this chapter. In both these examples, manoeuvre warfare worked even in a context in which urban areas were a significant feature of the tactical geography. The fight in Ukraine has often been focused on urban centres, not least because of their political significance and because of their position relative to lines of communication. The Kharkiv and Kursk offensives succeeded in doing what advocates of manoeuvre warfare have previously argued manoeuvre should do: they circumvented the need for urban warfare by causing the systemic collapse and withdrawal of defending troops.[40]

[36] Randy Noorman, 'The Russian way of war in Ukraine: A military approach nine decades in the making', Modern War Institute, 15 June 2023. https://mwi.westpoint.edu/the-russian-way-of-war-in-ukraine-a-military-approach-nine-decades-in-the-making/.

[37] Steve Maguire, 'Yes, manoeuvre is alive. Ukraine proves it', Wavell Room, 4 November 2022. https://wavellroom.com/2022/11/04/yes-manoeuvre-is-alive-ukraine-proves-it/.

[38] Pat Garrett and Frank Hoffman, 'Maneuver warfare is not dead, but it must evolve', *Naval Review Proceedings* 149/11/1,449, November 2023. https://www.usni.org/magazines/proceedings/2023/november/maneuver-warfare-not-dead-it-must-evolve.

[39] Alex Horobets, 'Assessing Ukraine's Kursk operation', *European Security & Defence*, 4 September 2024. https://euro-sd.com/2024/09/articles/40133/assessing-ukraines-kursk-operation/.

[40] Isabelle Khursudyan, Paul Sonne, Serhiy Morgunov, and Kamila Hrabchuk, 'Inside the Ukrainian counteroffensive that shocked Putin and reshaped the war', *The Washington Post*, 29 December 2022. https://www.washingtonpost.com/world/2022/12/29/ukraine-offensive-kharkiv-kherson-don

The profoundly attritional character of the fighting for towns such as Bakhmut in 2023, on the other hand, was because Russian troops 'failed to conduct basic maneuver warfare against Ukrainian forces'.[41] The Ukrainian defenders, so the argument goes, should themselves have utilized manoeuvre warfare to avoid attrition, given that, in a war between a Russian Goliath and a Ukrainian David, it seemed self-evident that attrition would favour the former.[42] At the same time, the employment of manoeuvre warfare in Ukraine also demonstrates that it is an operational strategy that isn't necessarily escalatory and that does have applications beyond limited wars.

Critics have argued that manoeuvre warfare is intrinsically risky to implement, particularly in cases where belligerents have nuclear weapons because it is an operational strategy designed to produce decisive victory through the systemic collapse of an adversary's forces. Manoeuvre warfare is therefore difficult to modulate and the high costs that its successful imposition might inflict on an enemy carry with it the risk of unintended escalation, especially in wars of national survival.[43] Yet, the war in Ukraine is an existential struggle, certainly for Ukraine, but also arguably for Putin, and successes in manoeuvre warfare have not led to uncontrolled escalation, even when, in the case of Ukraine's Kursk offensive, the attack was launched into Russian territory itself. Manoeuvre warfare is therefore a more flexible operational strategy than many allow.

A second defence of the relevance of manoeuvre warfare is that in Ukraine neither of the belligerents has adopted attrition as their first choice; nor does it seem to be their preferred option. At the end of 2023, General Valeri Zaluzhny assessed what his forces needed to fight the war effectively, including air superiority, better electronic warfare capabilities, and improved logistics. Significantly, in doing so, Zaluzhny stated that his aim was 'finding a way out of the positional form of warfare' and restoring manoeuvre.[44] Ukraine's failed offensive in the summer of 2023 was an attempt to end the attritional character of the war and to achieve breakthroughs and the restoration of manoeuvre.[45]

This has also been evident in the Russian approach. The initial Russian plan for the invasion of Ukraine was itself manoeuvrist in conception, relying on surprise, speed, and shock to conduct a *coup de main* against the political and military leadership of

etsk/; Patrick Bury, 'Ukraine war: Kursk offensive has taken the war into Russia and put Putin on the back foot—for now', *The Conversation*, 24 August 2024. https://theconversation.com/ukraine-war-kursk-offensive-has-taken-the-war-into-russia-and-put-putin-on-the-back-foot-for-now-236663.

[41] Cyril Tregub, 'War in Ukraine: How to help David to down Goliath', Friends of Europe, 5 June 2024. https://www.friendsofeurope.org/insights/war-in-ukraine-how-to-help-david-to-down-goliath/.

[42] Sean Bell, 'Ukraine must embrace the West's manoeuvre warfare instead of being dragged into traditional Russian-style combat', *Sky News*, 11 March 2023. https://news.sky.com/story/ukraine-must-embrace-the-wests-manoeuvre-warfare-instead-of-being-dragged-into-traditional-russian-style-combat-12830728.

[43] Carter Malkasian, 'Airland battle and modern warfare', *International Forum on War History: Proceedings*, September 2014: 117.

[44] Zaluzhny, 'Modern positional warfare'; 'War of attrition: An interview with Ukriane's commander-in-chief on the breakthrough he needs to beat Russia', *The Economist*, 4 November 2023. https://athenalab.org/wp-content/uploads/2023/11/ZALUZHNYI_FULL_VERSION-2.pdf.

[45] See Jack Watling, Oleksandr V. Danylyuk, and Nick Reynolds, 'Preliminary lessons from Ukraine's offensive operations, 2022–23', Special Report, Royal United Services Institute (RUSI), July 2024.

Ukraine. This plan came closer to success than might be appreciated. Subsequently, in operations such as those at Vuhledar in early 2023 in the south, and through Vovchansk towards Kharkiv beginning in May 2024, the clear intent was to try to achieve breakthroughs that would produce an alternative to attritional warfare. That both of these operations suffered heavy losses and degenerated into sustained attritional fights was not the intended outcome. Indeed, this observation leads to another important defence of manoeuvre warfare: context.

Advocates for manoeuvre warfare could claim that the limited impact of manoeuvre warfare in Ukraine stems from contextual factors that would not necessarily be replicated in other conflicts. Presented in this way, the problem in Ukraine isn't that manoeuvre warfare doesn't work: it's that Russian and Ukrainian forces can't make it work.

The Russian military is a case in point. It pursued from 2008 a modernization process designed to produce forces that were smaller, more professional, networked, integrated, agile, and mobile. Technologies, especially those relating to information and data, artificial intelligence, and precision firepower were prominent in this programme. Its doctrines, embodied in ideas such as Reconnaissance Strike and Reconnaissance Fire complexes, were indicative of a military more than capable of manoeuvre warfare approaches.[46] The western military district commander, Colonel-General Aleksandr Zhuravlev, argued in 2019 for the need to: 'devote particular attention to unorthodox thinking, departing from established stereotypes, and using nonstandard methods'.[47]

With hindsight though, Western assessments radically overestimated Russian capabilities for reasons relating to the political context of military assessments, practical difficulties in assessing non-material factors, and the complexity of strategy, especially in terms of its interactive nature.[48] There were evident critical structural weaknesses relating to issues of civil-military relations and the stultifying influence of a Soviet-style military culture in which individual initiative was discouraged.[49]

As it turns out, the war since 2022 has shown that the previous operations in Donbas and Syria were not meaningful tests of the Russian military as a whole. The operations involved the deployment of relatively small numbers of regular forces and niche capabilities, and the approaches adopted during those conflicts could not be

[46] Lester W. Grau and Charles K. Bartles, 'The Russian reconnaissance fire complex comes of age', University of Oxford Changing Character of War Project, May 2018. https://static1.squarespace.com/static/55faab67e4b0914105347194/t/5b17fd67562fa70b3ae0dd24/1528298869210/The+Russian+Reconnaissance+Fire+Complex+Comes+of+Age.pdf.

[47] Timothy Thomas, 'Russian lessons learned in Syria: An assessment', MITRE Center for Technology and National Security, June 2020, 6. https://www.mitre.org/sites/default/files/publications/pr-19-3483-russian-lessons-learned-in-syria.pdf.

[48] See, for example, Bettina Renz, 'Western estimates of Russian military capabilities and the invasion of Ukraine', *Problems of Post-Communism*, 12 September 2023. https://doi.org/10.1080/10758216.2023.2253359.

[49] See Alexander Golts, 'Russia's military failures in Ukraine: Causes and consequences', SCEEUS Report No. 4, 2023. https://sceeus.se/en/publications/russias-military-failures-in-ukraine-causes-and-consequences/; Philip Wasielewski, 'The roots of Russian military dysfunction', Foreign Policy Research Institute, 30 March 2023. https://www.fpri.org/article/2023/03/the-roots-of-russian-military-dysfunction/.

scaled up to the massive requirements of the Russo-Ukraine War. The Russian military instead demonstrated weaknesses in a whole range of areas that prevented it from employing manoeuvre warfare extensively: intelligence, planning, joint warfare, command and control, combined arms, training, and equipment. The inability of the Russian air force to achieve air superiority, despite its enormous superiority on paper over its Ukrainian adversary, was symptomatic of these problems and made manoeuvre warfare an even more challenging proposition.

Ukraine had also attempted pre-war processes of military reform. But, despite making many important advances, it was a military that still suffered from a range of problems, not least that the expansion in its size diluted the quality of its training and equipment.[50] Moreover, it too continued to suffer from the lingering impact of a Soviet military culture, as well as an inability to attain sufficient air control. The lack of manoeuvre warfare, it has been argued by many, is a function of the inadequacies of the militaries of both sides. North Atlantic Treaty Organization (NATO) forces would not suffer these limitations.

In addition, even if the modern battlefield is becoming increasingly transparent and lethal, manoeuvre warfare is developing in response to these challenges, reflected in the US and NATO concepts of multi-domain operations (MDO). MDO seeks to more tightly integrate military campaigns, moving beyond traditional joint operations towards an approach that might see the almost seamless fusion of capabilities across different domains (land, sea, air, space, cyber, human) and at different levels of war (tactical, operational, and strategic). The key to this is investments in digital architectures, including artificial intelligence, information technology, and data networking, that will allow the creation of a single 'system of systems'.[51]

MDO will allow those forces that can implement it effectively to drastically increase the tempo at which military operations can be conducted. It will increase the agility, flexibility, and survivability of friendly forces by allowing more dispersed and diverse forces to integrate their efforts against the enemy.[52] MDO is suffused with the spirit of manoeuvre warfare.[53] It rejects attritional approaches: its core defeat mechanism is dislocation of the enemy, achieved through generating 'simultaneous effects from all domains faster than the enemy'.[54] The goals of MDO are systemic paralysis of the adversary and 'shattering the coherence of his military system'.[55]

[50] See Deborah Sanders, 'Ukraine's third wave of military reform 2016–2022: Building a military able to defend Ukraine against the Russian invasion', *Defense & Security Analysis* 39/3 (2023): 312–28.

[51] TRADOC, 'The operational environment', 16; Ministry of Defence, 'Chief of the defence staff, General Sir Nick Carter delivers a speech'; US Government Executive, *Multi-doman battle.*

[52] US Army, 'Army futures command concept for maneuver in multi-domain operations', AFC Pam 71-20-1. US Army Training and Doctrine Command, 7 July 2020, 8. https://api.army.mil/e2/c/downloads/2021/01/20/2fbeccee/20200707-afc-71-20-1-maneuver-in-mdo-final-v16-dec-20.pdf.

[53] Dwight Phillipps, 'Multi-domain operations: Passing the torch', Hague Centre for Strategic Studies, November 2023, 3. https://hcss.nl/report/multi-domain-operations-passing-the-torch/ (accessed 12 November 2024).

[54] Tom Greenwood and Pat Savage, 'In search of a 21st century joint warfighting conflict', *War on the Rocks*, 12 September 2019. https://warontherocks.com/2019/09/in-search-of-a-21st-century-joint-warfighting-concept/.

[55] US Army, 'Army futures command concept', 10.

All of this reinforces a point that advocates of manoeuvre warfare have already made clear: it is a demanding mode of warfare that requires skilled and capable forces. Indeed, given the greater transparency and lethality evident on the Ukraine battlefields, challenging solutions reflected in MDO become necessary to make manoeuvre warfare work. The attritional character of the war in Ukraine therefore serves to reinforce the importance of manoeuvre warfare, rather than undermine it; the conflict illustrates the cost of an inability to conduct manoeuvre warfare effectively. Attrition is what belligerents are left with if they are unable to properly implement manoeuvre warfare as an operational strategy. More of an emphasis on manoeuvre warfare, not less, is required if the West is to avoid becoming mired in attritional conflicts that they are unsuited to fight.

Manoeuvre Warfare and History

The preceding discussion illustrates that commentators can take the same case study and yet come to completely different conclusions regarding the utility of manoeuvre warfare as an operational strategy. If this is frustrating it should also be of no surprise. History demonstrates that one of the perennial challenges in drawing useful lessons from the process of observing the wars of others is the extent to which that evidence can be used to support divergent positions.[56] The impact of various factors including imperfect information, variable contexts, politics, organizational culture, and wishful thinking make military 'lessons' highly malleable.[57]

The likely character of the First World War with hindsight might have been evident in conflicts such as the American Civil War (1861–5), the Franco-Prussian War (1870–1), and the Russo-Japanese War (1904–5). Yet commentators at the time could reach no consensus on the relevant lessons, with the evidence used to support contending perspectives on the implications of modern firepower and the relative weight that should be placed upon technological and human factors.[58]

However, if the evidence from the Russo-Ukraine War cannot be taken to provide a definitive conclusion on the utility of manoeuvre warfare, it does provide further support for the argument that placing too much emphasis on manoeuvre, at the expense of alternatives, is dangerous. The Ukraine war does this by reinforcing a range of historical observations regarding the role of attrition and the challenges of making manoeuvre warfare work.

The first of these issues becomes evident when one places the war in Ukraine into the context of previous conflicts. For critics of the historical evidence base for

[56] See, for example, Jonathan Bailey, 'Military history and the pathology of lessons learned: The Russo-Japanese War, a case study', in Williamson Murray and Richard Hart Sinnreich (eds), *The past as prologue* (Cambridge: Cambridge University Press, 2006), 170–94.

[57] See Thomas R. Mahnken (ed.), *Learning the lessons of modern war* (Stanford, CA: Stanford University Press, 2020).

[58] See Antulio Echevarria II, *Imagining future war: The West's technological revolution and visions of wars to come, 1880–1914* (Westport, CT: Praeger, 2007).

manoeuvre warfare, the war in Ukraine simply provides yet more evidence that modern warfare tends to involve a mix of manoeuvre and attrition and that the latter is often the predominant element. History shows that large-scale wars usually can't be ended by a single campaign: belligerents can generate the forces to man continuous defence-in-depth and to provide reserve forces and fall-back defences, and offensive operations have proven to be difficult and costly.[59]

This was the case in the Second World War. There were many campaigns in which the tenets of manoeuvre warfare were realized: France in 1940; Operation Barbarossa in 1941; and Operation Cobra in 1944. But these operations did not end the war as a whole, and these operations punctuated increasingly attritional fighting, as shown by battles in 1943 at the Mareth Line and Kursk; in 1944 in operations such as Epsom, Goodwood, Market Garden, and the Siegfried Line; and in 1944–5 along the Gothic Line.

For critics of manoeuvre warfare, the evidence from the past has never supported the idea that one can separate manoeuvre from attrition. Attrition and manoeuvre are not an 'either/or'—most campaigns involve both.[60] The proportion might vary with one more dominant at a given point in time, but even where campaigns have been labelled as manoeuvre warfare, they have involved important aspects of attrition.[61] This is because the capacity for manoeuvre has often depended upon prior attrition to wear down an adversary sufficiently for breakthroughs to be obtained.

In line with this, in Ukraine, as Gady and Kofman note: 'it does seem fairly clear that the war has seen prolonged phases in which the ability to manoeuvre has been earned chiefly through extensive attrition and the destruction of the enemy's capacity rather than through cognitive effects or effective employment of combined arms'.[62] But it is also the case that periods of manoeuvre often tend to end in a return to attrition because of logistics constraints, enemy adaptation, accumulating losses, and the need for operational pauses to regroup and reorganize. Ukraine's Kursk offensive made impressive early gains, but within five weeks it had become attritional in character, with casualty rates resembling other parts of the fighting.[63]

Linked to the previous point, a second general criticism is that viewing the character of the fighting in Ukraine through the lens of manoeuvre warfare misses the real sources of tactical and operational success in land warfare. For Stephen Biddle, for example, the balance between manoeuvre and attrition in land warfare is a function of the relative quality of force employment. By the term 'force employment', Biddle focuses on competence in 'modern system' warfare, something that he defines as the

[59] Kagan et al., *Ukraine and the problem of restoring manoeuvre*, 11.

[60] Garrett and Hoffman, 'Maneuver warfare is not dead'.

[61] Heather Venable, 'Paralysis in peer conflict? The material versus the mental in 100 years of military thinking', *War on the Rocks*, 1 December 2020. https://warontherocks.com/2020/12/paralysis-in-peer-conflict-the-material-versus-the-mental-in-100-years-of-military-thinking/; Edward N. Luttwak, *Strategy: The logic of war and peace* (Cambridge, MA: Belknap Harvard, 2001), 12; Tooke, 'Blending maneuver and attrition', 12.

[62] Gady and Kofman, 'Making attrition work'.

[63] Ivana Kottasova and Kostya Gak, 'Ukraine's Kursk offensive was seen as a major success, but it came at a huge cost', *CNN*, 13 September 2024. https://edition.cnn.com/2024/09/13/world/ukraine-kursk-offensive-cost/index.html.

'tightly interrelated complex of cover, concealment, dispersion, suppression, small-unit independent maneuver, and combined arms at the tactical level, and depth, reserves, and differential concentration at the operational level of war'.[64]

Where the gap in quality of force employment between the belligerents is very high, as it was between the Iraqis and coalition forces during the Gulf War in 1991, or between British and Italian forces in Operation Compass in 1940–1, then manoeuvre warfare becomes much more prevalent. But where the gap is much narrower, attritional warfare becomes much more likely, as it was at Alamein in 1942 or the early stages of the Normandy campaign in 1944. Land warfare is among other things an adversarial and relative activity, and, where the gap in quality of force employment is narrower, the constraints imposed by conditions such as decision-making disruption and dislocation are likely to be mutually imposed, making things difficult for both sides.[65]

Manoeuvre warfare when it occurs is one possible product of the interaction between more fundamental issues relating to relative competence. In Ukraine, we see this, for example, in the extent to which Ukrainian opportunities for manoeuvre warfare outcomes have been reduced by Russian military learning and adapting and how Russian mobilization, preparation, planning, and military conduct have shifted towards those designed to exploit attrition.[66] Successful offensive manoeuvre warfare against defence-in-depth requires, as Biddle argues, a skilled attacker and a compliant defender. The failure of Ukraine's summer 2023 offensive stemmed from the fact that the attacking forces simply didn't have the large relative advantage in competence and motivation necessary to achieve a decisive breakthrough and to restore manoeuvre.[67]

Underpinning aspects of the first two criticisms is a third that explains why attrition might become more likely as conflicts progress. This is the impact of attrition on the capacity of a military to conduct manoeuvre warfare. If the war in Ukraine underlines the fact that attrition is a regular feature of campaigns, even those that contain aspects of manoeuvre, then the capabilities for manoeuvre warfare developed in peacetime are likely to decline quite quickly in the face of casualties and necessary wartime expansion. Unless one wins very quickly, losses in trained leaders, specialists, rank and file, and equipment, combined with the problems of training and equipping replacements or new forces to the same level of competence mean that the capacity to meet the demands of manoeuvre warfare begin to decline, often very quickly.

[64] Stephen Biddle, *Military power: Explaining victory and defeat in modern war* (Princeton, NJ: Princeton University Press, 2004), 28.

[65] Michael Kofman, 'A bad romance: US operational concepts need to ditch their love affair with cognitive paralysis and make peace with attrition', Modern War Institute at West Point, 31 March 2021. https://mwi.usma.edu/a-bad-romance-us-operational-concepts-need-to-ditch-their-love-affair-with-cognitive-paralysis-and-make-peace-with-attrition/.

[66] Watling et al., 'Preliminary lessons', 27.

[67] Stephen Biddle, 'How Russia stopped Ukraine's momentum: A deep defence is hard to beat', *Foreign Affairs*, 29 January 2024. https://www.foreignaffairs.com/ukraine/how-russia-stopped-ukraines-momentum.

This was a feature evident in both world wars in which, during examples such as the Allied offensive in October and November 1918, and Operation Goodwood in 1944, the attacks were negatively impacted by mutually reinforcing problems of high casualty rates among officers and non-commissioned officers (NCOs), relatively limited training, gaps in essential skills, which together manifested in poor combined arms coordination, low levels of initiative, and limited gains.[68] These problems have afflicted both sides during the Russo-Ukrainian War. The problem facing the belligerents is not just that there were shortcomings in both sides' regular militaries at the moment the war began, it is also the often-catastrophic effect of losses since then, which has eroded the capabilities of both, as well as the dilution in quality caused by expansion in the size of their militaries. Towards the end of 2024, Russian forces had reportedly lost 90% of the forces that they had begun the war with.[69] Ukrainian losses have also been extremely high, with some units suffering 100% casualties in the first twelve months of the war.[70] Under these pressures, even NATO-trained Ukrainian forces have struggled with the demands of the fighting.[71]

The impact of attritional warfare has transformed the armies of both sides into very different organizations than those they began the war with. As Gady and Kofman note: 'Neither army looks like it did at the beginning of the war, each having become a variegated force composed of mobilised personnel, auxiliaries, foreign fighters and, in Russia's case, mercenaries, including ex-prison inmates'.[72] These new armies struggle with the basics of staff work and combined arms warfare, making the high standards demanded by manoeuvre warfare approaches largely unattainable except in limited circumstances.

The Russo-Ukrainian War also illustrates the challenge of implementing manoeuvre warfare's core defeat mechanisms. Manoeuvre warfare is predicated on attacking enemy intangibles through dislocation and disorientation. This approach relies on an ability to assess the state of the enemy's morale and cohesion. As intangibles, however, these are not easy to measure objectively and, as the fighting in Ukraine demonstrates, this often leads to wishful thinking. The initial Russian offensive plan of February 2022, for example, was based on the assumption, what some commentators have termed a 'bet', that surprise and speed would lead to the systemic collapse of Ukrainian resistance. But this assumption proved to be false and 'when the bet

[68] See Alex Vershinin, 'The attritional art of war: Lessons from the Russian war on Ukraine', *RUSI Journal*, 18 March 2024. https://rusi.org/explore-our-research/publications/commentary/attritional-art-war-lessons-russian-war-ukraine; Jonathan Boff, *Winning & losing on the Western front* (Cambridge: Cambridge University Press, 2012), 245–6; Biddle, *Military power*, 121–2.

[69] Pjotr Sauer, 'Ends justifies the means: High death toll fails to shift opinion on Ukraine war', *The Guardian*, 22 October 2024. https://www.theguardian.com/world/2024/oct/22/high-russian-death-toll-fails-shift-opinion-ukraine-war.

[70] Isabelle Kurshudyan, Paul Sonne, and Karen DeYoung, 'Ukraine short of skilled troops and munitions as losses, pessimism grow', *The Washington Post*, 13 March 2023. https://www.washingtonpost.com/world/2023/03/13/ukraine-casualties-pessimism-ammunition-shortage/.

[71] Eric Schmitt and Helene Cooper, 'Ukrainian troops trained by the West stumble in battle', *The New York Times*, 2 August 2023. https://www.nytimes.com/2023/08/02/us/politics/ukraine-troops-counteroffensive-training.html.

[72] Franz-Stefan Gady and Michael Kofman 'Ukraine's strategy of attrition', *Survival* 65/2 (28 March 2023): 14.

on speed failed, the plan was invalidated'.[73] Similarly, Ukraine's summer offensive of 2023 assumed that, though outnumbered and facing an enemy in prepared defensive positions, success could be achieved by shock action that would break Russian morale and cause a collapse in their resistance.[74]

As a result, the defeat mechanisms underpinning manoeuvre warfare became a kind of magic wand, the waving of which appeared to make uncomfortable material realities, such as inadequately trained troops and lack of resources, disappear. The difficulty in measuring the actual level of morale and cohesion in Russian forces also encouraged Ukraine to persevere in their failed summer offensive even after the lack early on of major gains, because it was hoped that Russian forces were near the brink of collapse and that another push would precipitate their final disintegration.[75] Because of the problems in objectively measuring progress in manoeuvre's core defeat mechanisms, therefore, the outcome was less realistic planning assumptions and more attrition.

Last, whether MDO really offers a new way to harness the forces of change in the service of manoeuvre warfare is also open to question. Despite the salience of MDO in contemporary military thinking, it remains a contested concept. For some critics, MDO is an empty idea marked by Babylonian terminological confusion, poor institutional fit, and immature technologies.[76] Its main defeat mechanism, cognitive paralysis caused by imposing simultaneous multiple dilemmas on the enemy, likewise has been labelled 'a faith-based enterprise' that would not work against peer adversaries.[77]

The war in Ukraine reinforces these points. Critics argue that the fighting there is more complex than envisaged and it is too amorphous to fit within the sort of MDO campaign envisaged by the West.[78] In addition, while space and cyber factors are significant in the Ukraine conflict, this doesn't necessarily create a new form of warfare or require a paradigm more than an evolved form of jointery or a more integrated approach.[79] All the fighting in Ukraine demonstrates is what we already should have known: combined arms warfare is becoming more complex but even more important; new technologies have continued the process of expanding the battlefield and encouraging dispersion; and the importance of mass, logistics, and adaptive dynamics shape situations in which it is 'cumulative or attritional effects that

[73] Watling et al., 'Preliminary lessons', 28.
[74] Ibid., 17.
[75] Ibid., 30.
[76] David Ellison and Tim Sweijs, 'Empty promises? A year inside the world of multi-domain operations, *War on the Rocks*, 22 January 2022. https://warontherocks.com/2024/01/empty-promises-a-year-inside-the-world-of-multi-domain-operations/.
[77] Kofman, 'A bad romance'.
[78] Benjamin Jensen, 'How US forces can adopt Ukraine's unconventional multidomain approach', *Defense News*, 22 March 2024. https://www.defensenews.com/opinion/2024/03/22/how-us-forces-can-adopt-ukraines-unconventional-multidomain-approach/.
[79] See Theodora Ogden et al, 'The role of the space domain in the Russia-Ukraine war', Centre for Emerging Technology and Security, 23 February 2024. https://cetas.turing.ac.uk/publications/role-space-domain-russia-ukraine-war.

create the conditions for maneuver through enemy exhaustion, rather than shock'—which is an idea that differs from MDO's current emphasis on creating "windows of opportunity"'.[80]

Going forward, the challenges of transparency and lethality may well get even more intense as the application of autonomous systems, already a feature of Ukrainian battlefields, increases, and new capabilities for autonomous swarming become possible.[81]

Conclusion: Manoeuvre Warfare, Risk, and Doctrinal Shock

The war in Ukraine currently sustains opposing views on manoeuvre warfare. Advocates argue that the Russo-Ukrainian conflict provides a stark example of the necessity for competence in manoeuvre warfare; the character of the war is a result of the inability of the belligerents to implement manoeuvre warfare properly the consequence of which is a more protracted and costly struggle. Critics argue, on the other hand, that the conflict shows that the theory of manoeuvre warfare is difficult to implement and needs to be thought of as an adjunct to, and not a replacement for, a capacity to fight attritional warfare. Where does this leave us?

First, nothing in the Russo-Ukrainian conflict suggests that manoeuvre warfare has become less challenging. Recurrent problems such as the difficulties posed by layered defence-in-depth, the planning and execution of combined arms warfare, and assessing key intangibles such as enemy morale and cohesion seem to have been multiplied by technological developments that have increased the transparency and lethality of the modern battlefield.

Second, Russian operations in Ukraine reinforce the point that attrition can work.[82] Historically, attrition 'has been consistently successful in overcoming the threat of imminent defeat or of resolving strategic and tactical dilemmas that preclude decisive victory'.[83] Attrition is the operational strategy that has defined the character of the war in Ukraine, and it is the method that has allowed Russia to take the initiative. As the war also illustrates, the opportunities for manoeuvre warfare may be conditional on preceding periods of attrition.

Third, and following on from the two preceding points, the war makes clear the risks entailed in not having a workable 'Plan B'; of not having a clear answer to the question: 'What if manoeuvre warfare fails to deliver?' Attrition is the main operational strategy that both sides in the Russo-Ukraine War have had to employ, even though it is clear that it is neither side's preferred option. Historically, land warfare has

[80] Phillipps, 'Multi-domain operations', 5.

[81] Kagan et al., *Ukraine and the problem of restoring maneuver*, 41; Chase Metcalfe, 'Is there anything to learn from the war in Ukraine?', War Room, 30 November 2023. https://warroom.armywarcollege.edu/articles/learning-lessons-1/#:~:text=William%20Owen%20argues%20in%20%E2%80%9CThe,noteworthy%E2%80%9D%20to%20an%20informed%20professional.

[82] William Freer, 'How Britain can better support Ukraine to fight a war of attrition', Council on Geostrategy (geostrategy.org.uk), 18 December 2023.

[83] Carter Malkasian, *A history of modern wars of attrition* (Westport, CT: Praeger, 2002), 8.

been influenced by the tension between idealized views of how it should be fought, shaped by an obsession with human intuition, skill, and decisiveness, and the realities, as a result of challenging political and material factors, of how it has usually had to be fought.[84] Both Russia in 2022 and Ukraine in 2023 based their planning on manoeuvre warfare assumptions: the assumption, for example, that speed and shock could substitute for mass and protracted operations. What the war demonstrates is that, if the idealized concept of manoeuvre warfare cannot be implemented in the way that is envisaged, then the messier material and political realities of attrition are likely to have to be confronted.

In these three points, the Russo-Ukraine War reinforces the risk associated with an overreliance on the efficacy of manoeuvre warfare. Even militaries that continue to preference manoeuvre warfare should consider the demands of attritional contingencies: currently, however, they do not. NATO doesn't have the doctrine or campaign design frameworks necessary to fight a war like that in Ukraine.[85] United Kingdom doctrine on land power addresses the issue of manoeuvre warfare but does not discuss attrition.[86] As Lawrence Freedman notes, 'The biggest surprises in war often lie in what happens after the first engagements'.[87] The Ukraine War reinforces the lesson of the costs of neglecting this point. It reinforces the point that, when manoeuvre warfare does not deliver the expected results, belligerents often undergo a form of 'doctrinal shock'. In this, commanders initially persevere with manoeuvre based on the logic that 'The doctrine is sound, but next time apply it better'.[88] Then, when forced to embrace attrition, they do so in an ad hoc and uncertain fashion because their organizations are conceptually, culturally, and materially unprepared for it.[89]

What an alternative doctrine and framework would look like is something that needs to be explored, with ideas advanced including a Cold War-style, deterrence-focused take on active defense, or a version of the French interwar methodical battle.[90] Western armies, of course, also have a history of attritional warfare which merits study by contemporary militaries. The Korean War (1950–3), for example, provides an illuminating case study of the challenges of manoeuvre warfare and the use of attrition in a limited and defensive role by numerically inferior forces.[91] Equally, thought needs

[84] Cathal J. Nolan, *The allure of battle: A history of how wars have been won and lost* (Oxford: Oxford University Press, 2017), 11.

[85] Kagan et al., *Ukraine and the problem of restoring maneuver*, 30.

[86] British Army, 'The tenets of land doctrine: The manoeuvrist approach', *UK land power*, 6th ed., Joint Doctrine Publication 0–20, section 3.14, October 2023. fttps://assets.publishing.service.gov.uk/media/6523da4aaea2d000132199d4/20231008-JDP_0_20_UK_Land_Power.pdf.

[87] Quoted in Walker Mills, 'A century of getting the future of war (mostly) wrong', Modern War Institute, 18 June 2018. https://mwi.westpoint.edu/century-getting-future-war-mostly-wrong/.

[88] John Shy, 'First battles in retrospect', in Charles E. Heller and William A. Stofft (eds), *America's first battles, 1776–1965* (Lawrence, KS: Kansas University Press, 1986), 333.

[89] Nolan, *The allure of battle*, 4–6.

[90] David Johnson, 'The army risks reasoning backwards in analyzing Ukraine', *War on the Rocks*, 14 June 2022. https://warontherocks.com/2022/06/the-army-risks-reasoning-backwards-in-analyzing-ukraine/; Gady, 'Manoeuvre versus attrition', 137–9, 143.

[91] See Malkasian, *A history of modern wars of attrition*.

to be given to the potential impact of nuclear weapons on manoeuvre warfare and their role in limited and attritional scenarios.[92]

Whatever the prospective solutions, the Ukraine war demonstrates the need at the moment to stop viewing attrition as an inferior and avoidable form of operational strategy and to begin to theorize and experiment on how it might be approached and implemented.[93] History demonstrates that some wars can indeed be won quickly under the right sorts of conditions.[94] But whether these conditions can be reduced to the implementation of a particular operational strategy, in this case manoeuvre warfare, is highly debatable. As the historian Cathal Nolan has argued: 'War is not reducible to one thing, be it a presumed tradition, a fighting style or even the idea of decisive battle. Enemies adjust. War evolves'.[95]

[92] Zachary Morris, 'Emerging US army doctrine: Dislocated with nuclear-armed adversaries and limited war', *Military Review*, January–February 2019. https://www.armyupress.army.mil/Journals/Military-Review/English-Edition-Archives/Jan-Feb-2019/Morris-Doctrine/.

[93] Kagan et al., *Ukraine and the problem of restoring maneuver*, 12–16.

[94] Nolan, *The allure of battle*, 11.

[95] Ibid.

3

The City as a Battlefield

(Re)Thinking Urban Warfare in the Wake of Ukraine

Mikael Weissmann and Douglas W. Winton

Introduction

> [W]ar today often takes place in cities and it likely will in the future as well. We have to be prepared for battle in dense urban environments.[1]

The long-held belief that future combat will occur in urban environments was definitively confirmed after the full-scale invasion of Ukraine in 2022.[2] While urban warfare is not a new phenomenon, emerging trends are increasingly pushing conflicts into urban areas on an unprecedented scale. Taken together, these developments emphasize the importance of cities in modern warfare and frame the analysis that follows. This chapter will examine the challenges and evolving expectations for urban operations on contemporary and future battlefields, with a focus on the lessons learned from the Russia-Ukraine War from February 2022 to February 2025.

Urban warfare is generally neither desired nor planned for, nor expected by conventional military forces. However, its historical frequency highlights the need to reassess our assumptions. For Western militaries, this reality has immediate practical implications: they must prepare not only to attack into urban terrain but also to defend.

[1] General Randy George, chief of staff of the United States Army, Kermit Roosevelt Lecture, Royal United Services Institute, 9 May 2024. https://www.rusi.org/research-event-recordings/recording-kermit-roosevelt-lecture-given-general-randy-george-chief-staff-united-states-army.

[2] Funding has been received from the Joint Advanced Staff and Command Programme Directorate (PL HOP), the Officers' Programme Directorate (PL OP) of the Swedish Defence University and the Swedish armed forces' research programme. The views and opinions presented in this chapter are those of the authors alone and do not necessarily represent those of the Swedish Defence University or the Swedish armed forces, nor the US Department of Defence or the US National Defense University. During the preparation of this work, a number of generative artificial intelligence (AI) and AI-assisted technologies have been used in the writing process, including Parsd, ATLAS.ti, DeepL, ChatGPT, and Grammarly. These tools were employed to support the writing process and to enhance and refine the text. The authors carefully reviewed, edited, and validated the content to ensure its accuracy and integrity, taking full responsibility for the final published work. The authors contributed equally to this chapter.

Mikael Weissmann and Douglas W. Winton, *The City as a Battlefield*. In: *Russia's War in Ukraine and Modern Warfare*. Edited by: Mikael Weissmann and Niklas Nilsson, Oxford University Press. © Mikael Weissmann and Douglas W. Winton (2026). DOI: 10.1093/9780198970903.003.0003

The canon of military history and strategic studies is rich with accounts of urban battles and campaigns.[3] Many of these have become iconic in the minds of scholars and military professionals alike. Historically, urban warfare has been approached with trepidation, leading military leaders to seek to isolate and bypass cities whenever possible. However, the war in Ukraine has starkly reminded us that this persistent challenge has not lessened despite advances in technology, societal norms, or demographic shifts. The fundamental truth is that militaries cannot simply wish away the complex problem of urban warfare. This enduring pattern makes Ukraine's recent experience particularly instructive. The war in Ukraine has highlighted these findings and introduced some new ones.

This war also underlines that a force outgunned or outmanned can boost its chances of success by defending in a city.[4] Sun Tzu's wisdom in seeking alternatives to beseiging cities remains sound, but responsible militaries must foresee that stratagems aimed at tempting adversaries out of urban areas will not always succeed. To cite General James Rainey, the commanding general of Army Futures Command: 'the probability of urban combat is indisputable, as much as we would like to bypass urban areas'.[5] These insights set the stage for the questions this chapter addresses.

Against this backdrop, this chapter critically re-examines pre-2022 perspectives on urban warfare in light of the war in Ukraine. It evaluates whether previous insights on urban warfare remain applicable, how they should be refined, and what new lessons have arisen. The analysis emphasizes key developments that require further study, adaptation, and incorporation into future strategies. To assist the reader, we next outline the chapter's structure.

This chapter is organized into three sections. The first section outlines the development of urban warfare in Ukraine by examining key examples of urban battles, focusing on the cases of Kyiv, Mariupol, Bakhmut, Avdiivka, and Sudzha.

The second section examines the relevance of our previous understanding of urban warfare, pinpointing areas needing further development and highlighting new insights gained from the conflict in Ukraine. It emphasizes the challenges faced by the armed forces of liberal democracies, particularly focusing on the civilian dimension and the implications of international humanitarian law (IHL), also known as the law of armed conflict (LOAC), in urban warfare—especially when confronting a peer or near-peer adversary that does not follow these legal norms.

[3] See for example Gregory Fremont-Barnes (ed.), *A history of modern urban operations* (Cham: Palgrave Macmillan, 2020); Alec Wahlman, *Storming the city: U.S. military performance in urban warfare from World War II to Vietnam* (Denton, TX: University of North Texas Press, 2015); Anthony Tucker-Jones, *Battle of the cities: Urban warfare on the eastern front, 1941–1945* (Barnsley, PA: Pen and Sword Military, 2023); Louis A. DiMarco, *Concrete hell: Urban warfare from Stalingrad to Iraq* (2012); Antony Beevor, *Stalingrad* (London: Penguin Books, 1999).

[4] Laurent Gisel et al., 'Urban warfare: An age-old problem in need of new solutions', Humanitarian Law and Policy, 27 April 2021. https://blogs.icrc.org/law-and-policy/2021/04/27/urban-warfare/.

[5] General James Rainey, commanding general of Army Futures Command, AUSA Warfighter Summit and Exposition, Fayetteville, NC, 27 July 2023. https://businessnc.com/four-star-army-general-says-technology-changing-war-strategies/.

The final section extracts lessons from the Ukraine conflict for future urban combat, proposing a tentative framework for urban warfare and illustrating how the North Atlantic Treaty Organization (NATO) Joint Function Framework can assist in preparing for upcoming urban battles. In short, the chapter progresses from case studies, to assessment, and then to a forward-looking framework.

Before turning to these sections, it is important to clarify how we use key terms. This chapter covers the full spectrum of urban warfare, from strategic to tactical levels. While aiming to present general lessons on urban warfare based on the Ukraine experience, this chapter concentrates on lessons relevant to NATO and its like-minded allies. It examines urban warfare from a liberal democratic state perspective and within the context of NATO capabilities and constraints, including the necessity to fully consider IHL, as well as to address the challenge of fighting an adversary that ignores the same constraints.

A core implication previewed here is the need to prepare for extended urban combats, which require intense firepower, strong logistics, and the ability to use irregular tactics in operational planning. It also involves developing two skills—both mastering high-tech, connected operations and conducting independent small-unit tactics—while maintaining the flexibility to transition fluidly between them.

Conceptual Framework and Terminology

Before analysing the complexity of urban warfare, it is important to clarify the scope and terminology used in this chapter. We cover the entire scope of urban warfare, from the strategic to the tactical level. The terms 'urban warfare', 'urban operations', 'urban battles', and 'urban combat' are closely related but carry distinct nuances in military contexts, broadly corresponding to the strategic, operational, and tactical levels of warfare.

Urban Warfare refers to the broader strategic conduct of military campaigns aimed at controlling cities, covering political, humanitarian, and logistical dimensions. It involves prolonged campaigns where cities become focal points due to their political, economic, or military value. **Urban Operations** denote the deployment of military forces within cities to carry out the full range of military missions, which may include combat operations, humanitarian assistance / disaster response, post-conflict stabilization, peace support, etc. **Urban Battles** are specific, standalone conflicts within a city, often serving as decisive moments in a larger campaign. These battles consist of multiple engagements and are characterized by combined arms and in contemporary settings by joint integration. **Urban Combat** describes the tactical execution of military operations in urban environments, focusing on immediate engagements marked by close-quarters fighting.

By establishing these distinct levels of analysis, this framework reveals critical gaps in military understanding and planning—particularly the tendency to conflate operational and tactical challenges or overlook the strategic-political dimensions unique to urban environments. This differentiation sets the stage for examining where current doctrines are insufficient and how the evolving lessons from Ukraine illuminate these deficiencies.

Battles in cities remain central to fighting between industrial militaries as much as it has been central to counter-insurgency operations. Global urbanization trends are decreasing the prevalence of politically significant areas that are non-urban, thereby suggesting a rising likelihood of urban conflicts. These strategic patterns set the scene for today's operational challenges.

Furthermore, we must shift our focus from planning and training for conflicts against irregular forces, where we hold a position of strength and technological superiority, to readying ourselves for confrontations with peer or near-peer adversaries possessing advanced technological capabilities. Although it is debatable how closely Russia's technology matches that of the West, it is clear that Russia operates at a higher competence level and has more robust access to advanced technologies than irregular adversaries, against whom most Western planning for urban operations has traditionally been directed.

NATO's Strategic Forecast Analysis in 2017 identified rapid urbanization as a factor that might shape its future security environment,[6] leading to the development of the concept for urban operations. A key finding from the wargame that validated the concept was the importance of resilience and agility.[7] Furthermore, as Anthony King has noted, the shrinking of militaries, especially among Western countries, increases the likelihood of combat occurring in cities.[8] These strands of analysis reinforce the strategic trendline described above.

Urban environments intensify everything we know about warfare. Regardless of the adversaries fighting, cities have elevated significance because the interactions between dense terrain, diverse human and societal interests (including the infrastructure that supports them), and political significance exponentially increase the complexity that confounds all military operations. The interplay of chance, passion, and reason intensifies when military forces clash in cities.

There is nothing about fighting in cities that fundamentally upsets how Clausewitz describes the nature of warfare as a violent clash of human will to achieve political objectives. What does change is the degree of friction generated by the urban system itself. Nonetheless, the complexity of cities compounds the leadership challenge commanders always face, given the inherent role of chance in war.

Cities' three-dimensional complexity intensifies the demands on intelligence professionals to collect, analyse, and assess the deluge of information. The compartmentalization of terrain impedes manoeuvre and forces the decentralization of units and

[6] Headquarters, Supreme Allied Commander Transformation, Strategic Foresight Analysis, 4 October 2017. https://www.act.nato.int/wp-content/uploads/2023/05/171004_sfa_2017_report_txt.pdf.

[7] Jozsef Bodnar and Sue Collins, 'Joint urban operations', *The Three Swords Magazine* 34 (2019): 93–6. https://www.jwc.nato.int/images/stories/_news_items_/2019/three-swords/NATOUrbanization_2035.pdf.

[8] Anthony King, *Urban warfare in the twenty-first century* (Cambridge: Polity Press, 2021).

thus complicates command and control (C2) and slows the tempo. The prevalence of ready-made fortifications increases the protection of defending forces and increases the demand for firepower. The combination of slow tempo and increased demand for firepower creates a surge in logistics requirements.

Taken together, these effects slow decision cycles while increasing resource consumption.

Nevertheless, the most fundamental aspect of cities is that people reside there, and their presence further amplifies each of these challenges. Given the central role of cities in driving human prosperity,[9] their strategic significance extends far beyond the geographic confines of the fighting. These effects are amplified in the modern information environment, where nearly everyone has the technological means to broadcast their experience globally and instantly across multiple platforms. Consequently, tactical actions can have strategic consequences within hours rather than weeks.

Urban Warfare in Ukraine: Five Illustrative Battles

Since Russia's full-scale invasion of Ukraine in February 2022, the war has been intermittently marked by dramatic offensives, stalled offensives, surprising counter-offensives, positional warfare, deep strikes against critical nodes, and the incorporation of new technologies and weapons. Unsurprisingly, both Russia and Ukraine have shrouded their battlefield losses; nonetheless, the human toll has been constant and mounting with some estimates of total loss of life over 500,000.[10]

Another common feature of the war, likely intensifying the casualties, has been battles in cities. These urban fights have amplified the civilian costs of this war with the United Nations reporting more than 10,000 civilian deaths and nearly 20,000 civilian casualties with the vast preponderance occurring in territory controlled by Ukraine and nearly 40% occurring during March 2022.[11]

Beyond the loss of life, the war and its destruction of civilian infrastructure have caused the displacement of more than fourteen million Ukrainians.[12]

[9] Jo Beall and David Adam, 'Cities, prosperity, and influence', British Council, March 2017. https://www.britishcouncil.org/sites/default/files/cities_prosperity_and_influence.pdf.

[10] Helene Cooper et al., 'Troop deaths and injuries in Ukraine war near 500,000, U.S. officials say', *The New York Times*, 18 August 2023. https://www.nytimes.com/2023/08/18/us/politics/ukraine-russia-war-casualties.html.

[11] United Nations Human Rights, 'Ukraine: Protection of civilians in armed conflict, December 2023 update', Office of the High Commissioner, 15 January 2024. https://ukraine.un.org/sites/default/files/2024-01/Ukraine%20-%20protection%20of%20civilians%20in%20armed%20conflict%20%28December%202023%29_ENG_0.pdf.

[12] Human Rights Watch, World Report 2023, 629. https://www.hrw.org/sites/default/files/media_2023/01/World_Report_2023_WEBSPREADS_0.pdf. See also Zaid bin Inam and Sarwat Rauf, 'Understanding urban warfare and its manifestation in Russi-Ukraine war', *Strategic Studies* 43/2 (Winter 2023); Institute of Strategic Studies Islamabad, 10 January 2024. https://issi.org.pk/wp-content/uploads/2024/01/Zaid_bin_Inam_and_Sarwat_Rauf_SS_No_2_2023.pdf.

Kyiv (February–April 2022)

When Russia invaded Ukraine on 24 February 2022, its forces advanced towards Kyiv, the capital, with the apparent goal of encircling the city and overthrowing the Ukrainian government. This opening move by the Russian forces underscores the enduring significance of cities, in particular capitals, as the symbols of political power and control. Facing a numerically and technologically superior adversary, the Ukrainian forces adopted a strategy of urban defence, leveraging the city's natural defensive barriers and constructing additional fortifications. Civilians participated in the defence effort by establishing checkpoints, making Molotov cocktails, and enlisting in territorial defence units. Despite initial advances, Russian forces encountered fierce resistance. Logistics issues, underestimation of Ukrainian resolve, United States (US) intelligence to Ukrainian forces, and strong international support for Ukraine hampered the Russian offensive.[13]

This defence of Kyiv during the Russian incursion in 2022 is a testament to the resoluteness and tactical ingenuity of Ukrainian forces. Despite the asymmetry in military might, Ukrainian defenders successfully stymied the multiple-axis advance on their capital, preventing the rapid takeover many had feared at the outset of hostilities. The Ukrainians' successful counter-attack of the Antonov Airfield, 20 kilometres north of Kyiv foiled Russian plans to use the airfield for inserting forces that could have reinforced the armoured assault.

A variety of Western-supplied weapons, including Javelins, next generation light anti-tank weapons (NLAW), and Stinger air defence missiles, were critical capabilities that enabled the armed forces of Ukraine to halt attacking formations and wreak havoc on their supply lines. The organic integration of Ukrainian defence forces and civilian guerrillas also stymied the Russian forces' efforts to enter the city from multiple axes. Within a week, the Russian assault into Kyiv had culminated and its limited offensive operations in the city's outskirts failed to achieve meaningful operational results despite the immense cost borne by the civilians. By the end of March, following a strategic stalemate and significant losses, Russian forces began to withdraw from the Kyiv region to refocus efforts in eastern Ukraine.[14]

[13] Adam Entous and Michael Schwirtz, 'The spy war: How the C.I.A. secretly helps Ukraine fight Putin', *The New York Times*, 25 February 2024. https://www.nytimes.com/2024/02/25/world/europe/cia-ukraine-intelligence-russia-war.html. See also Alexander J. Motyl, 'Russians Likely to Encounter Growing Guerrilla Warfare in Ukraine,' *Foreign Policy,* April 1, 2022. https://foreignpolicy-com/2022/04/01/ukraine-russia-war-guerrilla-partisans-civilian-militia/; Phil Thomas, 'Tactical Sustainment: Unraveling Kyiv's Urban Battlefield,' *Army Sustainment* (Spring 2024), April 23, 2024, https://www.army.mil/article/274906; and Marta Keep and Alyssa Demus, *Resisting Russia: Insights into Ukraine's Civilian-Based Actions During the First Four Months of the War in 2022* (Santa Monica, CA: RAND, 2023). https://www.rand.org/content/dam/rand/pubs/research_reports/RRA2000/RRA2034-1/RAND_RRA2034-1.pdf.

[14] Louis DiMarco, 'Urban combat lessons learned: The Battle of Kyiv, February–March 2022', *Strategein* 5 (Winter 2023): 1–16. https://doi.org/10.5281/zenodo.10407617. See also Sudarsan Raghavan et al., 'In Kyiv under siege, neighbors dig trenches and raise barriers to brace for Russian assault', *The Washington Post* 28 February 2022. https://www.washingtonpost.com/world/2022/02/28/ukraine-russia-kyiv-defense/.

Mariupol (February–May 2022)

The Battle of Mariupol stands as one of the most harrowing and prolonged engagements of the Russo-Ukrainian War in 2022, underscoring the stark brutality of urban combat and the immense human suffering that accompanies a siege in modern warfare. Mariupol, a key gateway for Russia to control for reliable land access to Crimea, suffered some of the most intense destruction and highest civilian casualties of the war. With Russia ultimately taking control of the city, the true number of civilian casualties remains likely unknowable, but United Nations estimates begin at 1,300 civilian deaths. Multiple failed attempts to provide secure evacuation corridors for the city's 400,000 civilian inhabitants after the fighting began, contributed to more than 2,500 civilian deaths.[15]

Within a week of Russia's full-scale invasion of Ukraine, the Russian armed forces, aided by the Donetsk People's Republic—People's Militia, had surrounded the city and commenced an artillery bombardment. The speed of the Russian encirclement was enabled by units advancing from Crimea towards Mariupol, overwhelming the Ukrainian defenders focused on attacks from the east. Russian bombardment of critical infrastructure throughout the city complicated Ukrainian efforts to mount a credible defence and provide for the basic needs of the city's residents. By the middle of March, civil life had ceased to function, and those remaining in Mariupol were fighting Russians or fighting to survive or both.[16]

By early April, Russian forces had advanced through the city under the cover of near-constant bombardment and forced Ukrainian defenders into a couple of small enclaves. Eventually all remaining defenders, primarily the Azov Regiment of the Ukrainian national guard, and many civilians collapsed into Azovstal, a steel plant complex. The industrial site offered many bunkers enabling Ukrainian survival, but the concentration also facilitated the Russian bombardment. On 1 May, the United Nations and the International Committee of the Red Cross brokered a deal to allow civilians in the plant to leave. Two weeks later, Ukrainian forces began to surrender, with the final resistance collapsing on 20 May. The Russians won control of the city by extracting an enormous price but paid with their own resources as well. The urban terrain enabled a small band of defenders to endure the bombardment longer than they would have otherwise and thus delay the attackers' control of the city.[17]

[15] 'Ukraine, ensure safe passage, aid for Mariupol civilians', Human Rights Watch, 21 March 2022. https://www.hrw.org/news/2022/03/21/ukraine-ensure-safe-passage-aid-mariupol-civilians.

[16] Shaun Walker et al., 'Mariupol: The ruin of a city', *The Guardian*, 23 February 2023. https://www.theguardian.com/world/ng-interactive/2023/feb/23/mariupol-the-ruin-of-a-city.

[17] Walker et al., 'Marioupol'; see also Michael Schwirtz, 'Last stat at Azovstal: Inside the siege that shaped the Ukraine war', *The New York Times*, 24 July 2022; Valerie Hopkins, Ivan Nechepurenko, and Marc Santora, 'The Ukrainian authorities declare an end to the combat mission in Mariupol after weeks of Russian siege', *The New York Times*, 16 May 2022. https://www.nytimes.com/2022/05/16/world/europe/azovstal-mariupol.html.

Bakhmut (May 2022–May 2023)

Much of the battle for Bakhmut is similar to the experiences in Mariupol, including the Russian reliance on artillery and complete destruction of infrastructure. One distinguishing characteristic was Russia's use of the Wagner Group which likely contributed to the duration of the battle and the accompanying high military casualties. Importantly, the fighting in and around Bakhmut was as much about the southern and northern flanks around the city as fighting for the city itself.

The battle for Bakhmut began as early as May 2022 as Ukrainian defenders collapsed from other positions around Donbas; however, the Russian assault on the city did not begin until later in the summer. The size of the Ukrainian commitment to defending Bakhmut, combined with the inexperience and ill discipline of the Russian forces, led to the battle being characterized by trench warfare during the autumn and winter months. Russian forces slowly but methodically overwhelmed Ukrainian defences and advanced to encircle the city, effectively destroying the city in the process in what many analysts described as a 'meat-grinder'. Through the spring of 2023, Ukrainian forces slowly moved their defences west, ceding more and more of the city to Russian control. In May 2023, Ukrainian forces began counter-offensive operations to roll back the Russian siege around Bakhmut; however, these localized offensives were unable to force Russian forces to abandon the city.[18]

Many criticized the Ukrainians for their persistent defence of Bakhmut as it drained scarce resources, especially artillery ammunition and lives, which would be desperately needed in the much-anticipated summer counter-offensive. However, the battle also forced the Russians to commit more resources than they had intended and probably heightened tension between the Wagner Group and the Russian armed forces. Ukrainian forces showed adaptability and resilience amidst ongoing Russian bombardment. They employed flexible defensive tactics, moving forces to respond to Russian pressure points, and used anti-tank weaponry that had proven effective earlier in the war. These defensive tactics were reinforced by targeted counter-attacks, which sometimes regained lost ground or inflicted notable personnel and equipment

[18] Lukas Milevski, 'The urban warfare debate, the Russian invasion of Ukraine, and the Bakhmut conundrum', Wavell Room, 3 November 2023. https://wavellroom.com/2023/11/03/the-bakhmut-conundrum/. See also David Axe, 'Russia is wasting its last good troops in a pointless attack on a worthless town', Forbes, 21 October 2022. https://www.forbes.com/sites/davidaxe/2022/10/21/russia-is-wasting-its-last-good-troops-in-a-pointless-attack-on-a-worthless-town/?sh=2998e191351d; Herb Scribner, 'Ukrainian city of Bakhmut now in "ruins" after Russian attack: Zelensky', Axios, 10 December 2022. https://www.axios.com/2022/12/10/volodymyr-zelensky-russia-bakhmut; 'Russia takes Bakhmut: Taking stock of the war's bloodiest battle so far', *The Kyiv Independent*, 21 May 2023. https://kyivindependent.com/russia-takes-bakhmut-taking-stock-of-the-wars-bloodiest-battle-so-far/; Chris York, 'Nearly 20,000 Wagner mercenaries died taking Bakhmut', *The Kyiv Independent*, 17 June 2024. https://kyivindependent.com/nearly-20-000-wagner-mercenaries-died-taking-bakhmut-media-investigation-confirms/; and Michael Koffman, 'The Russia-Ukraine War: Military operations and battlefield dynamics', in Hal Brands (ed.), *War in Ukraine: Conflict, strategy, and the return of a fractured world* (Baltimore, MD: Johns Hopkins University Press, 2024), 114–15. https://muse.jhu.edu/book/122782.

losses on the Russian forces. The battle likewise offered both sides the chance to test new tactics and technology.[19]

Avdiivka (October 2023–February 2024)

Seizing control of Avdiivka, an industrial city just 25 kilometres from Donetsk (controlled by pro-Russian separatists since 2014) was crucial to Russia's aims of dominating the entire Donbas and led the Ukrainian army to establish fortified defences in the city to prevent its fall.

The Russian strategy for capturing the city followed a familiar pattern, beginning with intense bombardments using artillery and air strikes from April 2022, which caused numerous civilian casualties and destroyed much civilian infrastructure. Throughout 2022 and early 2023, Ukrainian successes in thwarting Russian ground assaults into the city appeared to provoke Russia to intensify its efforts to destroy Avdiivka and make it indefensible. In autumn 2023, Russian attempts to wrest control of the city from Ukraine increased.

Despite Russia's significant manpower advantage, Ukrainian defenders in Avdiivka inflicted heavy casualties in prolonged fighting that escalated in late 2023. By early 2024, with Ukrainian forces deprived of Western (mainly US) aid, Russian forces had nearly encircled the city, forcing a Ukrainian withdrawal.[20]

Sudzha (August 2024–March 2025)

In August 2024, Ukraine launched a surprise cross-border offensive into Russia's Kursk Oblast to seize the initiative, disrupt Russian military operations, and gain leverage in ongoing battles across the broader Russo-Ukrainian front. The Ukrainian force of 11,000 troops leveraged deception, surprise, rapidly advancing small units, and electronic neutralization of Russian drones and communications, enabling them

[19] Michael Kofman and Rob Lee, 'Perseverance and adaptation: Ukraine's counteroffensive at three months', *War on the Rocks*, 4 September 2023. https://warontherocks.com/2023/09/perseverance-and-adaptation-ukraines-counteroffensive-at-three-months/; see also Andrew E. Kramer, 'The final blocks: Inside Ukraine's bloody stand for Bakhmut', *The New York Times*, 12 April 2023. https://www.nytimes.com/2023/04/12/world/europe/bakhmut-ukraine-russia.html.

[20] Michael Schwirtz, 'A Ukrainian town goes underground as Russia assaults the east', *The New York Times*, 20 April 2022. https://www.nytimes.com/2022/04/20/world/europe/avdiivka-ukraine-russia.html. See also Cash Boyle, 'Russian losses in battle for Avdiivka may be worst of 2023, says UK', *The Guardian*, 28 October 2023. https://www.theguardian.com/world/2023/oct/28/russian-losses-in-battle-for-avdiivka-may-be-worst-of-2023-says-uk; David Axe, 'Three thousand Russians were killed or wounded for every square mile they captured around Avdiivka', Forbes, 12 December 2023. https://www.forbes.com/sites/davidaxe/2023/12/12/three-thousand-russians-were-killed-or-wounded-for-every-square-mile-they-captured-around-avdiivka/?sh=2167dc534792; Shaun Walker, 'Ukrainian forces withdraw from Avdiivka to avoid encirclement, army chief says', *The Guardian*, 16 February 2024. https://www.theguardian.com/world/2024/feb/16/ukrainian-forces-give-up-some-positions-in-avdiivka-as-russian-assault-continues; and Jaroslav Lukiv, 'Avdiivka: Ukraine troops leave embattled eastern town', BBC, 17 February 2024. https://www.bbc.com/news/world-europe-68323366.

to advance as far as 35 kilometres, seizing control of 1,250 square kilometres and as many as ninety settlements. The Ukrainian force established its main command node and defences in Sudzha, a town of around 5,000 residents, just 9 kilometres from the Russia-Ukraine border.[21]

By September, Ukraine had invited the United Nations and the International Committee of the Red Cross to facilitate the evacuation of civilians from Sudzha and other population centres in Kursk.[22] However, when Russia reclaimed Suzha in March 2025, 2,000–3,000 civilians were reported as remaining.[23]

In the months following the Ukrainian attack into Kursk, the Russian armed forces amassed 78,000 troops augmented by 11,000 troops from North Korea and gradually rolled back Ukrainian forces' control of the region. The Ukrainians had also reinforced their initial invasion force with as many as 20,000 additional troops. Both sides suffered intense casualties during this fighting. A major contributor to Ukraine's inability to hold this ground was the lack of reliable communications because the Starlink satellite internet constellation did not work in Russia. The other major vulnerability was the reliance on one resupply artery on the highway from Sumy into Sudzha.[24]

While military operations in Sudzha are unique because Ukrainian armed forces initiated the offensive, the majority of the combat in the city reverted to the common scenario of Russian attacks which ultimately overwhelmed Ukrainian defences six months after the initial Ukrainian offensive. Early in March 2025 Russian forces had advanced through Ukrainian defences east and south of Sudzha leaving Ukraine's sole access to the city from the main highway leading the 9 kilometres west back to Sumy, Ukraine. The battle for Sudzha shares many features with other urban battles in the war, such as significant use of firepower and the destruction of normal civilian life.[25] One unique highlight of the Battle of Sudzha was how Russian forces achieved surprise in their assault into Sudzha by marching troops 10 miles through an abandoned gas pipeline.[26]

[21] Asami Terajima, 'The rise and fall of Ukraine's Kursk gambit', *Kyiv Independent*, 1 April 2025. https://kyivindependent.com/the-rise-and-fall-of-ukraines-kursk-gambit/. See also Olga Tokariuk, 'Ukraine's gamble in Kursk restores belief it can beat Russia', *Chatham House*, 19 August 2024. https://www.chathamhouse.org/2024/08/ukraines-gamble-kursk-restores-belief-it-can-beat-russia-it-requires-western-response; and Donald Hill, 'Don's weekly: Part 1 (Analysis of the Kursk operation)', Substack, 17 March 2025. https://substack.com/home/post/p-159237660?utm_campaign=post&utm_medium=web.

[22] 'Ukraine invites UN and ICRC to Russia's Kursk region,' *Reuters*, 16 September 2024. https://www.reuters.com/world/europe/ukraine-asks-un-icrc-join-humanitarian-effort-russias-kursk-region-2024-09-16/; see also 'Russia-Ukraine international armed conflict: Helping civilians cross frontlines and borders', International Committee of the Red Cross, 31 March 2025. https://www.icrc.org/en/article/russia-ukraine-international-armed-conflict-helping-civilians-cross-frontlines-and-borders.

[23] Anatoly Kurmanaev, 'Russia begins assault on Ukrainian-occupied town in Kursk region', *The New York Times*, 11 March 2025. https://www.nytimes.com/2025/03/11/world/europe/ukraine-sudzha-russian-forces.html.

[24] Angelica Evans, 'Ukraine's Kursk incursion: Six month assessment', Institute for the Study of War, 6 February 2025. https://understandingwar.org/backgrounder/ukraine%E2%80%99s-kursk-incursion-six-month-assessment.

[25] Ibid.

[26] Kurmanaev, 'Russia begins assault'.

Unlike previous defences in Mariupol and Bakhmut, the Ukrainians quickly recognized their vulnerability in Sudzha and ordered a withdrawal. The Russian assault on retreating Ukrainian forces was characterized by intense firepower and first-person view drones controlled by fibre-optic cables to avoid Ukrainian electronic countermeasures.[27] Although Ukrainian forces did not mount as determined a defence for Sudzha as they had in Bakhmut and Mariupol, the destruction of the city was just as complete.[28]

Emerging Insights from the Ukraine Experience

Russia's war in Ukraine has not invalidated previous wisdom on urban warfare; rather, it has reinforced and expanded upon it while highlighting new developments that warrant further attention. Building on this understanding, we draw on the insights presented in the chapter 'Urban Warfare: Challenges of Military Operations on Tomorrow's Battlefield',[29] re-evaluating the pre-2022 conclusions in light of Russia's war in Ukraine. We evaluate which findings are still relevant, identify emerging trends, and examine their broader implications.[30] Beyond Ukraine's particular situation, this analysis aims to extract lessons that improve our understanding of urban warfare in both current and future conflicts (See Table 3.1).

The full-scale invasion of Ukraine in 2022 has underscored the critical importance of *preparing and planning for your own wars, not others'*. The Ukraine experience has highlighted the importance of concentrating on the particular forms of urban warfare likely in one's own operational environment, especially as the threat from Russia has increased. Moreover, it is essential to *think beyond the traditional framing of asymmetrical warfare*, as the Russian invasion has demonstrated the urgent need to *prepare for contingencies against high-tech, peer, or near-peer adversaries.* When adversaries with symmetrical capabilities engage in urban combat, there is an increased need for finding asymmetrical approaches that rely on deception, exploiting the terrain, and targeting logistics.

[27] Jonathan Beale and Anastasiia Levchenko, 'Everything is finished: Ukrainian troops relive retreat from Kursk', *BBC News*, 17 March 2025. https://www.bbc.com/news/articles/c0q198zyppqo; see also 'Pushback in Kursk', Meduza, 11 March 2025. https://meduza.io/en/feature/2025/03/11/pushback-in-kursk.

[28] Daniel Bellamy, 'Ukraine loses Sudzha, the biggest town it had occupied in Russia's Kursk region', euronews, 16 March 2025. https://www.euronews.com/my-europe/2025/03/16/ukraine-loses-sudzha-the-biggest-town-it-had-occupied-in-russias-kursk-region; see also 'Russia hopes to continue its rapid territory gains in Kursk region', *Radio Free Europe / Radio Liberty*, 13 March 2025. https://www.rferl.org/a/russia-kursk-rapid-gains-ukraine/33346834.html.

[29] Mikael Weissmann, 'Urban warfare: Challenges of military operations on tomorrow's battlefield', in Mikael Weissmann and Niklas Nilsson (eds) *Advanced land warfare* (New York: Oxford University Press, 2023).

[30] Not all takeaways are directly applicable to the war in Ukraine, as some focus on scenarios other than full-scale conventional wars or engagements with irregular forces. However, these insights have not been disproved either.

Table 3.1 Challenges of military operations on tomorrow's battlefield

Key Takeaways in Light of Russia's War in Ukraine	
Multi-domain operations (MDO) are crucial for success	**Remain valid**—Warfare has occurred across the five domains as well as in the information environment, and the ability for MDO has been crucial for success.
Breakthroughs in technology are crucial for the future of urban warfare	**Remain valid**—Novel, emerging, and breakthrough technologies have been crucial on the urban battlefield; however, as expected, they have not been a panacea making the fighting in Ukrainian cities any less dirty than those of the past.
The demand for intelligence is paramount	**Remain valid**—Good intelligence has been essential for success in the urban battles in Ukraine.
Urbanization turns the future urban battlefield into a nightmare.	**Remain valid and enhanced**—Avoiding urban battles has not been possible and fog and friction have turned the urban battlefields into nightmarish hells.
The importance of the information environment cannot be underestimated	**Remain valid and enhanced**—The information environment and 'battle of narratives' have been crucial for Ukraine to sustain itself against a superior adversary.
Think beyond the asymmetrical warfare	**Remain valid and enhanced**—If anything, the Russian invasion has shown that there is a need to prepare for contingencies against high-tech, peer, or near-peer adversaries.
Plan for your own, not others' urban wars	**Remain valid and enhanced**—The Ukraine experience has reinforced the need to focus on preparing for the specific types of urban warfare expected in its own operating environment, especially as the threat from Russia has become more immediate.
All urban warfare will have a civilian dimension	**Remain valid and enhanced**—The presence of civilians has had a significant impact as both a constraining and enabling factor. Ukraine has also underscored the differing approaches to IHL and the LOAC between Western and Russian forces in relation to civilians and civilian infrastructure.

The Ukraine example also confirms that avoiding urban battles is not possible and that fog and friction turn the urban battlefield into a nightmare.

The critical need for multi-domain operations (MDO) remains valid. In Ukraine, warfare has occurred across the five domains and in the information environment, and *the ability to conduct MDO has been crucial for success*. The importance of the information environment is evident at first glance, as *it and the 'battle of narratives'* have been crucial for Ukraine to sustain itself against a superior adversary.

The technological and intelligence dimensions have both been essential features in urban warfare in Ukraine. *Novel, emerging, and breakthrough technologies have been crucial on the urban battlefield* but, as expected, they have not been a panacea making the fights in Ukrainian cities any less dirty than those of the past. *Good intelligence has been an essential component for success* in the urban battles.

The civilian dimension has emerged as one of the most prominent aspects of urban battlefields in Ukraine, with the presence of civilians significantly impacting the conflict as both a constraining and enabling factor. The war has also highlighted the *contrasting approaches to IHL* between Western and Russian forces, particularly concerning civilians and civilian infrastructure.

Reinforcing and Refining Previous Wisdom

It is evident that we do not need to discard what we know about urban warfare because of the war in Ukraine, but we must adapt and refine our understanding of the subject. Certain areas require special attention and will be discussed below. We will start by analysing how fog and friction generate chaotic battlefield conditions, the multi-domain aspects of urban warfare, the vital importance of information and intelligence, and the need for technological integration and adaptation. The civilian dimension, which is especially important, will be allocated its own section due to its significance and the realities of planning for possible conflicts with Russia.

Clearly, the Ukraine experience transcends the pre-2022 understanding, revealing areas that need further development, insight, and integration into future strategies—especially as these aspects are likely to be vital in future conflicts against peer or near-peer opponents. These include preparing for prolonged urban battles, which requires intense firepower and robust logistics; the necessity for being ready to deploy irregular tactics; and the importance of developing dual competencies—mastering both high-tech, connected operations and independent small-unit tactics, with the ability to shift seamlessly between them. Additionally, it is essential to recognize the reality of non-traditional warfare tactics that may challenge established rules of engagement.

Fog and Friction and the Nightmare of Urban Combat

> '[T]he horrific reality of urban combat' is a key observation that is driving how the United States Army is thinking about lessons from the Ukraine War.[31]

Any force fighting in a city must recognize that fog and friction, which Clausewitz believed permeated all combat, have an even greater influence in urban battles. Cities are complex, with many intricate and important details that are not readily apparent and amplify the uncertainty that hampers all military operations. As the size and

[31] General James E. Rainey, Commanding General of Army Futures Command, *Breaking Defense*, Interview, 10 October 2023. https://breakingdefense.com/2023/10/looking-through-the-front-windshield-afc-gen-rainey-talks-ukraine-force-changes/.

complexity of urban areas grow, military commanders and their staffs are increasingly challenged to understand the details and connections.

Moreover, cities tend to split military units into small, disconnected subunits, requiring them to operate with a high level of decentralization. Urban settings also offer concealment advantages, making thorough reconnaissance and surveillance vital for locating targets. These features together place a strain on human capacity to monitor and understand the many simultaneous actions on the battlefield. When commanders lack awareness of what is happening across the entire area of operations, it becomes extremely difficult to direct and control their units effectively.

This is a persistent reality of urban combat that ongoing investments in C2 systems, even when enhanced by artificial intelligence (AI), do not resolve. Generative AI can be highly valuable in planning and preparing for urban combat by speeding up military staff's understanding of the terrain. However, because urban combat continually alters the terrain, current generative AI systems will likely have decreasing utility in providing real-time assessments that influence decision making across the necessarily decentralized operations.

The features of a city that cause confusion also increase the friction that hampers military operations. Any military action results in changes to urban terrain, leading to greater friction during urban combat compared to fighting in open terrain. A building that acts as a defensive position or offers concealment one day may become rubble the next—by either chance or design—removing ambush sites, changing routes and passability, and creating new fields of vision. Furthermore, power and water supplies may be disrupted and then restored at unpredictable intervals.

Operations conducted within buildings often reveal unexpected internal structures and configurations that can confuse and complicate efforts. Essentially, the complexity of urban environments makes everything more difficult. The forced dispersion of units further increases the number of locations where forces must deal with the multiple challenges posed by urban terrain, thereby increasing friction. Although enhanced intelligence and surveillance systems may help to reduce this inherent confusion, they cannot fully alleviate the urban conditions that intensify the battlefield friction.

Multi-Domain Aspects

It is evident from the Ukraine experience that urban operations are, by necessity, a multi-domain undertaking, especially when facing peer-to-peer, high-tech, and highly qualified opponents. Today's urban battlefield underlines the importance of integrating MDO across all domains and environments.

Although land remains the primary theatre, success depends also on the air, space, cyberspace, and maritime domains, as well as the information environment. For instance, satellite-based intelligence improves situational awareness, cyber and electronic warfare disturb enemy communications and targeting, and aerial reconnaissance alongside precision air support offer vital advantages in a dense urban terrain.

Conventional ground operations are greatly augmented by intelligence from space and cyberspace, coupled with support from cyber and electronic warfare capabilities. Aerial reconnaissance and close air support further act as force multipliers.

In short, multi-domain coordination is vital for addressing the challenges of urban warfare. This coordination goes beyond purely military efforts. Nothing is more crucial than synchronizing military and non-military action in urban combat, where political, economic, and social factors directly influence operational outcomes. Successful urban operations ultimately depend on the ability to rapidly and effectively synchronize efforts across multiple domains.

Narratives and Strategic Communication

Narratives and strategic communication are crucial for maintaining public support and countering adversary narratives in urban warfare, where tactical and strategic factors are closely linked. In Ukraine, both Russian and Ukrainian forces have used media and information operations to shape public perception domestically and internationally of the conflict. During urban battles like the siege of Mariupol, both sides employed coordinated information campaigns and cyber operations to manage or steer narratives and influence audiences. Additionally, Western nations have played a vital role in shaping the discourse by actively spreading narratives in support of Ukraine and countering Russian propaganda. Notably, the West has also begun integrating intelligence analysis into strategic communications in an unprecedented way, using intelligence insights to strengthen public messages and policy decisions.

A key part of this effort involves the strategic use of intelligence for public release, balancing transparency with operational security to shape perceptions and prevent disinformation. Convincing narratives must be developed and disseminated proactively before and during urban operations, rather than only responding to adversary narratives. To achieve this, well-developed and tested frameworks for strategic communication and debunking are necessary to ensure the timely, accurate, and coordinated dissemination of information. By managing information proactively, military and political leaders can reinforce legitimacy, enhance resilience, and better disrupt, block, and debunk enemy propaganda.

Technical Integration and Adaptation

In March 2024, General James E. Rainey, commanding general of Army Futures Command, made a keynote presentation at the US Army Global Force Symposium in Huntsville, Alabama, saying that 'The amount of technical disruption in the character of war is unprecedented, and it just continues to go faster and faster ... Whatever you think you know this year, come back in 90 days, and you'll know something different'.[32] 'To give Soldiers what they need to win the nation's wars now and in the future,

[32] General James E. Rainey, commanding general of Army Futures Command, US Army Global Force Symposium, Huntsville, AL, 2 April 2024. https://www.army.mil/article/275040/army_futures_command_general_lays_out_continuous_transformation_plan.

the Army must continuously transform and adapt to advances in technology'.[33] This aligns with the experience of urban warfare in Ukraine, where technological integration and adaptation have been key factors in urban battles, influencing both tactical engagements and broader operational effectiveness.

The deployment of drones, electronic counter-measures, and advanced surveillance technologies has proven pivotal in Ukraine, enhancing both reconnaissance and targeting capabilities. The ongoing interaction between offensive and defensive technological progress—such as drone warfare, air defence systems, and innovative electronic counter-measures—highlights the need for continuous adaptation. Success in urban combat increasingly depends on the swift integration of emerging technologies, particularly the use of unmanned systems in the air, on the ground, and below the surface, while proactively anticipating and neutralizing enemy counter-measures. A notable example is the deployment of fibre-optic drones in Ukraine, which has sparked an ongoing contest between Russian and Ukrainian forces to detect and neutralize them.[34]

While drones have become a vital part of modern urban warfare, fundamentally changing battlefield capabilities and strategic decisions, their use in Ukraine has shown their versatility in reconnaissance, target identification, and precision strikes, improving situational awareness and operational efficiency. Ukrainian forces, for example, have utilized drones for live intelligence gathering, directing artillery fire, and carrying out kamikaze attacks against high-value targets.

Meanwhile, Russian forces have adapted by employing electronic warfare tools to disrupt drone operations, highlighting the ongoing technological and adaptability contest in urban combat. However, while drones offer significant advantages, they are not a universal solution. Their effectiveness is limited by factors such as signal jamming, limited endurance, and vulnerability.

The continuous development of drone warfare and electronic counter-measures emphasizes the need for sustained investment in technology and training units to effectively adapt and incorporate new capabilities. Armed forces must focus on research and development to maintain their operational edge.

Future urban combat will demand specialized autonomous systems capable of navigating complex environments, performing logistical support missions, and enhancing force protection. Innovations like AI-driven swarm drones, autonomous resupply vehicles, and resilient communication networks will be vital for sustaining urban operations. By investing in technologies designed to meet the unique challenges of urban warfare, military forces can ensure they remain agile and resilient in an increasingly contested battlespace.

[33] Ibid.

[34] Vikram Mittal, 'Ukraine and Russia battle to defeat "un-jammable" fiber-optic drones', Forbes, 19 January 2025. https://www.forbes.com/sites/vikrammittal/2025/01/19/ukraine-and-russia-battle-to-defeat-un-jammable-fiber-optic-drones/; David Hambling, 'Kalashnikov unveils "jam-proof" attack drone', Forbes, 9 May 2024. https://www.forbes.com/sites/davidhambling/2024/01/03/kalashnikov-unveils-jam-proof-attack-drone/.

Nevertheless, one should not believe that technology will eliminate either the need for or the conditions of urban battles. As General Rainey correctly observes,

> We are absolutely going to need to do things like fight in cities—close with and destroy enemies in the most horrific conditions; that's been a constant throughout the history of warfare. That last 500 meters of fighting when nothing is working, that's not going away. There is no technological solution that is going to offset that.[35]

Intelligence Is Paramount

The demand for intelligence is crucial given the increased complexity of urban warfare compared with operations in rural areas. Cities are not only interconnected and intricate hubs of human activity but also environments where the civilian population often exceeds the number of enemy combatants. As a result, gathering comprehensive and multi-faceted intelligence is vital—not only to track enemy movements but also to understand civilian dynamics, which can both facilitate and hinder military operations.

Future urban warfare is likely to be highly data driven, depending on large streams of intelligence from multiple sources, including signals intelligence (SIGINT), imagery intelligence (IMINT), open source intelligence (OSINT), social media intelligence (SOCMINT), and human intelligence (HUMINT) gathered on the ground. In this context, the importance of the local population must be emphasized, as it is both an invaluable source of intelligence—evident in the use of mobile apps in Ukraine to obtain otherwise inaccessible information—and a potential security threat, as it can inadvertently act as a channel for disinformation or collaborate with adversaries.

As discussed earlier, the deployment of drones and advanced surveillance technologies[36] has significantly enhanced situational awareness at tactical levels, providing front-line units with immediate real time intelligence.

Enduring Challenges

Many of the long-standing realities of urban warfare still apply in Ukraine, although they have evolved over time. Despite advances in technology, intelligence capabilities,

[35] General James E. Rainey, Commander, United States Army Futures Command, 2025 AUSA Global Force Symposium Day 3: Keynote Presentation. https://innovation.army.mil/Test-Page/videoid/956796/.

[36] See for example Jack Watling and Nick Reynolds, *Meatgrinder: Russian tactics in the second year of its invasion of Ukraine, RUSI Journal*, 19 May 2023. https://www.rusi.org/explore-our-research/publications/special-resources/meatgrinder-russian-tactics-second-year-its-invasion-ukraine; Federico Borsari and Gordon B. 'Skip' Davis Jr, *An urgent matter of drones* (Washington, DC: Center for European Policy Analysis (CEPA), 2023); Stacie Pettyjohn, 'Evolution not revolution: Drone warfare in Russia's 2022 invasion of Ukraine', Center for a New American Security (CNAS), February 2024.

and the development of urban environments, there has been no fundamental change in the nature of urban warfare. While these persistent challenges remain, the progress in technology, intelligence gathering, and urban infrastructure has created new opportunities and challenges that need to be addressed.

Units move **more slowly in urban terrain than they do in open terrain**. An enduring reality of urban battlefields is that the numerous buildings of multiple levels create myriad fortified fighting positions which offer **advantages to defenders**. These defenders often have increased incentives to retain their positions because of these advantages and the political importance of cities. These advantages are heightened by the limited opportunities for offensive manoeuvre along a high-speed corridor of approach.

The **buildings and narrow streets create blind spots and restrict** the ability to see in all directions, forcing units to deliberately clear each segment of their route. Some streets enable speedy movement, but they are usually easily observed, and units moving along them are susceptible to ambushes. This forces units to break through **walls or move subsurface to travel** without being observed. The complex terrain also **forces units to fragment and conduct decentralized operations, placing increased demands on C2 systems**. The myriad **natural fortified fighting positions** in a city incentivize advancing units to rely on substantial firepower to demolish known and suspected defensive positions, but the ensuing rubble further blocks and slows the advance. Units that try to manoeuvre without supporting fires will find themselves repeatedly, if not continually, exposed to well-placed ambushes and snipers.

Additionally, **the three-dimensional quality** of the terrain forces units to clear up stairwells of multi-story buildings and into the substructure of tunnels and sewers beneath the city. The requirement to **leave forces behind to hold cleared terrain multiplies manning requirements**, making urban operations more manpower intensive and slower than normal operations.

These factors slow units in cities, causing longer battles than in other terrains and increasing casualties. It also raises demand for supplies like ammunition, medical items, and spare parts. Challenges extend beyond merely military and technical aspects; urban warfare always includes a civilian component, and IHL is always a relevant factor.

The Civilian Dimension and the Importance of IHL

> We're a values-based military, and we abide by the law of armed conflict, and that's not going to change.[37]

Fighting among civilian populations frustrates militaries from liberal democracies that are committed to fighting in a way consistent with IHL, especially when their adversaries do not. The inherent challenge of fighting in consistence with

[37] Rainey, Interview, 10 October 2023.

IHL is intensified when combatants use civilians and protected sites as cover and concealment. For militaries fighting for liberal societies, having widespread non-combatants on the battlefield complicates military planning. Liberal societies expect professional militaries to prevent and mitigate civilian harm whenever possible.

These are noble and proper goals, but they limit military options and create opportunities for those who do not share such liberal values. Furthermore, a communications-savvy adversary can exploit the inevitable civilian harm that occurs in an urban battle to undermine the credibility of militaries fighting in accordance with liberal ideals. Such attacks on credibility can weaken domestic political will and international support for an operation, granting strategic advantages regardless of tactical success. When civilians remain on an urban battlefield, there are no foolproof methods to prevent civilian casualties. In the best cases, the civilian population evacuates, resulting in most casualties being among the armed forces. Even when people leave to avoid immediate injury or death, there is a significant human cost because residents depend on the city's infrastructure for vital services.

The requirement to provide those services for those who remain in active combat areas temporarily shifts from civilian to military authorities. Even with constrained and precise firepower, urban combat invariably destroys infrastructure, which is often needed to support suburbs and rural areas beyond the city. Eventually, many residents return; then, planning for the restoration of services and often temporary shelter becomes necessary. This responsibility is frequently shared between military and civilian authorities. It is impossible to eliminate the civil cost of an urban battle. This reality increases the complexity of conducting urban battles.

IHL requires military forces to operate in a manner such that anticipated harm to civilians is proportional with the value of the intended military objective. The inability to apply a universal formula or heuristic to this standard provides another layer of complexity when fighting in a city. Although this principle of proportionality does not establish a legal requirement that military forces prevent civilian casualties, NATO has made a commitment to take 'all feasible measures ... to avoid, minimize and mitigate harm to civilians'. This commitment advances its credibility and legitimacy.[38]

> This commitment also includes a pledge by NATO to facilitate access to basic needs for the civilian population and, if possible, to provide support to civilians in case of emergencies. NATO commanders on the ground are also required to limit the use of artillery, heavy weaponry, and airstrikes to avoid civilian casualties in densely populated urban terrains.[39]

[38] NATO Policy for the Protection of Civilians, 9 July 2016 (updated 24 June 2021). https://www.nato.int/cps/en/natohq/official_texts_133945.htm.

[39] Philippe Michel-Kleisbauer, 'Urban warfare', NATO Parliamentary Assembly, Science and Technology Committee, Sub-Committee on Technological Trends and Security, 20 November 2020. https://www.nato-pa.int/download-file?filename=%2Fsites%2Fdefault%2Ffiles%2F2020-10%2F040%20STCTTTS%2020%20E%20REV.%201%20-%20REVISED%20-%20URBAN%20WARFARE.pdf. See also NATO STANDARD, Allied Joint Doctrine, Allied Joint Publication-01, edition F version 1, December 2022. https://www.cimic-coe.org/resources/external-publications/ajp-01-edf-v1-f.pdf: 'NATO's approach to PoC is based on legal, moral and political imperatives and, consistent with applicable legal frameworks,

However, NATO's adversaries are unlikely to be bound by these same principles. The civilian casualties in urban battles in Ukraine have been significant, mainly due to Russian reliance on extensive firepower, as seen in Mariupol, Bakhmut, and Avdiivka. The tragedy of these urban conflicts underscores the importance of this renewed focus on reducing and responding to civilian harm during military operations. As Andrew Bell explains, commanders face a trilemma of using their units to accomplish the mission, ensuring sufficient force protection, and fulfilling the obligations of civilian safeguarding.[40] This protection of civilians extends to ensuring their access to humanitarian aid.[41] Implementing effective civilian protection heightens the need for commanders to coordinate with civil authorities and non-governmental organizations, further complicating the challenges of urban fighting.

Technological advances in surveillance and precision munitions foster unrealistic expectations of discreet combat operations that will minimize suffering. Technology-reliant militaries must recognize that improvements in precision can reduce but never eliminate collateral damage and harm to non-combatants. Guided munitions with large blast areas remain very destructive despite continuing improvements in weapons and munitions accuracy.

Furthermore, precision cannot eliminate errors in intelligence, targeting, programming, or munitions manufacture. High-explosive weapons are designed to cause damage and harm; once fired, their effects cannot be precisely determined before detonation. Technology will not resolve the problem of non-combatant harm in urban combat. Combat breaks things and kills people; these consequences are always more severe when the combat occurs in a city.

Discussion: Preparing for Urban Combat

The war in Ukraine has shown that urban warfare is evolving, highlighting key areas that need further development, refinement, and integration into future military strategies. These lessons are especially important in preparing operations against peer or near-peer adversaries, where urban combat will likely be central. Military forces must get ready for prolonged, high-intensity urban combat by balancing firepower, logistics, and information warfare while maintaining operational effectiveness and compliance with international norms.

Addressing these challenges requires not only doctrinal changes but also a fundamental shift in how urban operations are conceptualized, trained for, and

NATO recognizes that all feasible measures must be taken to avoid, minimize and mitigate harm to civilians.'

[40] Andrew Bell, 'Civilians, urban warfare, and US doctrine', *Parameters* 50/4 (Winter 2020): 33–45. https://press.armywarcollege.edu/cgi/viewcontent.cgi?article=2686&context=parameters.

[41] Harriet Norcross Eppel, 'Ukraine, urban warfare, and obstacles to humanitarian access: A predicament of public international law', *BYU Law Review* 49/3 (Winter 2024): 925–960. https://digitalcommons.law.byu.edu/lawreview/vol49/iss3/11/.

executed. The complexity of future urban battlefields necessitates an adaptable, multi-dimensional approach—one that integrates emerging technologies, innovative tactics, and a comprehensive understanding of the operational environment not only for achieving military success but also for ethical and humanitarian reasons, given its profound impact on civilians, economic stability, and the environment. We here propose a tentative framework for addressing the challenges of urban warfare (Table 3.2).

The framework is built around four enduring realities of cities that will impact urban combat operations: **1) the compartmented and three-dimensional terrain**; **2) prevalence of fortifications**; **3) presence of civilians**; and **4) infrastructure critical for military operations and civil life**. Each of these characteristics interacts with one another in a complex adaptive system, which creates unique challenges that military forces do not face when fighting in more open terrain or remote locations. Consequently, military forces that must operate in cities need to intensify their planning and preparation to wage a successful urban operation.

NATO doctrine establishes a Joint Functions Framework to 'assist commanders to integrate political, military and civilian actions through the operational domains' (Figure 3.1). This framework is intended to help military forces describe their capabilities, which should be modified based on the particular operational circumstances.[42] 'Using joint functions, commanders can determine force requirements. Common to joint operations at all levels are the functions: manoeuvre, fires, C2, intelligence, information, sustainment, force protection, and civil-military cooperation.'[43] This framework presents manoeuvre, fires, information, and civil-military cooperation (CIMIC) as the mechanisms to influence the abilities, attitudes, and behaviours of multiple audiences. These four functions are informed and directed by intelligence and C2 and supported by sustainment and force protection.

This framework offers a practical, if imperfect, way to consider the implications of applying military power in urban settings. Using this model also emphasizes a key finding of this study: urban combat is not an entirely different or new form of warfare; rather, the inherent complexity of cities combined with their strategic importance amplifies combat and often the significance of the outcomes.

The interaction of the four unique attributes of urban operations described above creates distinct opportunities and challenges across the joint functions. Therefore, military planners cannot approach urban operations in functional stovepipes;

[42] AJP-01 Allied Joint Doctrine, Edition F Version 1, December 2022. U.S joint doctrine and U.S. Army doctrine use a different model of joint functions or Army warfighting functions. The United Kingdom also uses a slightly different model. However, the functions in the NATO framework are generally recognized; the distinctions in the models do not undermine the generalizability of ideas presented here. (U.S joint doctrine (Joint Publication 3-0, *Joint Operations and Campaigns*, 18 June 2022; see Chapter III, 'Joint Functions'); U.S. Army doctrine (Field Manual 3-0, *Operations,* October 2022 (see Chapter 2, 'Applying and Generating Combat Power,'); AJP-01 Allied Joint Doctrine, Edition F Version 1 with UK national elements December 2022).

[43] Allied Joint Publication-3, Allied Joint Doctrine for the Conduct of Operations, edition C version 1, February 2019. https://www.coemed.org/files/stanags/01_AJP/AJP-3_EDC_V1_E_2490.pdf.

Table 3.2 Framework for addressing the challenges of urban warfare

Challenges (Urban Reality)	Impact	Effect	Priority Needs and Solutions
Compartmented and three-dimensional terrain	• Fragmented units and decentralized fighting • Increased fog and friction	• Complex C2; increased fratricide risk • Increased demand for manoeuvre forces • Slow tempo and prolonged battles • Increased demand for intelligence and fires	• Small units with dual competencies (combined arms at lower echelons) • Robust sustainment
Multiple fortifications	• Defence advantage intensifies	• Increased force protection • Slow tempo and prolonged battles • Increased demand for intelligence and fires	• Small units with dual competencies (combined arms at lower echelons) • Surge/reinforcing fires • Robust sustainment
Civilians	• Increased CIMIC demands • Increased IHL scrutiny • Increased importance of strategic narrative • Increased importance of HUMINT	• Restrictive rules for the use of force • Increased demand for manoeuvre forces • Slow tempo • Increased force protection risks	• Investment in developing reliable HUMINT sources • Robust CIMIC cells • Principles of military necessity, distinction, and proportionality understood and incorporated across the force
Civilian infrastructure	• Distracts from units desire to focus on achieving effects on the enemy	• Restrictive rules for the use of force • Increased demand for manoeuvre forces • Increased intelligence demands to assess critical infrastructure nodes • Limits on use of fires	• Robust CIMIC cells • Principles of military necessity, distinction, and proportionality understood and incorporated across the force

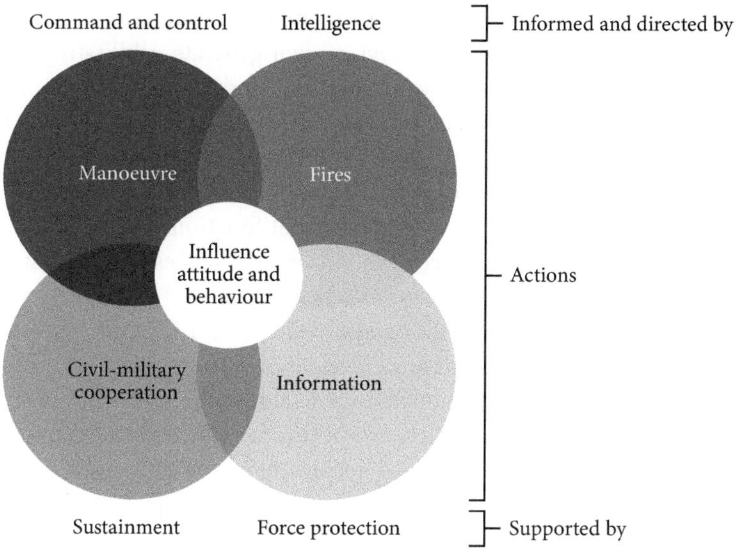

Fig. 3.1 The Joint Function Framework

Source: Allied Joint Function Framework, Allied Joint Publication-01, Allied Joint Doctrine, edition F version 1, December 2022, 106, figure 4.4.

instead, they must employ systems analysis to assess the interactions, which will differ by city and evolve over time within a given city or urban operation. Ultimately, planners must identify and address the trade-offs that best support the achievement of the specific political aims of the operation.

Compartmented and three-dimensional terrain and the accompanying increased fog and friction have their most direct effect on how units manoeuvre by forcing the fragmentation of units and accordingly a decentralized approach to C2. This fragmentation and decentralization of C2 then has cascading effects through all the other functions, which are challenged to marshal their effects and capabilities.

The compartmented and three-dimensional terrain prevents units from operating in mass formations that often create advantages in open terrain. Because units must operate in small elements, there is a need to create combined arms capabilities at lower echelons than would otherwise be ideal. This has an impact on unit training because it forces small units to have dual competencies in integrating high-level support and conducting independent small-unit tactics.

This fragmentation of units also prevents the direct visual coordination that often facilitates mass manoeuvre in open terrain, thereby increasing the risk of fratricide as units and leaders are often unsure of what is around the corner, in an adjacent building, beneath, or even in the next room. The compartmented nature of the terrain also tends to absorb more manoeuvre forces as commanders try to prevent the ceding of cleared terrain to the enemy.

A third significant impact of compartmented and three-dimensional terrain is the reduction of tempo. High-tempo manoeuvres often generate debilitating effects in open terrain by fragmenting and bewildering opposing forces. However, in cities, speed is often unattainable and rarely can be achieved on a scale sufficient to make a meaningful difference. The compartmented and three-dimensional terrain requires large numbers of fragmented manoeuvre units forced to operate with a slow tempo; this induces additional burdens across multiple joint functions.

The sustainment function faces particularly severe challenges in supporting urban operations. The distribution system cannot rely as effectively on a few well-placed nodes and must disperse in unfamiliar ways, which strain C2 and transportation assets. The slow tempo and increased force levels compound the issue by demanding more resources, especially ammunition and medical supplies.

The fires function must contend with a non-linear battlefield and targeting in three dimensions, as manoeuvre units strive to achieve effects with fires on the tops of buildings with varying heights and below the surface at multiple depths. The intelligence function also faces challenges due to the dispersed nature of operations and the conditions that differ across a city. The complex terrain undermines effective C2, which in turn amplifies the need for intelligence while making it harder to gather. Macro-level intelligence about the city that informs higher echelon commanders is unlikely to provide the detailed information needed by small, fragmented units, each addressing a unique tactical challenge.

The **prevalence of fortifications** is closely connected to the problem of compartmented and three-dimensional terrain. Most buildings that are a predominant feature of cities are built to endure for years, if not decades, and are designed to protect their inhabitants from multiple adverse conditions. Thus, forces fighting in cities have numerous fortified locations to fight from, offering increased access to static force protection measures.

This aspect is amplified in the modern era as cities have increased their reliance on subsurface infrastructure, and the technology required for underground construction has diffused widely. This is a primary reason why cities amplify the inherent advantages of defence. The widespread use of three-dimensional fortifications enhances the effects of a slow tempo and the need for additional forces to hold cleared areas.

These fortifications also create increased demands for fires, which in turn amplifies the challenges of sustainment. The need for fragmented units to operate across multiple fortifications intensifies the demands on intelligence to identify the most suitable fortifications and those actively used by enemy forces. The widespread presence of suitable fortifications makes locating the enemy akin to finding a needle in a haystack.

As described earlier in the chapter, the **presence of civilians** is the predominant feature of urban fighting that distinguishes it from warfare in open terrain. Most notably, the function of CIMIC is brought to the fore when military units operate in cities. Military units are typically poorly trained and ill suited to provide the functions of civil government, which are essential for sustaining and restoring normal civilian life, and require intense coordination with local governments and multiple international humanitarian support organizations.

The presence of civilians, the need to conduct operations in accordance with IHL, and the requirement to coordinate with civil support organizations increase planning challenges. These factors often lead to restrictive rules on the use of force, which can slow decision making in manoeuvre and fires functions. Protecting civilians and escorting them away from combat zones further intensifies the demand for additional manoeuvre forces and results in a slower tempo, creating cascading effects in C2, intelligence, and sustainment. There are also heightened force protection concerns, as civilians frequently shift between being combatants and non-combatants.

The presence of civilians in urban battlefields also elevates the importance of human intelligence over signals and imagery intelligence. Civilians increase the demands on the information function, as strategic messaging must consider the diverse viewpoints across international organizations and within the civilian population. The strategic narrative must incorporate the principles of military necessity, distinction, and proportionality, which in turn must be internalized by fragmented small-unit manoeuvre forces. The presence of civilians tends to heighten international awareness and scrutiny of military conduct more than would otherwise occur.

Closely related to the presence of civilians, **infrastructure critical for military operations and civil life** is another factor that intensifies planning requirements and influences the conduct of urban operations. Modern militaries have enormous demands for energy and transportation systems, which are typically found in and near cities to support civil life.

Using these resources to support operations or targeting them to disrupt the adversary will have an effect on the civilian population and increase CIMIC and information activities, as mentioned earlier. Cities also contain critical services such as medical, water, sewage, waste disposal, electricity, and communications, which are vital for civilian life. The military and civilian importance of infrastructure demands as much focus from units as the goal of exerting effect on the enemy within the city. Civil authorities must be integrated into military planning to ensure these essential systems remain functional. Any damage to these systems will raise scrutiny and complicate strategic messaging efforts. Identifying critical infrastructure nodes imposes new requirements on the intelligence function. The need to protect or preserve these nodes may increase the demand for manoeuvre forces, thereby heightening sustainment challenges, and it can limit the use of fires.

Conclusion

The Ukraine experience clearly shows that urban warfare continues to be a core aspect of modern conflict, with cities serving as strategic, political, and operational centres. Avoiding urban combat is often impractical, and the difficulties presented by dense, compartmentalized terrain require a change in doctrinal thinking. The battles in Kyiv, Mariupol, Bakhmut, and Avdiivka highlight the defender's advantage in urban settings, where well-prepared defensive positions and fortifications considerably slow offensive efforts. Yet, this war also highlights the myriad factors that frustrate efforts to turn tactical advantage into strategic advantage. These engagements also emphasize the logistical challenges and high attrition rates involved in extended urban battles, stressing the need for adaptability, resilience, and multi-domain coordination.

The Ukraine conflict highlights the changing role of technology in urban warfare. The widespread use of drones, electronic warfare, and sophisticated surveillance has changed tactical battles, but these tools are still supportive rather than game changing. Intelligence remains crucial, especially in managing the complex civilian aspects of urban warfare, where the balance between military needs and humanitarian concerns influences operational outcomes. The differing approaches to IHL between Western forces and adversaries such as Russia lead to additional strategic and political challenges. Future urban operations must integrate technological advancements while acknowledging the persistent human and physical complexities of cities.

4
Mission Command and the Russo-Ukraine War

Lessons for Future War

Chase Metcalf and John Nagl

Introduction

Russia's unprovoked expansion of its war on Ukraine in 2022 triggered the largest land war in Europe since World War Two.[1] Total casualties since 2022 exceed 1.4 million, and 650,000 or more troops are facing off along the front lines in Ukraine.[2] The scale of this war and its military operations differ markedly from those that most Western militaries have engaged in over the last two decades or more. This is especially true in terms of command and control (C2).

C2—defined as the exercise of authority and direction by a designated commander over assigned forces—is fundamental to the art and science of warfare.[3] Military operations are inherently a struggle between opposing forces and thus require continual adaptation. Mission command is an approach to C2 that empowers subordinate decision making and enables decentralized execution, allowing for the creation and exploitation of opportunities on the battlefield. Analysts and military professionals often view mission command as providing a competitive advantage for Western militaries and a key to defeating larger, less adaptive adversaries.

The Russo-Ukraine War provides a compelling case study for examining the principles of mission command and lessons for future warfare. The Russo-Ukraine War demonstrates that mission command—defined by disciplined initiative based on

[1] This chapter draws heavily on the research of the students and faculty who participated in the US Army War College Russia-Ukraine Integrated Research Project from academic year 2022–5, particularly the work of Cols Joe Junguzza, Jamon K. Junius, and Kelly Lelito who focused on the command and control (C2) warfighting function. The authors are grateful for their dedication and service. Dr Nagl is also grateful to the British army's Centre for Army Leadership, which invited him to give a lecture on Mission Command informed by lessons from the Ukraine war at Chatham House in November 2024.

[2] Eric Schmitt and Helene Cooper, 'Troop casualties in Ukraine war near 1.4 million, study finds', *The New York Times*, 3 June 2025. https://www.nytimes.com/2025/06/03/us/politics/russia-ukraine-troop-casualties.html.

[3] US Department of the Army, *ADP 6-0: Mission command—Command and control of army forces* (Washington, DC: Headquarters of the Army, July 2019), viii.

Chase Metcalf and John Nagl, *Mission Command and the Russo-Ukraine War*. In: *Russia's War in Ukraine and Modern Warfare*. Edited by: Mikael Weissmann and Niklas Nilsson, Oxford University Press. © Chase Metcalf and John Nagl (2026). DOI: 10.1093/9780198970903.003.0004

trust and shared understanding as well as risk acceptance—not only enables success in large-scale combat operations (LSCO) but is increasingly dependent on the integration of advanced technologies such as artificial intelligence (AI), autonomous systems, and cyber capabilities. Ukraine's partial adoption of mission command principles, supported by North Atlantic Treaty Organization (NATO) training and technological innovation, has enabled tactical agility and operational resilience. In contrast, Russia's rigid, centralized command structure and cultural resistance to disciplined initiative has led to battlefield failures. This contrast offers critical lessons for future warfare, particularly for militaries seeking to balance technological advantage with empowered leadership at all levels of command.

Command as a Warfighting Function—Framework for Analysis

Command is more than merely issuing orders—it is the cognitive and organizational process that synchronizes operations, empowers initiative, and enables a force to adapt to the chaos of combat. Leaders translate intent into action through command. Within the United States (US) army's doctrinal framework, C2 is a warfighting function that enables commanders to understand, visualize, describe, direct, lead, and assess operations.[4] It is particularly vital in LSCO, where complexity, scale, and contested environments demand leaders balance synchronization and flexibility.

Historically, mission command evolved from the Prussian concept of *Auftragstaktik*, which emphasized subordinate initiative within commander's intent.[5] The US military can trace the employment of mission command back to at least the Civil War, and it became doctrine in 2003 with the army's publication of 'Field Manual 6–0'.[6] It is guided by seven interdependent principles: competence, mutual trust, shared understanding, commander's intent, mission-type orders, disciplined initiative, and risk acceptance.[7]

At its core mission command is about empowering subordinates and enabling decentralized execution appropriate to the situation based on the above principles.[8] Tactical and technical competence, built through education and training, are the foundation of mission command and a prerequisite for trust.[9] Mutual trust, vertically and horizontally, is developed over time through working together and demonstrated

[4] Ibid., 2–12 to 2–16.

[5] Ibid., vii.

[6] Col. (ret.) Clinton J. Ancker III, 'The evolution of mission command in U.S. army doctrine, 1905 to present', *Military Review*, March–April 2013: 42–9.

[7] US Department of the Army, *ADP 6-0*, 1-6–1-14. NATO doctrine similarly emphasizes trust and mutual understanding as fundamental elements of mission command and unity of effort, timely and effective decision making, and decentralized execution as principles of mission command—see: NATO, AJP-01 edition F version 1, Allied Joint Doctrine, NATO Standardization Office, December 2022, 84–7.

[8] US Department of the Army, *ADP 6-0: Mission Command*, x.

[9] Ibid., 1–7.

competence.[10] Trust is essential in superiors' willingness to empower subordinates and subordinates' willingness to act with initiative. These two principles are the very foundation of mission command as a philosophy.

Shared understanding, commander's intent, and mission-type orders, then are central to enabling the disciplined initiative that is fundamental to mission command. Shared understanding of the environment and intent enable leaders at all levels to make rapid decisions and act in the absence of guidance. It both supports and derives from trust and requires clear communications.[11] Commander's intent contributes to shared understanding and more importantly provides the clear understanding of purpose that defines success.[12] Mission-type orders are based on commander's intent and tell subordinates what must be done without overly directing the how beyond what is necessary to synchronize activities.[13] These principles then enable subordinates to act with disciplined initiative to accomplish the mission without having to constantly seek permission or guidance.[14] Of course all of this requires commanders to define and accept risk. Prudent risk taking, and a willingness to underwrite that risk, to create and seize opportunities is built on the other principles and essential to decentralized execution and mission command.

These principles are more than abstract ideals—they are operational necessities in LSCO. In environments characterized by degraded communications, dispersed formations, and rapid tempo, centralized control becomes a liability. Mission command offers a framework for balancing synchronization with initiative, enabling subordinate leaders to act decisively within the commander's intent even when disconnected from higher headquarters.[15]

While mission command is often associated with decentralization, it is not synonymous with the abdication of control. As Colonel Joe Junguzza argues, effective command requires the ability to modulate levels of control based on operational context.[16] Complex operations requiring substantial levels of synchronization—such as a combined arms breach or a water crossing—demand tighter coordination and oversight, while fluid tactical engagements benefit from greater autonomy at lower echelons. This flexibility, and knowing when to tighten or loosen control, is a central aspect of the 'art' of command and optimizing effectiveness.

While mission command frequently works well at the tactical level, implementing it at the operational or strategic level presents distinct challenges. In addition

[10] Col. Jamon K. Junius, 'Mission command: An essential component for Ukraine's fight against Russia', in John Nagl and Katie Combe (eds), *A Call to Action: Lessons from Ukraine for the Future Force* (Carlisle Barracks, PA: USAWC Press, June 2024), 116; also see NATO, AJP-01, 85, on the importance of trust developed through shared experiences and a willingness to accept mistakes.

[11] US Department of the Army, *ADP 6-0: Mission Command*, 1-8–1-9.

[12] Junius, 'Mission command', 116–17.

[13] Junius, 'Mission command', 117.

[14] US Department of the Army, *ADP 6-0: Mission Command*, 1–13.

[15] Col. Joe Junguzza, 'Mission command as a key operational advantage in the Russo-Ukraine War', *Army War College Strategic Research Paper* (Carlisle, PA: Army War College, 2024), 3–4.

[16] Junguzza, 'Mission command as a key', 1–5.

to trust and initiative it requires capable staff and staff processes, interoperable digital systems, and well-trained leaders who can translate intent into synchronized actions across domains. Ukraine's experience during its 2023 counter-offensive illustrates this. Despite a culture of initiative and the technology to support mission command, battalion and brigade staffs struggled to synchronize fires, manoeuvre, and sustainment at the operational level in part due to lack of trained staff.[17]

The Russo-Ukraine War has illustrated the necessity and the difficulty of this balance. Ukrainian forces, particularly in the early phase of the war, demonstrated the ability to operate with decentralized initiative, enabling them to exploit Russian vulnerabilities through rapid, localized decision making. However, as operations scaled in complexity—think the summer 2023, the Ukrainian counter-offensive with its requirement for combined arms breaching at depth—Ukraine's ability to synchronize effects at echelon were found lacking, highlighting the need for trained staff at brigade and above.[18]

The war has also highlighted the necessity, and potential challenges, for mission command to evolve with technology. The proliferation and integration of advanced technologies, such as AI, autonomous systems, and cyber capabilities, are transforming the tempo and transparency of the battlefield. These technologies enable new levels of situational awareness, rapid decision making, and precision targeting that can accelerate the tempo and benefit from disciplined initiative. However, access to real-time information, predictive analytics, and the need to integrate capabilities from across multiple domains can amplify the temptation to micromanage, potentially undermining subordinate initiative.[19]

Ukraine's experience offers a cautionary but hopeful example. While the Ukrainian armed forces leveraged AI-enabled drones and digital command systems, such as Delta, to significant effect, they also faced challenges.[20] Throughout the war, Ukrainian junior leaders repeatedly demonstrated initiative in drone and cyber operations.[21] However, they would often fail to coordinate across echelons or overwhelm centralized coordination mechanisms, resulting in the misallocation of resources, such as fires.[22] This tension between tactical autonomy and strategic

[17] Dr Jack Watling and Nick Reynolds, 'Stormbreak: Fighting through Russian defences in Ukraine's 2023 offensive, Royal United Services Institute (RUSI), 4 September 2023, 21–2. https://www.rusi.org/explore-our-research/publications/special-resources/stormbreak-fighting-through-russian-defences-ukraines-2023-offensive; Jack Watling, Oleksandr V Danylyuk, and Nick Reynolds, 'Preliminary lessons from Ukraine's offensive operations, 2022–23', Royal United Services Institute (RUSI), 18 July 2024, 27–34. https://static.rusi.org/lessons-learned-ukraine-offensive-2022-23.pdf.

[18] Junguzza, 'Mission command as a key', 13–16; also see Watling and Reynolds, 'Stormbreak'.

[19] Col. Sean Madden, 'Mission command in the Ukraine-Russia war: Lessons from 2024', *Army War College Strategic Research Paper* (Carlisle, PA: Army War College, 2025), 2–3.

[20] Kateryna Bondar, 'Does Ukraine already have functional CJADC2 technology?', Center for Strategic International Studies, 11 December 2024. https://www.csis.org/analysis/does-ukraine-already-have-functional-cjadc2-technology.

[21] Madden, 'Mission command in the Ukraine-Russia war', 10–13, 17–20.

[22] Ibid., 10–14.

coherence underscores the importance of command systems that are technologically advanced yet culturally and procedurally aligned with the principles of mission command.

At its core, mission command is a human endeavour. No amount of technology can substitute for trust, competence, and shared understanding. Leaders must cultivate these attributes through professional military education (PME), realistic training, and leadership development. Perhaps most importantly, trust, both up and down the chain of command, must be earned through repeated interaction, demonstrated competence, and a culture that embraces disciplined initiative without fear of retribution.[23]

The contrast between Ukraine and Russia is stark. Ukraine's post-2014 reforms, supported by NATO training and a growing professional non-commissioned officer (NCO) corps, laid the groundwork for a more adaptive command culture.[24] In contrast, Russia's rigid hierarchy, lack of trust, and punitive approach to failure stifled initiative, leading to repeated operational failures and high casualty offensives that yielded limited territorial gains.[25]

Command involves far more than giving orders—it is an art that requires balancing authority, trust, decision making, and risk to create unity of effort. Command in war, amidst uncertainty and across the multiple domains of conflict, demands far more than doctrinal or technical knowledge; it requires judgement, adaptability, and shared understanding across the force. The war in Ukraine offers insights on command and contrasting command philosophies. The next section will show how Ukraine and Russia have pursued fundamentally different approaches to command and some of the implications for their battlefield performance.

Contrasting Approaches to Command: Ukraine vs Russia

The Russo-Ukraine War, since 2022, provides a stark contrast between how two militaries—sharing a common Soviet legacy—have evolved their command philosophies and practices. Ukraine's partial but evolving adoption of mission command has enabled tactical agility and operational adaptability, while Russia's rigid, hierarchical command model—rooted in its Soviet legacy—has generated consistent friction and underperformance. This section briefly examines some of the principal factors that shape each side's ability to implement mission command principles and how this has manifested broadly in the conflict.

[23] Junius, 'Mission command', chapter 7.

[24] Ibid.

[25] Junguzza, 'Mission command as a key', 5–13. See also Maciej Milczanowski, 'Mission command philosophy in the context of the war between Russia and Ukraine—Conclusions for Poland', Pulaski Policy Paper, Casimir Pulaski Foundation, 1 November 2024. https://pulaski.pl/en/mission-command-philosophy-in-the-context-of-the-war-between-russia-and-ukraine-conclusions-for-poland/.

Ukraine: Reform Through Necessity

Ukraine declared its independence from the Soviet Union in 1991 but remains encumbered to some extent by the Soviet's centralized command tradition. However, beginning in 2014—following Russia's annexation of Crimea and the war in Donbas—the Ukrainian military undertook serious military reform efforts to break from its Soviet legacy.[26] Driven by the Russian threat and supported by NATO's Defense Education Enhancement Program (DEEP), Ukraine prioritized PME, modernization of its officer and NCO corps, and a gradual embrace of Western command norms.[27]

Among the most significant reforms prior to 2022 was the growing prominence of mission command principles. In assessing Ukrainian reforms from 2016 to 2022, Deborah Sanders noted Ukraine's success in adopting a 'devolved system of command and control' that incorporated best practices consistent with the mission command approach.[28] Facilitating the adoption of a flatter, more dynamic command structure has been the promotion of a new generation of leaders, such as General Valeriy Zaluzhny, who championed pushing decision making and responsibility to the lowest level possible.[29]

These reforms manifested clearly in Ukraine's early wartime performance. During the defence of Kiyv in early 2022 and the Kharkiv offensive later that year, Ukrainian units demonstrated flexibility, speed, and an ability for localized decision making. These operations reflected not just tactical acumen but also a shift in command philosophy that rewarded initiative and adaptation over rigid adherence to orders.[30]

However, the Ukrainian approach did have limitations. Legacy Soviet attitudes remained embedded in some senior leadership, particularly at the operational level.[31] Ukraine also faces challenges in terms of scaling mission command from the tactical to the operational echelon. The challenge of scaling manifested primarily at brigade and battalion staffs, who struggled to synchronize manoeuvre, fires, and sustainment across larger formations due to a lack of training or resources.[32]

[26] Valeriy Akimenko, 'Ukraine's toughest fight: The challenge of military reform', Carnegie Endowment for International Peace, 22 February 2018. https://carnegieendowment.org/2018/02/22/ukraine-s-toughest-fight-challenge-of-military-reform-pub-75609; Major Stuart Gallagher, 'Mission Ukraine: U.S. Army leads multinational training group to counter Russian threat', Association of the United States Army, 19 May 2020. https://www.ausa.org/articles/mission-ukraine-us-army-leads-multinational-training-group-counter-russian-threat.

[27] Junius, 'Mission command', chapter 7. For a more in-depth discussion of the DEEP programme, see https://www.nato.int/cps/en/natohq/topics_139182.htm?utm.

[28] Deborah Sanders, 'Ukraine's third wave of military reform 2016–2022—Building a military able to defend Ukraine against the Russian invasion', *Defense and Security Analysis* 39/3 (June 2023): 312–28. https://doi.org/10.1080/14751798.2023.2201017.

[29] Ibid.

[30] Mykhaylo Zabrodski, Jack Watling, Oleksandr V. Danyluk, and Nick Reynolds, 'Preliminary lessons in conventional warfighting from Russia's invasion of Ukraine: February–July 2022', Royal United Services Institute (RUSI), 30 November 2022, 64. https://static.rusi.org/359-SR-Ukraine-Preliminary-Lessons-Feb-July-2022-web-final.pdf.

[31] Junius, 'Mission command', chapter 7.

[32] Junguzza, 'Mission command as a key', 13–15; Watling and Reynolds, 'Stormbreak'.

Despite these shortcomings, the Ukrainian command system proved adaptable and resilient. The growing adoption of mission command not only enabled Ukrainian forces early in the conflict but also facilitated the rapid integration of technologies such as AI, unmanned aerial vehicles (UAVs), and cyber capabilities, particularly in support of decentralized drone teams and predictive targeting as the conflict progressed.[33] The following section will explore these issues in more detail, but suffice it to say that Ukraine's implementation of mission command principles is at least partially responsible for allowing it to overcome the Russian advantage in mass.

Russia: Command and Control

In contrast, the Russian military entered the conflict with a command model firmly rooted in Soviet tradition. Despite some structural reforms initiated under Defence Minister Anatoly Serdyukov in the late 2000s, the Russian approach to command remained overwhelmingly centralized, hierarchical, and risk averse.[34] Consequently, Russian forces tend to maintain a strict adherence to orders with little to no flexibility to deviate.[35] Oleksiy Melynk, a Ukrainian officer who served a decade in the Soviet forces, described Russian officers as more concerned with the risk of being punished by a superior than with the risk of losing men or being killed themselves.[36]

This rigidity and inflexibility limit the opportunity for disciplined initiative and are a feature of the Russian system, thus making it highly resistant to change. This was evident from early in the war when Russian forces failed to adjust to Ukrainian resistance and followed the last orders given even when it degraded their tactical position.[37] Russian officer's willingness to blindly follow orders, regardless of battlefield realities, such as in the Battles of Kiyv and Mykolaiv, directly contributed to heavy casualties.[38] In several cases, senior officers were killed at the front because they felt compelled to micromanage tactical execution, a symptom of their subordinate's inability—or lack of authority—to act independently.[39]

Low levels of training and a lack of trust further undermined the Russian approach. Hastily mobilized formations, including prison units and conscripts, lack cohesion

[33] Madden, 'Mission command in the Ukraine-Russia war', 4–6, 10–15.

[34] Junguzza, 'Mission command as a key', 5–13; Cols Joe Junguzza and Kelly Lelito, 'What national culture teaches us about mission command', *Small Wars Journal*, 4 March 2024. https://smallwarsjournal.com/2024/03/04/what-national-culture-teaches-us-about-mission-command/. See also Philip Wasielewski, 'The roots of Russian military dysfunction', Foreign Policy Research Institute, 31 March 2023. https://www.fpri.org/article/2023/03/the-roots-of-russian-military-dysfunction/.

[35] Ben Hall, 'Military briefing: Ukraine's battlefield agility pays off', *Financial Times*, 26 May 2022. https://www.ft.com/content/9618df65-3551-4d52-ad79-494db908d53b.

[36] Ibid.

[37] Junius, 'Mission command', chapter 7; see also Victoriano Vicente et al., 'Command and control insights from Ukraine', Finabel—The European Land Force Commander Organization, October 2024. https://finabel.org/wp-content/uploads/2024/10/IF-PDFs.pdfxx.pdf.

[38] Junguzza, 'Mission command as a key', 2.

[39] Meredith Deliso, 'Why Russia has suffered the loss of an "extraordinary" number of generals', *ABC News*, 8 May 2022. go.com.

and basic battlefield competence.[40] In response, Russian leadership has relied even more heavily on coercive discipline, including documented cases of torture pits, execution for desertion, and severe punishments for battlefield withdrawal.[41] Importantly, analysis by the Center for Naval Analysis highlights that Russian thinkers and military leaders continue to emphasize the need to restore manoeuvre but are placing little emphasis on developing new operational concepts for doing so.[42] These measures certainly exacerbated the trust deficit between commanders and troops, contributing to a command culture that is reactive, brittle, and unable to exploit fleeting opportunities on the battlefield.

Culture and the Command Divide

At the heart of this contrast lies a fundamental difference in organizational culture. Ukraine, though still bearing the institutional scars of its Soviet legacy, especially at the operational level, is committed to building a professional, empowered force modelled on NATO norms. Ukraine's leaders have understood that effective command in LSCO depends on trust, competence, and adaptability—not simply control.

Russia, by contrast, has clung to an authoritarian command model that emphasizes compliance over creativity and subordination over judgement. Even when advanced technologies have been introduced, they have been layered atop a brittle structure that was incapable of effectively exploiting them. The result has been a disconnect between capability and command—one that has proved fatal in many engagements and has led to substantively high casualty rates. These organizational characteristics are deeply rooted in broader national cultures that shape how authority, initiative, and trust are viewed across societies.

National culture plays an outsized role in enabling or inhibiting mission command. Mission command—rooted in commanders' intent, trust, and shared understanding—emerges more readily in societies with strong cultural norms supportive of individual initiative and autonomy.[43] Colonels Junguzza and Lelito highlight how democracies with their greater emphasis on personal agency, self-reliance, and civic responsibility develop stronger foundations for subordinates to exercise initiative within commander's intent. Conversely, authoritarian regimes that prioritize centralized authority, conformity, and control, tend to prioritize the collective over the individual and inhibit subordinates initiative. These cultural foundations inevitably impact on military organizational culture and thus battlefield outcomes.

[40] Junguzza, 'Mission command as a key', 9–12.
[41] Ibid., 10–14.
[42] Michael Petersen, Paul Schwartz, and Gabriela Iveliz Rosa-Hernandez, *Russian concepts of future warfare based on lessons from the Ukraine war* (Arlington, VA: CNA Corporation, July 2025). https://www.cna.org/reports/2025/08/Russian-Concepts-of-Future-Warfare-Based-on-Lessons-from-the-Ukraine-War.pdf.
[43] Junguzza and Lelito, 'What national culture teaches us'.

Ukraine's embrace of mission command and Russia's entrenched centralization have had real operational implications. To better understand how these contrasting approaches have shaped battlefield outcomes, the following section examines three key case studies: the battle for Kiyv and Kharkiv which highlights the role of shared understanding and trust in decentralized initiative; the battle of Avdiivka which highlights the strategic implications of risk acceptance and force preservation; and the role of AI and technology in enabling decentralized initiative. These case studies illustrate the practical application of mission command principles as well as the institutional conditions that have enabled or inhibited their use in LSCO.

Command in Action: Case Studies from the Russo-Ukraine War (2022–Present)

The operational realities of the Russo-Ukraine War, and particularly the integration of increasingly advanced technologies, offer an important lens through which to examine the application—and limits—of mission command. The following three case studies illustrate how the principles of shared understanding and trust, risk acceptance, and integration of AI and technology—all core aspects of mission command—have shaped battlefield outcomes. They also highlight the interplay between leadership philosophy, organizational culture, and technology in large-scale war.

Early War—Battle of Kiyv and Kharkiv—Shared Understanding and Trust

The defence of Kiyv and the Ukrainian counter-offensive in Kharkiv during the early months of Russia's expanded invasion of Ukraine offer compelling examples of the importance of shared understanding and trust in enabling the disciplined initiative and decentralized execution that is essential to mission command.

Shared understanding, as defined in US army doctrine, is a common understanding of 'the operational environment, an operations purpose, problems, and approaches to solving problems.'[44] It further highlights that shared understanding and the flow of information to the lowest level is the basis for unity of effort and subordinate initiative.[45]

In the defence of Kiyv, Ukrainian commanders and forces demonstrated a remarkable effort to align their actions with the broader strategic intent, even in the face of rapidly evolving threats and degraded communications.[46] This was particularly evident in Ukraine's ability to leverage terrain, civilian support, and decentralized

[44] US Department of the Army, *ADP 6-0: Mission Command*, 1–8.

[45] Ibid.

[46] Zabrodskyi et al., *Preliminary lessons in conventional warfighting*, 28; Simon Shuster and Vera Bergengruen, 'How Ukraine and Valeriy Zaluzhny turned the tide of the war', *TIME*, 26 September 2022. https://time.com/6216213/ukraine-military-valeriy-zaluzhny/.

defensive positions to blunt Russia's initial attack on Kiyv.[47] Shared understanding of the operational environment and approach enabled Ukrainian forces to act on and exploit Russian vulnerabilities, even when out of communication with higher headquarters through rapid, localized decision making. This differed markedly from Russian forces, who proved unable or unwilling to act when the situation they faced in Ukraine differed significantly from expectations, or they lacked detailed instructions or a clear understanding of broader intent.[48]

In Kharkiv, shared understanding enabled a different kind of success. During the September 2022 counter-offensive, Ukrainian forces executed a rapid and coordinated manoeuvre that took advantage of Russia's degraded position and lack of forces to retake thousands of kilometres of Ukrainian territory.[49] This operation, enabled by a coordinated deception effort, required precise timing, coordination across multiple brigades, and the ability to exploit fleeting opportunities—all of which depended on a shared understanding of the operational plan and flexibility to adapt it in real time.

Mutual trust is another key element of mission command. Trust allows commanders to delegate authority and empowers subordinates to act decisively without fear of reprisal. Trust is developed over time as people work together and is essential for promoting initiative.[50] Trust is essential, both vertically, up and down the chain of command, and horizontally, between adjacent units and across the joint force.[51] This type of trust is essential when units must coordinate fires, manoeuvre, and logistics on a dynamic and fast-moving battlefield. In both Kyiv and Kharkiv, Ukrainian commanders demonstrated trust in their subordinate leaders, and those leaders, in turn, demonstrated initiative.

In contrast, the Russians demonstrated a lack of trust that certainly contributed to the Ukrainians' ability to achieve operational surprise and the Russians' poor operational performance. The lack of trust endemic in Russian forces is due to several factors, including fear of punishment, a lack of discipline, a culture of lying, and vast differences in unit training and competence—especially between Russian regular forces and those of the Donetsk People's Republic.[52] The fact that Russian leadership appeared surprised by the tactical situation, that Russian soldiers panicked when Ukrainian forces launched the Kharkiv counter-offensive, and that hundreds of thousands of men fled Ukraine when Putin called for mobilization to check the Ukrainian offensive is indicative of this lack of trust.[53]

47 Sarah Cahlan, Joyce Sohyun Lee, Meg Kelly, and Artur Galocha, 'Inside the battle for Kyiv: How Ukraine repelled the Russian onslaught', *The Washington Post*, 24 December 2022. https://www.washingtonpost.com/national-security/interactive/2022/kyiv-battle-ukraine-survival/.

48 Zabrodskyi et al., *Preliminary lessons in conventional warfighting*, 26.

49 Michael Kofman, 'The Russia-Ukraine War: Military operations and battlefield dynamics', in Hal Brands (ed.), *War in Ukraine: Conflict, strategy, and the return of a fractured world* (Baltimore, MD: Johns Hopkins University Press, 2024). https://muse.jhu.edu/pub/1/oa_edited_volume/chapter/3881920.

50 Junius, 'Mission command', 1.

51 Junguzza, 'Mission command as a key', 4.

52 Ibid., 6–9; Siobhán O'Grady and Anastacia Galouchka, 'Ukraine's offensive in Kharkiv, Kherson and Donetsk: How Kyiv's forces retook occupied territory', *The Washington Post*, 29 December 2022. https://www.washingtonpost.com/world/2022/12/29/ukraine-offensive-kharkiv-kherson-donetsk/.

53 Ibid.

Shared understanding and mutual trust are essential and reinforcing elements of mission command, which is at the core of a decentralized initiative. Combined, these three elements can have real battlefield implications, as illustrated by the battles of Kyiv and Kharkiv. Ukraine's former defence minister, Oleksiy Reznikov, attributed much of Ukraine's early success to having decision-making abilities at every level of the army, which allowed Ukraine to act faster than the Russians.[54] Meanwhile, Jeffrey Edmonds, a former Russian director for the US National Security Council, attributed Ukraine's success to its ability to conduct distributed operations based on superior 'distributed tactical-level operations'.[55] Ukraine's overperformance and Russia's underperformance in the early stages of this conflict are largely due to these factors.

Battle of Avdiivka (2023–2024) and Kursk (2024–2025): Risk Acceptance

The battle for Avdiivka, a fortified Ukrainian stronghold near Donetsk, revealed starkly different approaches to risk acceptance, a crucial element of mission command. Ukrainian commanders, facing encirclement and sustained Russian pressure, made the difficult decision to conduct a phased withdrawal.[56] This decision was significant, as it created a propaganda victory for the Russians and could have allowed them to use Avdiivka as a base for future offensive operations.[57] However, it was necessary, given Russia's quantitative advantages in artillery and manpower, as well as threats to Ukrainian supply lines.[58]

Ukrainian forces initially saw Avdiivka as an opportunity to attrit Russian forces. Fighting from defensive positions, Ukrainian forces used mines, first-person view drones, and fires to 'chew up' Russian forces, creating what Peter Zeihan estimated to be a 5:1 to 10:1 ratio in favour of the Ukrainians.[59] Ukrainian commanders likely hoped to leverage Avdiivka, as Bakhmut before it, to inflict casualties on Russia and erode Russian morale both at the front and in Russia.[60] However, Russian mass and ability to hold Ukrainian supply lines at risk eventually forced Ukraine to withdraw as

[54] Shuster and Bergengruen, 'How Ukraine and Valeriy Zaluzhny turned the tide of the war'.

[55] Ibid.

[56] Alex Babenko, Hanna Arhirova, Susia Blann, and Lori Hinnant, 'Russia takes control of Avdiivka in eastern Ukraine after Kyiv withdraws troops', *AP News*, 18 February 2024. https://apnews.com/article/russia-ukraine-war-avdiivka-2e827b4cae4698b3f6b80a421447fab8.

[57] Ryan N. Forte, 'The Battle of Avdiivka and its lessons on withdrawal under pressure', Modern War Institute, 12 March 2024. https://mwi.westpoint.edu/the-battle-of-avdiivka-and-its-lessons-on-withdrawal-under-pressure/.

[58] John Hardie, 'Battle of Avdiivka: A preliminary analysis', *FDD's Long War Journal*, 22 March 2024. https://www.longwarjournal.org/archives/2024/03/battle-of-avdiivka-a-preliminary-analysis.php.

[59] Ibid.; Peter Zeihan, 'The Ukraine war & the battle of Avdiivka', *Zeihan on Geopolitics*, 5 January 2024. https://zeihan.com/the-ukraine-war-the-battle-of-avdiivka/.

[60] James Horncastle, 'An unstable foundation: Russian morale problems in the Russo-Ukrainian war', Forum for Ukrainian Studies, 15 March 2024. https://ukrainian-studies.ca/2024/03/15/an-unstable-foundation-russian-morale-problems-in-the-russo-ukrainian-war/.

they prioritized force preservation and manoeuvre flexibility. This approach reflected a mature understanding of risk—not as something to be avoided, but as something to be managed in pursuit of long-term operational advantage.[61]

Conversely, Russia's approach to risk in Avdiivka demonstrated their willingness to sacrifice soldiers for limited operational gains. Russian commanders in Avdiivka launched repeated assaults, often with poorly trained forces or units of prisoners, in an attempt to overwhelm Ukrainian defences through attrition. These 'meat assaults', as they came to be known, resulted in staggering casualties—as high as 70% for some units—with minimal territorial gains.[62] This willingness to accept high losses in exchange for limited territorial gains reflects a deeply ingrained cultural preference for control and mass over adaptability and initiative.

This divergence in risk acceptance had strategic consequences. Ukraine's withdrawal, although executed somewhat haphazardly, imposed significant costs on Russia while preserving its combat power for future operations. Inversely, it achieved short-term operational gains but at a significant cost, with inevitable impacts on unit cohesion, morale, and operational tempo.[63] Ultimately, risk acceptance within mission command is about empowering leaders to make timely decisions under conditions of uncertainty that balance risk to force with mission accomplishment.

AI and Technology: Enabling Disciplined Initiative

Since the beginning of the war and the Battle of Kiyv, Ukrainian forces have consistently demonstrated a level of disciplined initiative that has helped mitigate Russia's advantage in terms of mass and firepower. From the early decision to move forces out of garrison to the use of hit-and-run tactics and willingness to manoeuvre in the absence of orders, Ukrainian forces have leveraged disciplined initiative and decentralized execution to deny Russia operational or strategic success.[64]

As the war has evolved, Ukrainian forces have effectively leveraged AI and advanced technologies to foster and enable this advantage. The use of AI models and predictive analytics to aid Ukrainian commanders in responding to Russian offensives or to enhance the targeting of Russian forces is but one example.[65] Systems like Delta and Kropyva enable Ukrainian forces to process large quantities of data from multiple sources and share them rapidly across systems, thus empowering Ukrainians to act faster than the more centralized Russian forces.[66] Another example of technology enabling disciplined initiative and decentralized operations is Ukraine's

[61] Junius, 'Mission command', chapter 7.
[62] Junguzza, 'Mission command as a key', 10–12.
[63] Ibid., 10–13.
[64] Cahlan et al., 'Inside the battle for Kyiv'.
[65] Samuel Bendett, 'Roles and implications of AI in the Russian-Ukrainian conflict', *Russia Matters*, 20 July 2023. https://www.russiamatters.org/analysis/roles-and-implications-ai-russian-ukrainian-conflict/.
[66] Madden, 'Mission command in the Ukraine-Russia war', 10–11. For more on DELTA and KROPYVA, see Bondar, 'Does Ukraine already have functional CJADC2 technology?'.

'Army of Drone bonus' and 'Brave 1 marketplace', which rewards soldiers with points for uploading videos of drones destroying Russian forces and the ability then to convert those points into new equipment for their unit.[67] These are just a few of the examples of Ukraine using technology to enable and empower their forces.

Ukraine has also seen technology allow civilians to support the war effort through disciplined initiative and decentralized operations. For instance, Ukraine's civilian 'IT Army' has collaborated with Ukraine's intelligence services in multiple attacks with economic losses estimated to exceed $1 billion as of early 2024.[68] Ukraine has also seen volunteer efforts, such as the North Atlantic Fella Organization, conduct effective information operations, and other groups step forward to fund or procure equipment directly for Ukrainian units.[69] These efforts highlight the ability of technology to enable initiative and decentralized operations.

In contrast, Russian forces, despite possessing similar technological capabilities, maintained a very centralized approach. At its core, an unwillingness to decentralize decision making and a failure to communicate accurate information limit the Russian force's ability to execute decentralized initiative.[70] As Colonel Joe Junguzza highlights, Russian leadership and forces lack trust up and down the chain of command, suffer from low levels of training, and, thus, their units suffer from low levels of cohesion, which limits disciplined initiative.[71] Even in places where Russia has demonstrated some degree of disciplined initiative—such as in drone operations early in the war—Russian efforts to centralize control through institutional reforms are likely to limit Russian forces in the future.[72] As a result, Russian forces primarily rely on mass and attrition rather than disciplined initiative or decentralized operations to achieve limited battlefield gains at a significant cost.

Command continues to grow more complex with the need to integrate capabilities from across multiple domains while avoiding detection and destruction on an increasingly transparent battlefield. The Russo-Ukraine War highlights this tension and the reality that large, static command posts, with their robust C2 capabilities and electromagnetic signatures, are no longer suited for their purpose.[73] With both sides in the conflict increasingly reliant on autonomous systems and engaged in a battle for electromagnetic superiority, this complexity is only going to grow.[74] However this

[67] Veronika Melkozerova, 'Points for kills: How Ukraine is using video game incentives to slay more Russians, *POLITICO*, 29 April 2025. https://www.politico.eu/article/ukraines-army-have-video-game-like-digital-weapons-store-deadly-realistic/.

[68] 'Ukraine's decentralized warfare: The battle of the common person, *Asia Times*, 11 November 2024. https://asiatimes.com/2024/11/ukraines-decentralized-warfare-the-battle-of-the-common-person/.

[69] Ibid.

[70] Wasielewski, 'The roots of Russian military dysfunction'.

[71] Junguzza, 'Mission command as a key', 5–13.

[72] Kateryna Stepanenko, 'Russian efforts to centralize drone units may degrade Russian drone operations', *Institute for the Study of War*, 13 December 2024. https://www.understandingwar.org/backgrounder/russian-efforts-centralize-drone-units-may-degrade-russian-drone-operations.

[73] Nagl and Combe, *A Call to Action*, xxiii.

[74] Samuel Bendett and David Kirichenko, 'Battlefield drones and the accelerating autonomous arms race in Ukraine', Modern War Institute, 10 January 2025. https://mwi.westpoint.edu/battlefield-drones-and-the-accelerating-autonomous-arms-race-in-ukraine/.

manifests in the future, the integration of AI and other advanced technologies will remain key to battlefield outcomes in Ukraine.

Lessons for the Future of Warfare

The Russo-Ukraine War has reaffirmed the enduring relevance of mission command in LSCO, while also revealing some of the institutional, cultural, and technological conditions necessary for its success. The war has validated the role of decentralized initiative based on shared understanding and trust, along with risk acceptance, while exposing the challenges of scaling these principles across echelons and integrating them with emerging technologies. Looking forward to warfare in an increasingly complex battlespace, the lessons from Ukraine offer both cautionary and instructive insights.

Culture Shapes Command More Than Doctrine

Perhaps the most enduring insight from the war is that culture, rather than doctrine, determines whether and how mission command is practiced. Ukraine's democratic ethos, reform-oriented leadership, and NATO alignment has fostered a command climate that values initiative and adaptability. In contrast, Russia's authoritarian model reinforces rigidity, fear, and a lack of trust, fostering a command culture that prioritizes obedience and control over initiative. This is consistent with Colonels Junguzza and Lelito's argument that national culture profoundly influences command behaviour and, thus, how forces implement mission command.[75]

This means that institutionalizing mission command requires more than doctrinal updates—it demands cultural transformation. Militaries must train and educate leaders to not only issue mission-type orders but also to create environments where subordinates feel empowered to act within the leaders' intent. This includes accepting mistakes will be made, rewarding initiative, and fostering a climate of mutual respect and accountability.

Mission Command—More Than a Slogan

One of the clearest takeaways from the Russo-Ukraine War is that mission command must be more than a slogan. Militaries must deliberately cultivate the cultural, institutional, and technical systems that support mission command. Ukraine's ability to empower leaders, adapt under pressure, and integrate emerging technologies has been the result of years of reform, NATO-supported PME, and a growing culture

[75] Junguzza and Lelito, 'What national culture teaches us'.

of trust and initiative.[76] Ukraine's willingness to empower its leaders at every echelon was crucial to the Ukrainian force's ability to act more quickly and thus blunt Russian forces early in the conflict, demonstrating the power of mission command. Conversely, Russian failures largely stemmed from a command culture that prioritized control and obedience over initiative despite a theoretical commitment to a more professional and dynamic force based on the Serdyukov reforms. Ultimately, mission command requires trained staff, the systems and processes to integrate and coordinate multi-domain capabilities, and clear roles and responsibilities at echelons of command to synchronize effects in time, space, and purpose.

Of course mission command, no matter how well implemented, cannot substitute completely for shortfalls in other areas such as logistics, mass, or firepower. Ukraine's 2023 counter-offensive illustrates this limitation. Despite a demonstrated ability to employ mission command principles and tactical agility Ukrainian forces have struggled to achieve operational breakthroughs due to a lack of trained staff, insufficient artillery, limited air support, and logistical constraints.[77] As this example highlights, mission command is important but it must be more than a slogan and still requires proper resourcing. Therefore, militaries must not view mission command as a replacement for shortfalls in the other warfighting functions but as a complement that maximizes their effectiveness.

Trust and Shared Understanding—Cornerstones of Disciplined Initiative

Inherent in every Ukrainian success—from Kyiv to Kharkiv and beyond—two principles consistently emerge: trust and shared understanding. These attributes have enabled Ukrainian decentralized execution, rapid adaptation, and resilience in the face of uncertainty. ADP 6-0 emphasizes that shared understanding is the foundation for unity of effort and disciplined initiative.[78] It further highlights the critical and interdependent relationship between trust, shared understanding, and the disciplined initiative that is essential to decentralized execution and mission command. Trust is best built ahead of conflict through training and shared experience, and reinforced through a command culture before and during combat. Ukraine's ability to maintain operational coherence and act in the face of surprise and contested communications is testament to the power of trust and shared understanding.

In contrast, Russian forces have consistently demonstrated a lack of trust—both vertically and horizontally. Poor training, a lack of transparency between leaders and subordinates, coercive discipline, and a culture of obedience have all exacerbated this trust deficit, which in turn has discouraged initiative.[79] The result has been a

[76] Junius, 'Mission command', 117–19.
[77] Watling et al., 'Preliminary lessons from Ukraine's offensive operations', 19–26.
[78] US Department of the Army, *ADP 6-0: Mission Command*, 1-8.
[79] Junguzza, 'Mission command as a key', 5–13, 16–20.

brittle command structure that has failed to adapt to battlefield realities, resulting in repeated tactical and operational failures.

The lesson here is that shared understanding and trust are essential to a force's ability to leverage disciplined initiative. Trust is built pre-conflict through training, shared experience, and a willingness to underwrite legitimate mistakes. Shared understanding requires the sharing of information and clarity of purpose. Together, these principles serve as a foundation for disciplined initiative and the decentralized execution that is at the very core of mission command.

Risk Acceptance—Strategic Imperative

Mission command, with its emphasis on disciplined initiative and decentralized execution, requires commanders to define and accept risk—not avoid it. Command in warfare inherently requires trade-offs between the risk to forces or mission and the potential benefits of action. The greatest opportunity can come from costly or high-risk options.[80] For instance, Ukraine's defence of Bakhmut or Avdiivka allowed them to impose significant costs on the Russians, but at substantial risk to their forces. Conversely, Russia's willingness to incur massive casualties through frontal assaults for limited territorial gains has undermined trust and unit cohesion, demonstrating a potential failure to balance operational risk with battlefield outcomes. Importantly, these two examples also reinforce the fact that risk is often far from cost free—something especially true in LSCO—and inherently involves uncertainty. Thus, militaries must train leaders to make decisions under conditions of uncertainty and be willing to accept prudent risk in pursuit of mission success.

This is best done prior to conflict by leaders cultivating and regularly encouraging subordinates at all levels to embrace prudent risk as part of executing disciplined initiative. Doing this requires them to overcome their natural risk aversion based on the fear of subordinates making mistakes, career risk derived from others' mistakes, and the discomfort of feeling out of control.[81] Failure to do so will inevitably result in a force that is risk averse and thus unable or unwilling to act with the disciplined initiative that is essential to the decentralized execution that mission command seeks to enable.

Mission Command—Technology Enabled but Human Centred

Ukraine's use of AI-enabled targeting, predictive analytics, and digital command systems, such as Delta and Kropyva, illustrates how technology can enhance mission

[80] US Department of the Army, *ADP 6-0: Mission Command*, 1–13.

[81] L. Burton Brender, 'The problem of mission command', RealClearDefense, 2 September 2016. https://www.realcleardefense.com/articles/2016/09/02/the_problem_of_mission_command_110008.html.

command when aligned with culture.[82] These tools have accelerated decision making, improved situational awareness, and enabled decentralized or distributed operations. The technologies have provided Ukrainian leaders with unprecedented battlefield awareness and empowered junior leaders to make informed and timely decisions that have offset Russian advantages in mass.

However, they have also highlighted challenges. Technology can create information overload, potentially paralysing commanders and staffs, or, conversely, it can tempt commanders and staffs to micromanage due to the level of situational awareness available. Technology can also create a dependency that risks becoming a vulnerability when disrupted. Ultimately, technology serves as an enabler for the human decision making required in command. Ukraine's lack of trained staff capable of integrating capabilities and synchronizing activities—despite functioning C2 systems—played a substantive role in the failure of its 2023 summer offensive.[83]

The lesson is clear: technology must be integrated in ways that support, rather than supplant, human judgement and disciplined initiative. Technology should enhance commanders' decision-making capabilities without becoming a single point of failure. This requires investment in systems as well as in training leaders and staffs to use them in ways that encourage initiative and adaptability while retaining the ability to operate in degraded technological environments.

Command in the Transparent, Contested, and AI-Enabled Battlespace

The war in Ukraine has also revealed the growing complexity of command in an increasingly transparent and technologically shaped battlespace. The proliferation of sensors and the increasing availability of data, combined with AI, are shaping the battle for an information advantage.[84] This makes the collection, processing, and sharing of data increasingly important on the modern battlefield. This means it is more important than ever that militaries consider the interoperability of their systems, both within their forces as well as with those of allies and partners.

The battle for information advantage, along with the proliferation of autonomous systems, cyber and electronic warfare capabilities, and more traditional capabilities, require commanders and staffs to possess the capacity and bandwidth to plan and execute multi-domain operations. However, large, static command posts with robust electromagnetic signatures are increasingly vulnerable to detection and destruction.[85] Thus, even as staffs are required to deal with greater complexity and the integration

[82] Madden, 'Mission command in the Ukraine-Russia war', 2–6, 10–11.

[83] Watling et al., 'Preliminary lessons from Ukraine's offensive operations', 32.

[84] Mike Hackett and John Nagl, 'A long hard year: Ukraine war lessons learned 2023', *Parameters* 54/3 (2024): 41–52. doi: 10.55540/0031-1723.3302.

[85] Nagl and Combe, 'Introduction', *A Call to Action*, xxiii–xxiv; see also Milford 'Beags' Beagle, Jason C. Slider, and Matthew R. Arrol, 'The graveyard of command posts: What Chornobaivka should teach us about command and control in large-scale combat operations', *Military Review*, Army University Press,

of more domains, they must do so while reducing their electromagnetic signature and moving more frequently to improve their survivability.

The proliferation and scale of this challenge continues to grow. The rapid expansion of cheap and attritable systems, the employment of AI-powered drones, and the constant struggle for technological advantage is driving both sides to adapt and innovate.[86] This adaptation cycle will continue as both sides evolve their approach to command. One data point that highlights the scale of this challenge is that the combatants employed between 1.5 and four million drones each in 2024, up from approximately 2,000 in 2022.[87] This pace and scale of change, in some ways, resembles that of World War One, where the combatants had to learn how to employ airplanes, tanks, chemical weapons, and improved artillery and machine guns.[88] As this cycle continues in Ukraine and future wars, the integration of AI and other advanced technologies will remain a critical enabler of command and central to battlefield outcomes.

Mission Command Is a Strategic Advantage—But One That Must Be Earned

Ultimately, the war in Ukraine reinforces the point that mission command can be a strategic advantage, but it is not a given. Ukraine has leveraged its superior ability to leverage mission command to prevent the larger and more powerful, but more centralized, Russian military from achieving its strategic objectives. Early in the war, this manifested in Russia's defeat in the Battle of Kyiv and Kharkiv. Today, Russia's high casualty rate for incremental gains reflects this. Ukraine's ability to do this is, in large part, a result of its training, education, and leadership culture, which is supportive of mission command principles.[89]

However, it is important to remember that while mission command, and command in general, are critical to success in war they are not the only elements.

Militaries preparing for future wars should consider if they have institutionalized mission command as a core aspect of their approach to warfare. This means embedding it in doctrine, reinforcing it through training and education, and living it through leadership. Most importantly, it means embracing a culture that encourages disciplined initiative, accepts risk, and underwrites legitimate mistakes.

March 2023. https://www.armyupress.army.mil/journals/military-review/online-exclusive/2023-ole/the-graveyard-of-command-posts/.

[86] Bendett and Kirichenko, 'Battlefield drones'.

[87] Kateryna Bondar and Samuel Bendett, 'The Russia-Ukraine drone war: Innovation on the frontlines and beyond', Center for Strategic and International Studies, 28 May 2025. https://www.csis.org/analysis/russia-ukraine-drone-war-innovation-frontlines-and-beyond.

[88] Stephan Bolton and John Nagl, 'Innovation and adaptation: Lessons from the Russia-Ukraine war, 2024', *Army War College Strategic Research Paper* (Carlisle, PA: Army War College, 2025), 17.

[89] Junius, 'Mission command', 117–19.

Conclusion

Command is a central and unifying aspect of warfighting. The war in Ukraine illustrates that adopting the principles of mission command can enhance battlefield outcomes, even against larger adversaries. At its core, mission command is about unleashing human judgement in the face of complexity, friction, and uncertainty—an imperative that grows more urgent as the battlefield becomes increasingly more transparent, multi-domain, and technologically advanced.

Ukraine's success in leveraging elements of mission command is not accidental. It reflects years of reform, training, education, and cultural change that have embraced the importance of disciplined initiative and decentralized execution at multiple levels. This cultural shift, combined with the integration of emerging technologies such as AI, has enabled an adaptive command system capable of offsetting Russian advantages in mass. Ukrainian units have been able to act more rapidly and with greater initiative than their Russian counterparts, who have proved unable to act in the absence of clear orders due to fear of punishment. This reinforces the view that trust, shared understanding, and risk acceptance are more than theoretical principles; they are practical enablers of battlefield success.[90]

Conversely, Russian forces' adherence to a centralized, hierarchical command culture, characterized by fear, rigidity, and punitive control, has proved brittle in the face of a dynamic battlefield. Their failure to cultivate trust or empower subordinates has limited Russian forces' ability to respond to unexpected battlefield developments and undermined unit cohesion, further limiting their operational effectiveness.[91] Despite fielding large numbers and advanced capabilities, the Russian command culture has proved unable to capitalize on opportunities or mitigate vulnerabilities. This contrast highlights a key lesson of the war: a command culture that suppresses initiative and prioritizes obedience will likely prove relatively inflexible, thereby incurring a high cost for battlefield gains.

While Ukraine's success illustrates the value of mission command, its experience also highlights some of the challenges of implementing at scale or in certain conditions. Ukraine's failed summer 2023 offensive highlights that operations requiring a high level of synchronization and control require well-trained staff—a resource that may not be readily available in all situations.[92] Ukraine's experience also highlights that mission command alone may not be sufficient to defeat a larger enemy, offering a cautionary tale for those who might underresource their military in the belief that mission command will prevail. This all reveals that mission command is more than just pushing decisions downward; it requires the staff capacity, training, and command systems, along with the resources, to integrate actions into a cohesive and operationally sufficient whole.

[90] Junius, 'Mission command', chapter 7.
[91] Junguzza, 'Mission command as a key', 5–13.
[92] Ibid., 13–16.

The character of warfare is changing with technology, making the battlefield more complex while also providing commanders and staffs with tools to manage that complexity. As AI, autonomy, cyber effects, and space-based capabilities proliferate, the battlefield will only become faster and less forgiving of delay or indecision. Military forces that can integrate these technologies while preserving human initiative will have an advantage. Doing so requires a command philosophy that prizes adaptability over control and a leadership culture that underwrites disciplined initiative rather than punishing deviation.

Ultimately, mission command should be seen as a strategic asset. If embraced and institutionalized, it enables the resiliency, agility, and lethality needed for modern LSCO. To fully harness the benefits of mission command, a military must go beyond doctrine. This means investing in training and education, fostering a culture of trust and risk acceptance, supported by technology that enables rapid and informed decision making founded on the commander's intent and mission-type orders. If this is done well, it empowers leaders at all levels to execute with the disciplined initiative and decentralized execution that is necessary on the complex and lethal battlefields of the future.

5

Russian Military Logistics and Its Impact on Russia's War with Ukraine

Christopher Kinsey and Ronald Ti

Introduction

A seminal work that links strategic success with logistics is Thomas Kane's 2001 book *Military Logistics and Strategic Performance.*[1] This book has stood the test of time with its most enduring achievement being that of giving due importance to logistics in the shaping of strategic success. A theme that runs throughout this work is that recognition of logistics capabilities by commanders enables strategic advantage. The dictum that logistics will continue to be a major factor in shaping future military operations is stressed throughout *Military Logistics and Strategic Performance.* It is this interdependency of operational logistics with strategic planning that has been most emphasized in the Russo-Ukrainian War as the discussion below makes clear.

When Russia launched on the 24th February 2022 what it called a 'special military operation' along multiple axes not only to seize the Ukrainian capital of Kyiv, replacing the government with one friendly to Moscow, but also to expand occupied territory into eastern Ukraine. However, those who were responsible for planning the operation did so based on fundamentally flawed assumptions about the strength and quality of the Ukrainian armed forces (UAF), the resilience of the government, and the actual level of Ukrainian public support for a Russian invasion.[2] Within days, the operation had encountered determined resistance from local military units and the country's special forces. Indeed, according to a Royal United Services Institute (RUSI) report, 'by day 3 it was apparent to Russian commanders that their plan had gone seriously wrong'.[3] Conducting offensive operations along multiple axes extending over a 1,000-kilometre front line would be a considerable challenge for any first-world military force, including the United States (US), but even more so for a military

[1] Thomas Kane, *Military logistics and strategic performance* (Abingdon: Routledge Press, 2001).
[2] Ronald Ti and Christopher Kinsey, 'Lessons from the Russo-Ukraine conflict: The primacy of logistics over strategy', *Defence Studies* 23/3 (2023): 381–98, at 5; also Robert Dalsjö, Michael Jonsson, and Johan Norberg, 'A brutal examination: Russian military capability in light of the Ukraine war', *Survival* 64/3 (2022): 7–28. doi: https://doi.org/10.1080/00396338.2022.2078044.
[3] Mykhaylo Zabrodskyi et al., 'Preliminary lessons in conventional warfighting from Russia's invasion of Ukraine: February–July 2022', Special Report, Royal United Services Institute (RUSI), 2022, 29.

Christopher Kinsey and Ronald Ti, *Russian Military Logistics and Its Impact on Russia's War with Ukraine*. In: *Russia's War in Ukraine and Modern Warfare*. Edited by: Mikael Weissmann and Niklas Nilsson, Oxford University Press.
 DOI: 10.1093/9780198970903.003.0005

that was less capable, particularly one handicapped by the lack of adequate defence logistical support.[4]

It is not surprising therefore that the plan to transition from attempting the *coup de main* to deliberate combat operations to defeat the UAF quickly fell apart as Russian military commanders struggled to control the fluidity of the front line especially due to the poor state of their defence logistics support. However, this is not to suggest that poor operational planning and the lack of central command were not also to blame as these would have been key factors in the unravelling of the operation. Nevertheless, while operations can survive poor planning and command, the chances of success diminish substantially in the absence of capable defence logistics support.

As detailed further below, many of the operational and tactical challenges that faced Russian forces in the early months of the invasion should have been foreseen by the Russian high command and thus mitigated. However, this would have required much greater attention to solving the many logistics challenges inherent in the operation from the outset. Instead, by downplaying or, worse still, ignoring the centrality of defence logistics to military strategy, Russian commanders condemned the initial phases of the operation to failure before it had even begun.

As Kinsey and Uttley remind us, defence logistics is 'a key enabler of the execution of campaign plans for military operations'.[5] Russian commanders not only ignored this vital principle, but also failed to understand or simply ignored the fact that 'defence logistics reflects the inherent dilemma confronting strategy formulation and execution in microcosm'.[6] These authors continue that 'the application of strategy against an intelligent, reactive, and adaptive adversary [which the UAF surely is] does not tolerate simple formulas or principles for success',[7] as was shown by the failure of Russia's initial formula to defeat the UAF. Such a catastrophic failure, moreover, raises key questions about why greater consideration was not given by the Russian high command to the challenges posed by defence logistics. Consequently, the fault of this catastrophic failure lay not at the door of operational commanders, though their decisions probably made the logistics situation on the ground worse, but was the outcome of more fundamental problems with the Russian way of supplying war. As noted in this chapter, issues of defence structure, doctrine, and command philosophies were collectively and ultimately responsible for Russia's defence logistics failure during the initial phase of the 2022 invasion.

Clearly, the Russian high command failed to generate a logistics system able to interface optimally with ground forces[8] and provide them with the capabilities and capacities necessary to guarantee success. This chapter investigates the reasons

[4] Ti and Kinsey, 'Lessons from the Russo-Ukraine conflict', 6.

[5] Matthew Uttley and Christopher Kinsey, 'The role of logistics in war', in Julian Lindley-French and Yves Boyer (eds), *The Oxford handbook of war* (Oxford: Oxford University Press, 2012), 402.

[6] Ibid., 402.

[7] Ibid.

[8] The chapter is focused on defence logistics in the land domain and support to ground forces. This is not to suggest that defence logistics support to the air and maritime domains is not important, but that we do not have sufficient space to discuss the other two domains in a short chapter.

behind this failure through an examination of the structural, doctrinal, and command philosophies that underpin Russia's defence logistics.

The chapter is divided into three sections that cover each of these issues, ending with the conclusion in the final section. The first section—structure—examines challenges associated with the 'push' system of delivering force projection and sustainment.

The second section—doctrine—analyses the impact of Russian military doctrine on its defence logistics system during the initial stage of the invasion. This section is concerned with the concept of 'active defence' and how it shapes Russia's military behaviour in both general and specific terms.[9] The third section—command philosophies—considers the philosophical foundation that shapes Russian military command ethos, and one that is opposed to individual decision making and initiative. It will be shown in this section how the command philosophy has a profound impact on logistics. The final section is the conclusion.

Finally, a key part of any operational plan is essential logistics operational planning. This must also be successfully completed, resulting in a practical logistics system that will sustain the capabilities required to attain the requisite tempo and power of operation needed for victory. The next section will summarize key logistics concepts important for understanding the discussion that will then follow in the first, second, and third sections.

'Push' and 'Pull' Logistics

Logistics systems can be categorized as 'pull' or 'push' systems, with these terms describing the location of the drivers determining sustainment levels. 'Pull' systems literally pull supplies and sustainment to the 'front line' based on demand signals generated by forward consumption. On the other hand, 'push' systems, as the name implies, push material and resources forward according to pre-determined usage rates. These are often determined by higher command levels whose physical distance from the front line frequently hinders timely information flow.[10]

The consequence of this is that actual logistics demand from forward echelons is unable to be communicated rearwards in a timely manner often resulting in a critical mismatch between pre-determined supplies and front-line unit requirements. Military logistics systems utilize a combination of these two categories with no one system purely a 'push' or 'pull' system. The Russian military operates in a similar manner and utilizes a combination of both. When a military logistics system is deployed in war, in whatever combination of 'push' or 'pull' modalities, it is exposed to considerable material stress. In the case of the Russian logistics pull system, this proved

[9] Michael Kofman et al., *Russian military strategy: Core tenets and operational concepts* (Washington, DC: Center for Naval Analysis, 2021), 10–14. https://www.cna.org/centers/cna/sppp/rsp/russian-military-strategy.

[10] Ti and Kinsey, 'Lessons from the Russo-Ukraine conflict', 9.

inadequate in Ukraine, leaving the 'push' system as the principal method to deliver logistics support to the special military operation.

A critical issue affecting Russian 'pull' logistics is the comparative lack of logistics enterprise resource management systems in the military logistics system that are now standard in Western armies. Such systems are the hallmarks of modern commercial logistics, and their relative paucity in the Russian military indicates logistics that does not reflect current supply chain management 'best practice'. Furthermore, the reason the chapter focuses on the push system is because this is the Russian military's primary method of delivering operational support. This is often the only viable system that can be employed by a highly centralized military command system dominant in Russia's armed forces. Later chapter sections will discuss the direct effects of this command philosophy on logistics.

Structural Challenges to Russian Defence Logistics

Russian's invasion of Ukraine could be described as an 'expeditionary' operation in that it has involved the projection and sustainment of force outside of the country. As Kinsey and Ti explain, such operations must rely on expeditionary logistics, which is the planning and movement of manpower and equipment and its sustainment from afar.[11] This distance may be hundreds or even thousands of kilometres from the force's homebase. This was the case with North Atlantic Treaty Organization (NATO's) intervention in Afghanistan, 2001, and the US / United Kingdom (UK) / multi-national intervention in Iraq, 2003, while another characteristic is that this movement and sustainment is performed in a hostile environment dominated by potential or actual violence.

Such are the challenges confronting military logisticians tasked with preparing and engaging in support of expeditionary operations that they have sought to optimize their actions through the *principles of logistics*[12] —foresight, efficiency, cooperation, simplicity, and agility. Nevertheless, as the following sections explain, optimizing the *principles of logistics* to help achieve operational success is not easily attained under the push system adopted by Russian military logisticians.

As described previously, push logistics literally 'pushes forward' the material and resources needed by the combat arms of the operation based on pre-determined usage schedules, often originating from centrally driven higher level planning sections who control the process.[13] The system works best in an operational environment where demand and consumption rates are stable and predictable, while its major disadvantage in supporting expeditionary operations is its inflexibility. Such inflexibility can quickly transform into a critical vulnerability in a rapidly changing

[11] Ibid., 6.
[12] Uttley and Kinsey, 'The role of logistics in war', 407–14.
[13] Ronald Ti, 'Brief: Russian military logistics', International Centre for Defence and Security, Series No. 3, June 2022, 2.

and unpredictable operational environment.[14] Nor can this inflexibility be easily overcome through the adoption of the *principles of logistics* mentioned above. Instead, the push system works against the *principles of logistics* as discussed below, while the inflexibility, itself, is one of the outcomes of the system.

The two *principles of logistics* that are most vulnerable to marginalization by the push system are 'foresight' and 'agility'. This is not to suggest that the other *principles of logistics* are not ignored, and thus their impact weakened by the push system, but that they are not as susceptible to marginalization by the system as are 'foresight' and 'agility'. It is definitely the case that uncertainty dominates the military logistician's environment.[15] This is even more so with expeditionary operations.

As previously noted, Russian commanders underestimated both the capability of the UAF but especially the will to resist of the Ukrainian population. If they had not, they would have in all probability organized their military logistics differently. Nor does it appear that the Ukrainian military were considered to represent a serious threat to their operation. Such was the thinking of the Russian high command that bringing 'foresight' to bear on the operation's planning was simply ignored or dismissed. This, in turn, would have meant that planning was immediately constrained by ideas, which would not have been the case if the 'foresight' principle had been employed. Moreover, there is little evidence that the Russian high command gave any consideration to gathering logistical intelligence, which is critical if 'foresight' is to be brought to bear on the planning process.

As Uttley and Kinsey explain, 'logisticians require information to anticipate and analyse likely future operations and predict the necessary requirements in terms of manpower, equipment and services to avoid the constraints under which the logistical systems work'.[16] In the case of the Russian high command, collecting this information was either not considered important, and therefore ignored, or simply disregarded as a result of the push system pre-determining usage levels of equipment, manpower, and services as well as scheduling their delivery. In effect, the push system proved to be ineffective as it encouraged Russian logisticians to ignore using their 'foresight' to think ahead about how resources might be delivered in the different theatres of operation if the system failed.

The other aspect of 'foresight' that the push system is unable to respond to effectively is the requirement for logisticians to work closely with all stakeholders to guarantee provisions and services are delivered to the right place in the right quantities and at the right time. For example, this might involve utilizing private sector capabilities, which can often be harnessed at short notice.[17] But this would require greater cooperation from both parties; something that is very difficult under the

[14] Ibid.

[15] Uttley and Kinsey, 'The role of logistics in war', 407.

[16] Ibid., 408.

[17] An example of a military successfully utilizing the market at short notice is the UK Ministry of Defence (MoD) during Operation TELIC (the 2003 invasion of Iraq and subsequent occupation). Even though 'the MoD was prevented from engaging with primary contractors early enough to allow all the required supplies to be delivered on time because of the need to maintain operational security' contractors were still

push system because it is centrally driven by high-level planning sections that are often far removed from the operational space, have little if any immediate understanding/knowledge of what is happening on the front line, lack the appropriate connections to the private sector that is a common feature of Western militaries and goes against the Russia's command philosophy that discourages individual initiative being applied by lower echelon logisticians.[18] In effect, the push system marginalizes or, worse, ignores the utility of other non-state stakeholder strategies designed to mitigate logistical challenges.

As mentioned earlier, the other principle that is vulnerable to the push system is agility. Of all social phenomena, war must rate very high in terms of the uncertainty that it entails. As the Ukraine war has shown, the supply chain environment that Russian forces are operating in is dominated by considerable uncertainty that benefits its asymmetric adversaries. Furthermore, operational success in Ukraine necessitates that logistical support is provided at the right time, in the right place, and in the right quantity to allow fighting units to continue to perform their mission. To achieve this in an environment that is subject to unpredictable threats means logisticians must adopt an agile mindset, able to think 'outside the logistical box' quickly to adapt logistical plans that are then able to meet the challenges these unpredictable threats pose.[19]

Agility refers to nimble and swift action and is usually associated with the physical environment. In the case of logistics, though, it relates to a state of mind that can quickly adapt plans to fit operational requirements. This, in turn, allows a commander to immediately respond to unforeseen events, while ensuring his force continues to be effective and flexible in overcoming unexpected challenges.[20] However, the problem with the push system is that it does not encourage logisticians to be resourceful, but instead to rely on established logistical systems. Nor should it, given that underpinning push logistics is a centralized command system that does not encourage, or even tolerate, independent thinking in its military staff, and this is a fundamental requirement for acting resourcefully. Nor are logisticians who operate under this type of command system necessarily educated to act resourcefully, another hurdle to agile logistics. Critically for field commanders, the failure by logisticians to think agilely in revising plans can quickly place limits on the military supply chain and thereby limit the response by fighting units to unforeseen events.

able to provide the MoD with a surge capability that was needed for the force to be deployed to the theatre of operation. See Christopher Kinsey, *Private contractors and the reconstruction of Iraq: Transforming military logistics* (Oxford: Routledge Press, 2009), 102–5.

[18] The ratio of military personnel to contractors in Iraq was eventually 1:1. Indeed, both the US and UK militaries are heavily reliant on contractor support for their expeditionary operations, without which these operations would not be possible. This support has also enabled them to draw on innovative ideas to help them manage fluid and dynamic operational environments successfully. For a discussion on the impact contractor support has had on the operational environment, see Christopher Kinsey and Malcolm Hugh Patterson, *Contractors & war: The transformation of US expeditionary operations* (Stanford, CA: Stanford University Press, 2012).

[19] Uttley and Kinsey, 'The role of logistics in war', 413.

[20] Ibid., 413.

Finally, what we are not arguing is that 'foresight' and 'agility' principles are totally absent from Russian defence logistics, but that they are not sufficiently embedded into the military logistical structure to have any significant impact on military operations. Indeed, while individual Russian military logisticians may have displayed these principles during the initial stage of Russia's invasion of Ukraine, without them being widely utilized their impact was always going to be limited and certainly not enough to change the initial military failures the invasion experienced. Furthermore, as mentioned earlier, the speed with which Russia's pull system fell apart suggests the argument does hold merit. Nevertheless, we do not attribute this failure only to the lack of 'foresight' and 'agility'. The next section now turns to another important factor contributing to the failure of Russian defence logistics in Ukraine: Russian military doctrine.

Russian Military Doctrine and Its Impact on Delivering Defence Logistics

Having previously discussed the structural challenges that faced Russian military logisticians at the start of the country's 'special military operation' this section focuses on Russia's military doctrine and its impact on the delivery of force projection and sustainment. This section will discuss the concept of 'active defence' and how it shapes Russia's military behaviour in both general and specific terms and what this meant during the initial stage of its operation in Ukraine.

Milevski comments that: 'modern Russian military doctrine is defensive, reflecting both the reality of its railways and the military's perception of Russia's geopolitical situation and imperatives'.[21] The first point to note is that the doctrine of active defence applies to both military and non-military means. 'It conceptualises what the Russian military should do to deter a war before it begins and the general tenets for how it would fight a war against a militarily superior opponent'.[22] The doctrine 'essentially focuses on installing wartime deterrence by denial by degrading the opponent's ability to employ his military effectively through the exploitation of asymmetric responses, resilient air defence, and ultimately seizing the strategic initiative'.[23] The second point to note is that the 24 February 2022 offensive should not be considered an operation under the 'active defence' doctrine. This is because the doctrine seeks to deter the threat of a strategic conventional offensive—initially through using indirect means—a threat which the Ukrainian military evidently did not pose at the time of Russia's invasion.[24]

[21] Lukas Milevski, 'Russian logistics and forward urban defence in the Baltic states', *Military Review* (Washington, DC: US Army University Press, 2022), 3. https://www.armyupress.army.mil/journals/military-review/online-exclusive/2022-ole/milevski/ (accessed 5th February 2024).

[22] Kofman et al., *Russian military strategy*, 10.

[23] Milevski, 'Russian logistics', 3.

[24] Kofman et al., *Russian military strategy*, ii.

On the other hand, the invasion clearly incorporated elements of 'active defence' doctrine. As some examples, Russia's show of force during the build-up to the invasion, its attempt to shape the psychology of Ukrainians beforehand, and the creation of a network of sleeper agents in major Ukraine cities to be activated both before and immediately after the invasion,[25] are all part of the non-military subthreshold activity that the doctrine mentions. However, not all key elements of 'active defence' doctrine were present, particularly in the operational space; this was manifested in key deficiencies such as where fragmentation of command showed the lack of an overall commander to oversee the entire operation and poor logistics planning.

According to Milevski, Russian military theorists anticipate the future battlefield to be fragmented, occupied by fewer troops compared to the two world wars, and lacking continuous front lines.[26] Furthermore, a future where the battlefield is fragmented will place a premium on manoeuvre. This also highlights the vulnerability of logistics, as demonstrated in Ukraine, with Russian forces not benefiting as significantly from manoeuvre warfare as they might have hoped. Milevski makes another, more important, point about Russian military logistics and the Ukraine war and that is that the function of railways is not as a means to invading a hostile territory but to sustain that invading force.[27]

This is not to diminish the role played by the railways in the invasion. The performance of the Russian army in Ukraine demonstrates how critical the railways have been to its penetration of the country. Without the railways, it is doubtful Russia would have been able to take and hold the amount of territory they now control. However, fixed lines of communication based on railways present logisticians with practical challenges delivering support in very fluid operational environments and that are dominated by manoeuvre, unless the motorized units supporting fighting units are operating in close geographical proximity to the railway heads.

Unfortunately, while this may be the case with some parts of the front line, there are significant areas where it does not apply, placing Russian fighting units at a considerable disadvantage logistically. The idea of 'active defence' as reflected in Russia's use of its railway system as part of its defensive doctrine does not sit well with, or even apply to, an offensive operation of the type being conducted in Ukraine. However, its effects on Russian operational thought are deeply ingrained, making it challenging for commanders and planners to conceptualize thinking outside its parameters.

The doctrine also places great emphasis on 'the use of standoff weaponry, [as well as] precision and massed fire … to maximise the survivability of Russian units and preserve the force'.[28] Moreover, this is where the doctrine has had limited success

[25] Russia's use of the Wagner Group to hunt down and assassinate Ukraine's President Volodymryr Zelensky and members of his cabinet is an example of non-military means that make up 'active defence' doctrine. See Louise Nordstrom, 'Wagner Group's bloody year in Ukraine: From murder squad to cannon fodder', *France 24*, 2 February 2023. https://www.france24.com/en/europe/20230222-wagner-groups-bloody-year-in-ukraine-from-murder-squad-to-cannon-fodder (accessed 5 February 2024).

[26] Milevski, 'Russian logistics', 3.

[27] Ibid.

[28] Kofman et al., *Russian military strategy*, 13.

with its use of long-range artillery strikes to inflict casualties and cause disorganization among Ukrainian fighting units. A strategy of massed fire, though, relies first on having enough stockpiles of munitions to maintain a relentless concentration of fire and, second, the means to replenish exhausted stocks of shells quickly. It also involves being able to move large quantities of shells from railheads to where they are needed using trucks. Replenishing stocks of shells has been particularly difficult for Russia, with the country apparently having to rely on North Korea for munitions.[29] The principal challenge has been maintaining sufficient replenishment of munitions.

According to Kofman et al., the '"active" within active defence envisions demonstrations or limited use of force, to convince opponents that the cost of aggression would exceed desired benefits'. If the doctrine of 'active defence' was behind the decision to invade Ukraine, it is unclear who the invasion was supposed to deter. Furthermore, the invasion can hardly be called a limited use of force and therefore does not fit neatly within the boundaries of the doctrine. The Russian high command might have perceived a threat to Russia's security from the West but if this is the case, then in line with 'active defence' doctrine surely Russia would initially address this perceived threat through the use of indirect means, instead of invading the country.

Alternatively, there may be other motive(s) behind the decision to invade Ukraine. This may also be why defence logistics was not part of the discussion behind the invasion. For example, the purpose of the invasion may have had less to do with the doctrine of 'active defence', and more about replacing a government desirous of stronger ties with the West with a supportive government in the Kremlin's 'near abroad' well within its sphere of influence. In conclusion, it would appear that no serious deliberation was given to the doctrine of 'active defence' in the planning of the invasion, or if it was, it only applied to that part of the strategy occupied with the use of standoff weaponry and mass fire to increase the survivability of Russian units and to protect Russia's forces in the country. The next section examines the impact of Russian military command philosophies and how they restrict defence logistics initiatives.

Russian Military Command Philosophies and How They Restrict Russian Operational Logistics

Russian command philosophy could be summarized in two words: central control. The command structure is pyramidal with essential command decisions made at the highest echelons. This is in contrast to the widely practised philosophy of mission command which is universally taught, practised, and promoted throughout NATO militaries.[30] In essence, mission command devolves much decision making

[29] Pjotr Sauer, 'Evidence mounts of North Korean arms to Russia in threat for Ukraine', *The Guardian*, 27 October 2023. https://www.theguardian.com/world/2023/oct/27/north-korean-arms-supply-russia-war-ukraine-munition-shipments (accessed 9 February 2024).

[30] NATO HQ, Allied Joint Doctrine for Land Operations, AJP 3.2, NATO, 2022, 41–4.

to subordinate levels of command. A key element is to clearly identify the overall commander's intent to subordinate commanders.

Mission command inculcates the subordinate commander with an understanding of the mission, enabling that subordinate with the situational awareness required to understand the assigned mission, but leaving specific details of execution to that subordinate. The result is that instead of all command decisions being made centrally, these can also be made peripherally, obviating the need to refer every decision to higher command levels. The major advantage of mission command over centralized command philosophies is that it is a command philosophy that assumes a non-linear battlespace that is both unpredictable and emergent in nature.[31]

The major deficiency of centralized command systems is that it is not always possible to refer all major decisions to higher command. In the 'heat of battle' and the 'fog of war' when communications are impaired or non-existent, the lack of decision making at the 'sharp end' means that initiative is restricted (or even stopped). In logistics, this is reflected in a push-based system where peripheral units cannot generate demand signals and supplies are pushed forward.

Given these circumstances it is clear how centralized military command systems and centralized 'push' driven logistics have contributed to Russian operational logistics problems in the current conflict. First, centralized command systems as typified by push logistics systems do not reflect actual demands at the front line, or 'sharp end'. This was one of the reasons pictures of stalled Russian military logistics convoys were shown at the commencement of the war.

Second, operational delays at the front line were not accounted for by a logistics system which had already been programmed to push supplies forward, particularly in the absence of real-time information about actual supply requirements. This situation has been exacerbated by the lack of a modern enterprise resource management system such as those presently in use by Western militaries. The need for centrally located commanders to make critical decisions has led directly to conspicuously high casualty rates among senior Russian commanders who have been killed in the forward edge of the battle area. By contrast, it would be unusual for Western military commanders of equivalent seniority to be physically present due to necessity.

Lastly, it appears that the Russian military lacks any higher echelon organization equivalent to a joint logistics command or even a joint logistics support group as developed within NATO during the past decade.[32]

Conclusion

This chapter has sought to explain the impact of Russian military logistics on its war with Ukriane. In doing so, the chapter argues that the lack of adequate defence

[31] Brian Cole, 'Clausewitz's wondrous yet paradoxical trinity: The nature of war as a complex adaptive system', *Joint Force Quarterly* 96 (2020): 42–9.

[32] NATO, Allied Joint Doctrine for the JLSG, AJP 4.6, edition C, §, Nato, 2018.

logistics support is at the root of many operational and tactical setbacks that have come to characterize the early stages of the 'special military operation'. As the chapter explains, many of the operational and tactical challenges that the Russian military faced in the early stages of the invasions, and in certain cases still faces three years into the war, should have been foreseen and mitigated by Russia's military high command. Instead, as a result of either downplaying or simply ignoring defence logistics, the initial stages of the 'special military operation' in particular were already precarious even before Russian troops crossed over the border into Ukraine.

At the heart of this failure are structural challenges that have undermined Russia's military's ability to supply and sustain its 'special military operation' to the degree needed to guarantee operational and tactical success. In particular, Russia's military has not been able to optimize the *principles of logistics*, particularly foresight and agility, to its advantage on the battlefield. Instead, its operations have lacked the flexibility needed to rapidly adapt to the changing operational and tactical environments that is essential to achieving operational and tactical success. Compounding this inflexibility is Russia's reliance on a logistics system that is centrally controlled and that pushes supplies forward according to pre-determined usage rates that are often decided by higher command levels with little or no knowledge of what is actually required by front-line units.

The final question posed in this chapter is, what are the most important lessons from the Ukraine conflict that can be learnt by NATO in the area of operational logistics? The first is that logisticians operate in a battlespace that is transparent, where there is nowhere to hide, and where logistics assets are targets at a considerable distance behind any notional 'front line'. The implication of this is that lacking any 'rear areas' or similar 'safe spaces', logisticians must think more like combat soldiers. They need to pay heed to camouflage, concealment, managing their electronic signature, and a host of tactical issues, which did not hitherto require consideration during the Cold War. The second lesson, which leads directly from the first lesson, is how to manage the exponential rise in the use of uncrewed aerial systems (UAS). This is a critical issue given the surge in the use of 'first-person view' (FPV) drones controlled by operators with vision goggles, which are having a very significant impact on the battlefield. The very recent introduction and expansion of UAS guided by fibre-optic cables has rendered jamming devices ineffective as these units do not require radio signals to control the drone. The lethality of such units will also increase exponentially as AI guidance models are developed and further refined, removing humans from the kill chain.

The implication is that given the now transparent battlespace, the use of these units extends the availability, reach, and lethality of adversaries, especially in targeting what were reasonably considered to be 'rear areas' out of reach. Lastly the third lesson is a more enduring one, and it is that warfare continues to be as unpredictable as it has always been, and that this unpredictability will continue to confound the best efforts of operational logisticians. This last lesson points to why logisticians must conduct their own threat assessment of the battlespace and always be aware of what and where the next threat may come from.

Postscript: 2025

This note provides an updated summary of the current 2025 operational logistics situation as this volume goes to print. The overall front line has become static over the past two years with attritional warfare occurring across static lines of defence across the entire front line.[33] While the overall logistics picture described in this chapter has remained fundamentally unchanged, logistics resupply movement in what constitutes the 'last logistics mile' in the respective front-line areas of both adversaries has become more restrained and more manoeuvre restricted. This has been due primarily to the massive increase in use of small unmanned aerial systems (sUAS), particularly FPV drones. As a result of the increasing range of sUAS, the risk areas for strikes from these weapons have been pushed further rearwards from each adversary's forward line of own troops (FLOT). This has necessitated most movements occurring at night, particularly casualty movements, which in turn has increased the time interval between the point of wounding and first surgery, increasing both mortality and morbidity.[34] The other major changes in the logistics picture are the increasing use of UAS (despite their current obvious load limitations[35]) to drop small quantities of supplies to the front line,[36] as well as the increasing dispersal of sustainment assets, with one report noting how Ukrainian resupply is distributed among fleets of pickup trucks, further dispersing and disaggregating the logistics supply transport chain. What has also increased is the risk to non-military transport and the civilian personnel who operate these commercial vehicles closer to the front line and, in 2025, increasingly including the Ukrainian railway network.[37] The success or failure of the Russian 'special military operation' continues to hang on the integrity of logistics and the imperative to maintain sustainment in the face of attritional, positional warfare.

[33] Jack Watling and Nick Reynolds, 'Tactical developments during the third year of the Russo-Ukrainian war', Special Report, Royal United Services Institute (RUSI), 2025. https://static.rusi.org/tactical-developments-third-year-russo-ukrainian-war-february-2205.pdf (accessed 6 October 2025).

[34] Ibid., 15–16; and Terry Rauch, 'Weekly review of medical readiness across the NATO AOR', *Aspen Medical USA*, September 75 (2025): 6.

[35] See JOUAV: Unmanned Aircraft System, 'How much weight can a drone carry? (Comprehensive 2026 guide)'. https://www.jouav.com/blog/how-much-weight-can-a-drone-carry.html#:~:text=Agricultural%20Drone&text=Payload%20capacities%20can%20range%20from,50%20kg%20of%20dry%20material (accessed 6 October 2025).

[36] Pieter De Nijs, 'Lines of sustainment: Evaluating Ukrainian military logistics in the Russo-Ukrainian war', *Info Flash: FINABEL*, July 2025, 1–16 at 3. https://finabel.org/lines-of-sustainment-evaluating-ukrainian-military-logistics-in-the-russo-ukrainian-war/; and Watling and Reynolds, 'Tactical developments', 17.

[37] See 'Russian strike hits passenger train in Ukraine causing casualties', Reuters, 4 October 2025. https://www.reuters.com/business/aerospace-defense/russian-strike-hits-passenger-train-ukraine-causing-casualties-2025-10-04/ (accessed 6 October 2025).

6
Health Services in War

Martin C. M. Bricknell

Introduction

This chapter will review emerging observations for health service support to armed forces from the Russo-Ukrainian war after the invasion by Russia in 2022. It will use the concepts described in the chapter on 'Military Health Services Supporting the Land Component in the Twenty-first Century' in *Advanced Land Warfare: Tactics and Operations*.[1] While many of the predictions on the impact of advances in land warfare on medical services have come to fruition, the rate of organizational and technological change during this war have provided additional signposts to future issues and challenges for the health service support function during conflict. The content of this chapter is based on reviewing open-source publications (primarily the *Ukrainian Journal of Military Medicine*[2] and *Current Aspects in Military Medicine*[3]) since 2022 identified using the search string 'lessons military medicine Ukraine' in Google and Google Scholar. This information was supplemented by reviewing notes and slides from presentations given by Ukrainian military medical services officers at the World Congress of Military Medicine in 2022 and 2024,[4] and at the International Conference on Disaster and Military Medicine in 2021, 2022, 2023, 2024, and 2025.[5] It is inevitable that most of this information concerns the armed forces of Ukraine (AFU) medical experience as there is little information available from open sources on the Russian Federation military experience.

This chapter covers five themes. The first discusses the realities of combat in a peer-on-peer conflict and the implication for health service support in terms of the

[1] Martin C. M. Bricknell, 'Military health services supporting the land component in the twenty-first century', in Mikael Weissmann and Niklas Nilsson (eds), *Advanced land warfare: Tactics and operations* (Oxford: Oxford University Press, 2023; online edn, Oxford Academic, 20 April 2023). https://doi.org/10.1093/oso/9780192857422.003.0011 (accessed 17 December 2024).

[2] The *Ukrainian Journal of Military Medicine* is an open-access journal published four times per year by the Ukrainian Military Medical Academy (UMMA). https://ujmm.org.ua/index.php/journal.

[3] *Current Aspects in Military Medicine* is an open-access journal jointly published once per year by the Ukrainian National Military Medical Clinical Centre and the UMMA. The website is at: https://camm-journal.com/index.php/journal.

[4] World Congress of Military Medicine 2022. https://icmmworldcongress2021.org/index.php; World Congress of Military Medicine 2024. https://icmm2024australia.com.au/.

[5] DiMiMed. https://events.military-medicine.com/dimimed/.

Martin C. M. Bricknell, *Health Services in War*. In: *Russia's War in Ukraine and Modern Warfare*.
Edited by: Mikael Weissmann and Niklas Nilsson, Oxford University Press.
DOI: 10.1093/9780198970903.003.0006

volume of casualties and the threats to the national medical system. The next theme examines tactical medical care on the battlefield covering pre-hospital care, tactical medical evacuation, and forward surgical support. The theme of operational organization of medical services discusses the distribution and reception of casualties into the civil-military health system, recovery and rehabilitation, and battleshock and mental health. The third theme, long-term strategic issues, looks at the medical assessment of recruits, medical retention standards, military disability pensions, long-term care of the war injured, military healthcare education, and finally, antimicrobial resistance. The final section considers the implications of the Ukrainian experience for military health systems across the globe as the threat of war increases due to the tensions across the global international system.

The Realities of Combat

In an unprecedented post on X on 8 December 2024, President Zelinsky announced that Ukraine had lost 43,000 soldiers and had 370,000 wounded and in need of medical assistance.[6] Although he suggested that Ukrainian data showed that approximately half of the soldiers wounded in action are later returning to the battlefield, this means that 185,000 military personnel were unfit for further action. President Zelinsky compared the Ukrainian numbers with an estimate of Russian losses of 198,000 killed and more than 550,000 wounded. While the Russian losses are far larger, the Ukrainian losses as a proportion of the adult population will have extensive and long-term impacts on economic output, health and social support, and the psyche of the whole country.

It is widely accepted that the modern battlefield has become almost transparent with ubiquitous sensors able to continuously detect and track military targets. The consumption of ammunition by both sides has almost exceeded North Atlantic Treaty Organization (NATO) and Russian production capacity and the front lines mirror the battlefields of World War One. Precision weapons, most particularly unmanned aerial systems (drones), have increased the likelihood of direct hits on vehicles and strong points. Beyond the increased lethality on the battlefield, there has been a substantial number of Russian operational-level attacks on hospitals, healthcare infrastructure, ambulances, and wider civilian infrastructure in complete disregard of the Geneva Conventions and other laws of war.[7] The war has had a devastating impact on the health of the Ukrainian population and the Ukrainian health system.[8] Unlike for

[6] President Zelinsky (@ZelenskyyUa), 'Yesterday, I visited President @EmmanuelMacron at the Élysée Palace and had a good meeting with President @realDonaldTrump'. Post on X, 8 December 2024. https://x.com/ZelenskyyUa/status/1865709519352873408.

[7] Luke Taylor, 'Ukraine: Nearly 2000 attacks on health facilities have been launched since invasion, says WHO', *British Medical Journal* 386 (2024): q386. doi: https://doi.org/10.1136/bmj.q1833.

[8] Martin Bricknell and George Bundy, 'The devastating impact of war on health and health systems', in Zeno Leoni, Maeve Ryan, and Gesine Weber (eds), *War in Ukraine—One Year On* (London: Kings College London, 2023). https://www.kcl.ac.uk/warstudies/assets/war-in-ukraine-one-year-on.pdf.

Russia, the whole Ukrainian population is at risk; armed forces, militia, local defence forces, and non-combatant civilians. In additional to direct attacks, the demand for trauma care has displaced care in the whole hospital system reducing cancer services, gynaecological services, and professional education programmes, and led to emigration of health professionals.[9] This has also occurred in the digital environment alongside the physical environment with multiple cyber attacks against healthcare information systems in Ukraine and its allied countries.[10]

Overall, as the battlefield has become more intense and more lethal, medical care for casualties has also become more complicated. An analysis of casualties between February and April 2022 of those casualties who survived to be treated in the National Military Medical Centre, Kiev found that the majority of wounded, 75%, had penetrating explosive injuries with the blast causing 16%, bullets 8%, and burns 1%. In common with the high-intensity wars of the twentieth century, mortars, artillery, and rockets cause more casualties than bullets. Patients with limb injuries made up 89% of the total received at the hospital compared with 65.7% in the period 2014–21. The reduction in other sites of injury is a possible indicator of a lower chance of survival for the wounded in that period after the invasion due to pressures on the medical system.[11] First-person view (FPV) drones have become the primary cause of battle casualties with survivors having proportionally more head, neck, and upper limb injuries that casualties from earlier in the war.

By the time of writing this chapter, the war has entered its fourth year with both sides still seeking operational-level success against their enemy. Both sides have had to conscript, recruit, and train new forces with limited opportunities for rotation and recovery of current forces. This, alongside continual losses, will have had a long-term impact on morale, motivation, and mental health of all military personnel and wider society. All these factors have contributed to placing intense demands on the military health systems of both countries both in the numbers of casualties to be treated and evacuated, and in the complexity and risk associated with fulfilling this mission.

The AFU have been fighting this war alone. Many countries have provided physical assistance with finance, weapons, ammunition, and other technologies. Though no international military forces have been involved in fighting, many NATO countries have provided training assistance to Ukrainian military personnel in third countries

[9] Ubydul Haque, Moeen Hamid Bukhari, Nancy Fiedler, Shanshan Wang, Oleksii Korzh, Juan Espinoza, Miraj Ahmad, Irina Holovanova, Tetyana Chumachenko, Olga Marchak, Dmytro Chumachenko, Osman Ulvi, Ifthekar Sikder, Hanna Hubenko, and Emily S. Barrett, 'A comparison of Ukrainian hospital services and functions before and during the Russia-Ukraine war', *JAMA Health Forum* 5 (2024): e240901. doi:10.1001/jamahealthforum.2024.0901.

[10] Rose Bernard, Gemma Bowsher, and Richard Sullivan, 'Cyber security and the unexplored threat to global health: A call for global norms', *Global Security: Health, Science and Policy* 5/1 (2020): 134–41. https://doi.org/10.1080/23779497.2020.1865182.

[11] Anatoliy Kazmirchuk, Yurii Yarmoliuk, Igor Lurin, Rostislav Gybalo, Olexandr Burianov, Serhii Derkach, and Kostiantyn Karpenko, 'Ukraine's experience with management of combat casualties using NATO's four-tier "changing as needed" healthcare system', *World Journal of Surgery* 46/12 (2022): 2858–62. doi:10.1007/s00268-022-06718-3.

and donations of medical equipment and supplies. This has included first aid training, combat medic training, and other technical medical education. International personnel with experience of caring for combat casualties have deployed into Ukraine with civilian non-government organizations (NGOs) to provide medical education programmes, medical equipment, and materiel, and to contribute to hands-on clinical treatment of Ukrainian casualties, both military and civilian. Some NGOs have provided specific support to the AFU medical services both in an advisory role and to augment the AFU treatment capacity. This experience is beginning to be reported in the academic literature alongside papers written in the Ukrainian military medical literature.

Tactical Medical Care

There are three health service support 'instruments of care' to be performed in the combat zone of tactical operations: pre-hospital care (emergency care and routine sick care), tactical medical evacuation, and initial surgical care (often called damage-control resuscitation and surgery). The intensity of the combat zone has particularly affected the ability to meet the NATO medical timelines of '10–1–2(2)+2'.[12] The US Joint Trauma System (JTS)'s Committee on tactical combat casualty care (TCCC) concepts and guidelines are widely accepted as the standard for medical care in the combat zone.[13] These have been taught to many AFU medical personnel with most of the material being translated into Ukrainian and presented in digital form.[14] This same model of pre-hospital emergency first aid has been taught to Ukrainian civilians.[15] Quinn et al. published a comprehensive review of pre-hospital lessons from the war in Ukraine involving authors from Ukraine and the US.[16] This paper identified ten lessons on the following topics: scene safety, triage, patient assessment, tourniquets, pain management, immobilization for transport, warming and blood products, infection control, all age groups, and clinical governance.

[12] This is defined as ten minutes to first aid, one hour to advanced resuscitation, two hours to damage-control resuscitation and surgery (which is assumed to take two hours), and an additional two hours for initial medical evacuation (tactical evacuation care, or TACEVAC) for further resuscitative, diagnostic, surgical, and specialist care capabilities necessary to stabilize the patient for strategic evacuation.

[13] Frank K. Butler, 'Two decades of saving lives on the battlefield: Tactical combat casualty care Turns 20', *Military Medicine* 182/3–4 (2017): e1563–e1568. https://doi.org/10.7205/MILMED-D-16-00214.

[14] D. V. Karamyshev, V. M. Zhdan, V. M. Dvornyk, L. P. Hordiienko, and Zh. P. Kundii, 'Universally applicable approaches to the tactical level of aid and medical support for the personnel of the armed forces of Ukraine', *Світ медицини та біології* 4/82 (2022): 74–79. http://repository.pdmu.edu.ua/handle/123456789/20119.

[15] V. Sulyma, U. Kuz, Y. Filiak, L. Tverdokhlib, K. Hafiychuk, and V. Sarancha, 'VYSHKIL—training of survival: A new approach to train civilians first aid skills', *Pain, Joints, Spine* 13/2 (2023): 77–84. https://doi.org/10.22141/pjs.13.2.2023.369.

[16] John Quinn, Serhii I. Panasenko, Yaroslav Leshchenko, Konstantyn Gumeniuk, Anna Onderková, David Stewart, A. J. Gimpelson, Mykola Buriachyk, Manuel Martinez, Tracey A. Parnell, Leonid Brain, Luke Sciulli, and John B. Holcomb, 'Prehospital lessons from the war in Ukraine: Damage control resuscitation and surgery experiences from point of injury to role 2', *Military Medicine* 189 (2024): 17–29. doi:10.1093/milmed/usad253.

Scene safety reflects the increased risk and lethality of the battlefield and the need for suitable protective equipment (helmets, body armour, eye protection) for military and civilian health personnel in the combat zone. Triage is the process for prioritizing casualties according to the severity of their injuries. Almost all casualty events cause several casualties which may overwhelm forward medical positions and may reoccur multiple times per day. This places significant challenges for medical personnel to triage patients for both treatment and evacuation, including the expectation that some patients may wait a considerable time before evacuation necessitating prolonged field care (PFC) measures. Patient assessment has to be both correct at first presentation and multiply repeated in order to detect any patients who are deteriorating.

Military personnel, medics, and doctors need specific training and equipment for the realities of battlefield medical care. Alongside training in TCCC, widespread distribution of individual first aid kits (IFAKs) is needed according to a standard packing list.[17] The reintroduction of the tourniquet into military pre-hospital care during the wars in Iraq and Afghanistan has been credited with saving many battle casualties. The versions in use by NATO forces are expensive and so local Ukrainian designs have become widely adopted. Although they can undoubtably save lives, bad designs, improper application, and prolonged use can damage limbs by cutting the blood supply to healthy limbs. A review of the use of tourniquets in sixty-nine patients showed that tourniquet use was appropriate in 24.6% of cases and that further training, including the use of local pressure bandages, is needed to ensure proper control of bleeding and fewer complications from tourniquet use.[18] It will be necessary to refocus TCCC training to ensure that those using tourniquets are much more discerning when delays in evacuation are inevitable.[19]

Tactical medical evacuation from forward areas is almost invariably conducted at night due to the threat from drones and indirect fire. Therefore most patients will not meet the 'golden hour' from wounding to surgery and so PFC measures will need to be implemented, sometimes over several days! PFC aims to stabilize, delay, or reverse the physiological deterioration of a seriously ill or injured patient through effective patient care supported by timely and appropriate medical interventions.[20] Medical

[17] V. V. Chorna, M. V. Matviichuk, V. M. Podolian, Y. M. Ivashkevych, V. M. Syvak, V. V. Slobodian, and M. S. Ivashkevych, 'Current issues of providing individual first aid kit of military personnel and civilians of Ukraine', *Ukrainian Journal of Military Medicine* 3/2 (2022): 92–101. https://doi.org/10.46847/ujmm.2022.2(3)-092.

[18] Vladyslav Yatsun, 'Application of hemostatic tourniquet on wounded extremities in modern "trench" warfare: The view of a vascular surgeon', *Military Medicine* 189 (2024): 332–6. doi:10.1093/milmed/usac208.

[19] Frank Butler, John B. Holcomb, Warren Dorlac, Jennifer Gurney, Kenji Inaba, Lenworth Jacobs, Bob Mabry, Mike Meoli, Harold Montgomery, Mel Otten, Stacy Shackelford, Matthew Tadlock, Justin Wilson, Kostiantyn Humeniuk, Oleksandr Linchevskyy, and Oleksandr Danyliuk, 'Who needs a tourniquet? And who does not? Lessons learned from a review of tourniquet use in the Russo-Ukrainian war', *Journal of Trauma and Acute Care Surgery* 97/2S (2024): S45–S54. doi: 10.1097/TA.0000000000004395.

[20] Mike Smith, K. Johnston, and Richard Withnall, 'Systematic approach to delivering prolonged field care in a prehospital care environment', *BMJ Military Health* 167 (2021): 93–8. http://dx.doi.org/10.1136/jramc-2019-001224.

personnel in casualty collection points or forward 'stabilization' points should be proficient with PFC techniques. Pre-hospital pain management is complicated by the lack of suitable pre-hospital opioid drugs in Ukraine. Patients need to have their wounds dressed and immobilized alongside rewarming if exposed to low temperatures and administration of blood or other fluids to replace blood loss. Antibiotic regimes should be started at the first point of medical care as wound surgery and debridement is likely to be delayed by many hours, even if this carries the risk of antibiotic resistance emerging in wound bacteria. Military medical units also need to anticipate casualties among militia volunteers and non-combatant civilians within their area of operations in the combat zone, alongside responding to non-trauma medical emergencies and co-morbidities in trauma patients.[21]

Almost no forward medical evacuation can be undertaken by helicopter due to the surface-to-air and air-to-air threat. The first stage of evacuation is almost invariably a hand carry by fellow fighters. Ground medical evacuation is conducted by armoured vehicles, fitted ambulances, general vehicles converted to ambulances, and vehicles of opportunity. Formal tasking and patient regulation are limited by the availability of communications and relatively ad hoc local command arrangements. There are emerging examples of successful use of remote control ground vehicles to evacuate casualties. This experience from Ukraine is a radical departure from the low-volume, high-speed forward medical evacuation of field medical care during the NATO mission in Afghanistan.

Epstein et al. have published the experience of the Global Surgical and Medical Support Group, an NGO with a strong cadre of ex-military medical personnel.[22] Their paper provides further evidence of differences in the tactical environment of the war in Ukraine compared to the NATO experience in Afghanistan. The paper focuses on the lessons for forward surgery, so called 'Role 2' medical care. Building on the observation of the importance of triage in the pre-hospital environment, surgical teams have had to frequently apply 'MASCAL' (mass casualty) in the assessment of patients for surgery. This means that some severely injured patients may be given compassionate end-of-life care rather than intensive resuscitation and evacuation in order to protect the capacity of the whole medical system. Time pressure and capacity may also mean that some patients with relatively minor wounds will not receive wound surgery at this level even if this results in deeper infection because of wound contamination.

[21] Viktor Semenov, S. Kumar, M. Y. Zak, O. Kuryata, A. Murach, I. Syrovatko, N. Krupchak, C. Cortellino, and G. Trombatore, 'Types of morbidity presenting in frontline-bordering regions of Ukraine: The experience of deploying additional mobile medical units', *BMJ Military Health* 170/e2 (2024): e110–e115. https://doi.org/10.1136/military-2023-002365.

[22] Aaron Epstein, Robert Lim, Jay Johannigman, Charles J. Fox, Kenji Inaba, Gary A. Vercruysse, Richard W. Thomas, Matthew J. Martin, Gumeniuk Konstantyn, and Steven D. Schwaitzberg, 'Putting medical boots on the ground: Lessons from the war in Ukraine and applications for future conflict with near-peer adversaries', *Journal of the American College of Surgeons* 237/2 (2023): 364–73. doi:10.1097/XCS.0000000000000707.

Delays to medical evacuation occur across the whole medical system and so some patients may receive more than one 'damage-control surgery' procedure alongside prolonged care in a field intensive care bed before further evacuation. Forward surgical teams may face significant challenges in the resupply of blood and other medical consumables when they hold patients for longer than planned for in medical doctrine. The techniques of damage-control resuscitation and damage-control surgery to save life pending further surgical operations to repair damage have been confirmed as key tenets of war surgery. However, it is also important to ensure clinical personnel are aware of other principles of war surgery such as 'second-look surgery', delayed primary suture, and avoidance of open fixation through contaminated wound tracks.[23] The personnel in these forward surgical teams also need to be competent to look after themselves in an austere field environment and to be able set up their facility in protected temporary accommodation such as underground carparks or basements.

It is important not to forget the responsibility of the military medical services to prevent disease and treat the sick and injured who do not require further evacuation. This requires a system for territorial medical support that matches arrangements for garrison healthcare in peace. This can also be a hub for the management of battleshock cases and for rotating medical staff from the front line.[24]

Operational Organization of Medical Services

The Ukrainian military health system has been transitioning from a Soviet model to a NATO model since the Russian invasion of the Donbas in 2014. This included the unification of medical services for the AFU under a single command and a step-by-step increase in its capacity with gradual adaptation to the system of medical support of the armed forces of NATO member states.[25] While the NATO standardization documents provided a valuable reference point, it became clear that these principles needed to be adapted for the specific circumstances of the AFU acknowledging that Ukraine has been fighting a war of national survival on its own territory.[26] NATO concepts of military Role 3 (in-theatre hospitalization) and military Role 4

[23] Sevan Gerard, Anna Onderková, Michael Meoli, Yaroslav Leshchenko, Madelyn Banks, C. Beato, Denys Surkov, and John Quinn, 'Large scale combat operations and far forward damage control resuscitation: Observations of Ukraine', *European Military Medical Services* 2024: 18–25. https://military-medicine.com/article/4272-large-scale-combat-operations-and-far-forward-damage-control-resuscitation-observations-of-ukraine.html.

[24] O. O. Mykyta, 'The system of territorial defense forces medical support of the armed forces of Ukraine', *Ukrainian Journal of Military Medicine* 3/3 (2022): 64–70. https://doi.org/10.46847/ujmm.2022.3(3)-064.

[25] V. O. Zhakhovsky, V. G. Livinskyi, T. M. Ostaschenko, O. M. Ivanko, and N. O. Vasiuk, 'Medical support of the armed forces of Ukraine in the context of public health care transformation: Regulatory and legal aspects', *Ukrainian Journal of Military Medicine* 2/4 (2021): 5–13. https://doi.org/10.46847/ujmm.2021.4(2)-005.

[26] V. G. Livinskiy, V. O. Zhakhovskyi, A. V. Shvets, and O. M. Ivanko, 'Standardization of medical support in the armed forces of NATO member states', *Ukrainian Journal of Military Medicine* 3/2 (2022): 5–18. https://doi.org/10.46847/ujmm.2022.2(3)-005.

(out-of-theatre hospitalization) have been adjusted to reflect Ukrainian needs.[27] This included the requirement for greater cooperation with the civilian healthcare system by increasing inter-ministry coordination of the activities of the medical services of the UAF and the civilian healthcare system of the state in a single medical space defined as 'a system of organizing medical care, which unites all medical resources of the state by joint management and defined financing mechanisms and ensures accessibility, quality and effectiveness of all types of medical care for all citizens of Ukraine, including military personnel'.[28] The introduction of martial law in 2022 provided the legal authority for full civil-military integration of health service support operations and authority for military personnel to access civilian medical services.[29]

In Ukraine, Role 3, or hospital-level care is a civil-military effort as all patients are 'funnelled' from the combat zone to area receiving facilities in the relatively safe 'territorial hospital bases' of the Ministry of Health. These might be reinforced with military medical personnel to provide care for military patients in separate departments of civilian hospitals but sharing key capabilities such as imaging and operating theatres.[30] This is very different from NATO medical doctrine which envisages a purely military chain of care within a theatre of operations. Thus provision of integrated health service support to national populations within the zone of military operations is an operational-level civil-military task that requires collaboration between both parties at both regional and national level.

Many countries experienced a similar requirement to integrate national health services during the COVID-19 pandemic health emergency. Medical evacuation from the hospital bases to the interior of Ukraine has primarily been undertaken by ambulance and rail since 2014, with very limited use of aircraft.[31] Trains have been used to move large numbers of casualties from eastern hospitals to hospitals in the west of the country staffed by Ukrainian medical personnel and international volunteers.[32] Military casualties are received in the 'rear area' under the supervision of the National Military Medical Centre and distributed to facilities focusing on care

[27] Dmytro Karamyshev, Liudmyla Hordiienko, and Mykola Lytvynenko, 'Management of the medical and evacuation system development of the armed forces of Ukraine according to NATO standards', *Public Administration and Regional Development* 20 (2023): 445–70. https://doi.org/10.34132/pard2023.20.08.

[28] V. O. Zhahovskiy, V. G. Livinskyi, S. O. Petruk, and O. V. Zhakhovska, 'Functional and organizational model of medical supply of the defense forces on the basis of a uniform medical space', *Ukrainian Journal of Military Medicine* 3/3 (2022): 35–47. https://doi.org/10.46847/ujmm.2022.3(3)-035.

[29] D. V. Karamyshev, V. M. Zhdan, V. M. Dvornyk, L. P. Hordiienko, and Zh. P. Kundii, 'Institutional principles of civil-military cooperation regarding medical support of the armed forces of Ukraine', *Bulletin of Problems in Biology and Medicine* 4/167 (2022): 66–74. doi: 10.29254/2077-4214-2022-4-167-66-75 (in English).

[30] A. V. Verba, T. M. Ostashchenko, I. B. Plis, S. S. Tuz, and I. V. Sobko, 'Experience of using consolidated medical units as part of a forward hospital base in conditions of high intensity armed conflict', *Ukrainian Journal of Military Medicine* 4/1 (2023): 13–20. https://doi.org/10.46847/ujmm.2023.1(4)-013.

[31] V. O. Zhakhovsky and V. G. Livinsky, 'Six years of war: Experience and lessons of medical support of the armed forces of Ukraine (first message)', *Current Aspects of Military Medicine* 27/1 (2020): 25–38. https://doi.org/10.32751/2310-4910-2020-27-03.

[32] Stig Walravens, Albina Zharkova, Anja De Weggheleire, Marie Burton, Jean-Clément Cabrol, and James S. Lee, 'Characteristics of medical evacuation by train in Ukraine', *JAMA Netw Open* 6/6 (2023): e2319726. doi: 10.1001/jamanetworkopen.2023.19726.

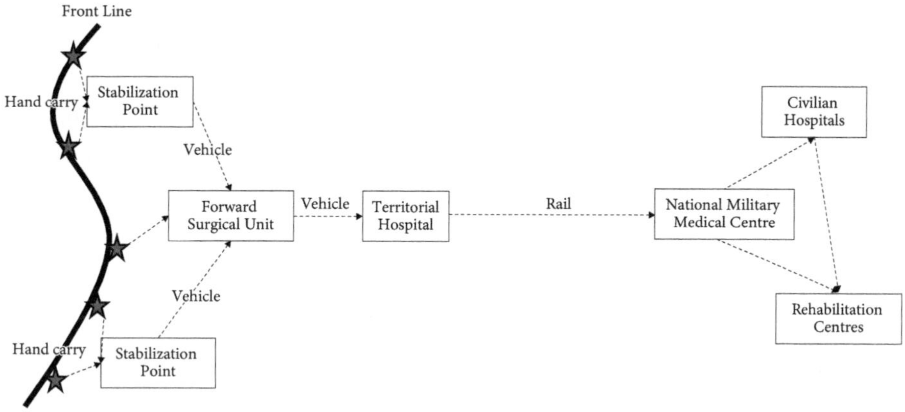

Fig. 6.1 Ukrainian scheme of medical evacuation

of military patients. A limited number of civilian and military trauma patients have been transferred to other European countries for hyper-specialist treatment under the European Union (EU) Civil Protection Mechanism.[33] A schematic summary of the Ukrainian scheme of medical evacuation is shown at Figure 6.1.[34]

In addition to creating a 'single medical space', it has been recognized that clinical care for the casualties of war should be standardized with common clinical protocols for the management of trauma cases and an agreed approach to data collection in registries to monitor the clinical performance of the whole medical system, civilian and military.[35] However, this has proved difficult to implement with many key features of a holistic clinical governance system remaining absent.

This chapter concentrates on the impact of war on the physical health of land forces. However, it is important not to forget the mental and social health consequences of war even if there is a delay between exposure to the violence of war and the full manifestation of these conditions. There is a considerable volume of academic literature on the mental and social health impact of the wars in Iraq and Afghanistan on NATO forces. The experiences of war were already creating a significant demand on mental health services for military personnel prior to 2022 with a predictable association between combat intensity and type and rate of mental health

[33] Anders Holtan, Fredrik Ottesen, Carl Christopher Hoel, Petter Iversen, Steinar Olsen, and Marianne Thorbjørnsen, 'Medical evacuation of patients from war-torn Ukraine to Norwegian hospitals', *Tidsskrift for Den Norske Legeforening* 143/8 (2023). https://doi.org/10.4045/tidsskr.23.0286.

[34] VP Pechyborsch, VM Yakimets, IV Ogorodniychuk, VV Yakimets, and OV Pechyborsch, 'Territorial defense and its medical support—today's challenge', *Ukrainian Journal of Military Medicine* 3, 1 (2022): 5–18. https://doi.org/10.46847/ujmm.2022.1(3)-005

[35] V. G. Livinskyi, V. O. Zhakhovsky, A. V. Shvets, and O. M. Ivanko, 'Standardization of medical aid in the health care system of Ukraine', *Ukrainian Journal of Military Medicine* 3/4 (2022): 26–40. https://doi.org/10.46847/ujmm.2022.4(3)-026.

admissions.[36] It is also necessary to consider how to mitigate the impact of war on the psychological and social health of families of military personnel deployed to the combat zone, as studies of Ukrainian provision suggested that more might be done to match the range of services that support service families in the US.[37] The intensity and duration of the current war means that Ukraine will face similar challenges in supporting their armed forces and veterans for many years.

Long-Term Strategic Issues

The health service support system also provides advice to the human resources function to the AFU. Both Ukraine and Russia have introduced conscription to mobilize sufficient personnel to meet the needs of the armed forces. This required a massive expansion of the capacity for medical assessment of recruits and a review of the medical and wider health standards for employment in the armed forces.[38] The detection of corruption in the system for granting medical exemption from mobilization has demonstrated that there also needs to be system of oversight to ensure that it is fair and valid.[39] In addition, the large number of war injured will have also put pressure on the system for medical assessment for ill-health pensions and an adjustment to medical retention standards to enable injured personnel to stay in service and contribute to the armed forces even if not fully fit for front-line service. The most visible example is the successful employment of soldiers with prosthetic lower limbs in combat roles.

Finally, there will be an ongoing requirement to provide medical advice to the military pension system in the assessment of disability and financial support for injured veterans for the remainder of their lives. This might include providing long-term rehabilitation services to veterans to support physical and mental recovery even after they have left military service.[40] It will be necessary to maintain the strategic relationships between the Ministry of Defence, the Ministry of Health, and the National

[36] Andriy Haydabrus, Mikel Santana-Santana, Yuriy Lazarenko, and Lydia Giménez-Llort, 'Current war in Ukraine: Lessons from the impact of war on combatants' mental health during the last decade', *International Journal of Environmental Research and Public Health* 19/17 (2022): 10536. https://doi.org/10.3390/ijerph191710536.

[37] Anatolii M. Hrynzovskyi, Serhii V. Bielai, Aleksandr M. Kernickyi, Vladimir I. Pasichnik, Vladimir S. Vasischev, and Aleksandr V. Minko, 'Medical, social and psychological aspects of assisting the families of the military personnel of Ukraine who performed combat tasks in extreme conditions', *Wiadomosci lekarskie* 75/1–2 (2022): 310–17. https://wiadlek.pl/wp-content/uploads/archive/2022/01/WLek202201228.pdf.

[38] A. G. Kyrychenko, S. V. Abramov, V. O. Voloshin, G. I. Titov, Y. M. Bylo, and I. Y. Khaniukova, 'Modern approaches to the organization of directing military personnel to pass medical and social examination under the martial law', *Ukrainian Journal of Military Medicine* 4/2 (2023): 5–12. https://doi.org/10.46847/ujmm.2023.2(4)-005.

[39] 'Ukraine's Zelenskiy decries corruption in military medical exemptions', *Reuters*, 31 August 2023, 12:28 a.m. GMT+1. https://www.reuters.com/world/europe/ukraines-zelenskiy-decries-corruption-military-medical-exemptions-2023-08-30/.

[40] A. V. Shvets, O. V. Horishna, Y. M. Deputat, O. V. Rychka, A. Y. Zhaldak, and A. Y. Kikh, 'Prognostic assessment of the need for medical rehabilitation among military officers of the armed forces of Ukraine based on the structure of their combat trauma', *Ukrainian Journal of Military Medicine* 3/3 (2022): 110–17. https://doi.org/10.46847/ujmm.2022.3(3)-110.

Academy of Medical Sciences to continue collaboration across the 'one medical space' to ensure the health of the population in case of future mobilization for war and the professional education of all health professionals to care for war casualties.[41]

Alongside supporting the overall recruitment process for the AFU, the Ukrainian military medical services have also faced a challenge in recruiting healthcare personnel. At the start of the war there were insufficient numbers of active-duty hospital-based healthcare personnel (surgeons, anaesthetists, specialist nurses, medical technicians) to sustain intensive forward surgical support for a campaign. Those personnel who deployed during the first phases of combat also needed to be withdrawn from the front line for physical and psychological recovery. This was resolved by rotating military personnel from rear-based roles and employing civilian volunteers or conscripts in the military medical services.

Linked to this issue and the challenge of standardization of medical care from the previous theme, Ukrainian lessons from the first six years of the war against Russia since the invasion of the Donbas region in 2014 highlighted the importance of training in wartime medical care to all of the country's healthcare personnel. This also showed the value of the Ukrainian Military Medical Academy (UMMA) as the focus for research and teaching in these subjects. The UMMA has played a key role in analysing the lessons from the transformation of the AFU medical system and increasing the number of presentations at international conferences and academic publications by Ukrainian authors.[42]

The adoption of distance learning in medical education maintained the output of the UMMA during the COVID crisis.[43] This has been extended into a formal evaluation of distance-learning education to develop the methods and tools to institutionalize distance learning in military medical topics,[44] though there are practical challenges with learning during war including internet outages, distractions by attacks and air raids, and students volunteering to support the emergency services.[45] In spite of these constraints, three times as many medical personnel underwent retraining and advanced training by distance learning in 2022

[41] O. M. Tkalenko, 'Justification of the functional and organizational model of preserving the health and professional capacity among military personnel in war and peace (post-war) times', *Ukrainian Journal of Military Medicine* 4/4 (2023): 5–16. https://doi.org/10.46847/ujmm.2023.4(4)-005.

[42] T. M. Ostashchenko, V. L. Savytskyi, and A. V. Shvets, 'Establishment and development of military medical science: today's challenges', *Ukrainian Journal of Military Medicine* ¾ (2022): 5–25. https://doi.org/10.46847/ujmm.2022.4(3)-005.

[43] N. Kozak, O. Rudynskyi, and D. Kozak, 'Pilot study applying distance learning during continuous professional development of military doctors in Ukrainian Military Medical Academy', *Current Aspects of Military Medicine* 28/1 (2021):10–23. https://doi.org/10.32751/2310-4910-2021-28-1-01.

[44] V. L. Savytskyi, T. V. Kutz, and N. M. Sydorova, 'Evaluation of the organization and quality of medical distance learning in crisis situations: Rationale and design of the DILEMMA survey', *Current Aspects of Military Medicine* 29 (2022): 29–38. https://doi.org/10.32751/2310-4910-2022-29-3.

[45] N. D. Kozak, G. Zaychenko, N. Gorchakova, A. Doroshenko, and D. Kozak, 'Recent challenges in higher medical, military and medical education: Sustainability trends in the prism of COVID-19 pandemic and Russian full-scale invasion 2022–2023', *Current Aspects of Military Medicine* 30/2 (2023): 23–30. https://doi.org/10.32751/2310-4910-2023-30-2-02.

compared with 2020.[46] Videoconferencing has also enabled international speakers to contribute to professional education and conferences for Ukrainian military medical personnel.[47]

International allies have not been able to formally deploy many of their military medical personnel in Ukraine to assist with the care of AFU casualties. However there has been a sustained programme of international assistance and partnering to provide educational materials, directed training, and financial support. An example is the translation of the complete suite of US military medical educational material on deployed medicine into Ukrainian.[48] The US and other allies have run partnering programmes in rehabilitation medicine by investment into military rehabilitation centres in Lviv and Irpin, outward visits of Ukrainian doctors to US military rehabilitation centres, and inward visits to Ukraine.[49]

This war has been fought under the shadow of chemical, biological, radiological, and nuclear (CBRN) threats as a result of known Russian military capabilities in this field. The most public risk has been the radiological consequences of accidental or deliberate damage to Ukrainian nuclear power facilities. There is also increasing evidence of Russian tactical use of riot control agents and incapacitants to force AFU personnel to leave underground bunkers. It remains necessary for the AFU to maintain their own CBRN defensive capabilities. While not directly a biological weapon, the substantial volume of contaminated war wounds and the widespread use of antibiotics has led to a significant increase in anti-microbial resistance among endemic bacteria in Ukraine. There is evidence of these strains being exported across Europe. This is likely to increase the pace at which anti-microbial resistance becomes an immediate threat to global health security.[50]

Wider Lessons for Health Service Support in War

This chapter has shown that many of the issues that were forecast to be important for the future of health service support to land warfare have transpired through the experience of the war in Ukraine. These observations are beginning to be converted into recommendations for other military medical services. There has been a steady stream of papers in international military medical journals explaining how

[46] N. D. Kozak, O. V. Rudynckyi, and D. O. Kozak, 'Features of the educational process organization at the faculty of retraining and advanced training of the Ukrainian Military Medical Academy in the conditions of martial law', *Current Aspects of Military Medicine* 30/1 (2023): 38–47. https://doi.org/10.32751/2310-4910-2023-30-1-03.

[47] S. O. Korol, V. Y. Bilyi, K. V. Gumeniuk, R. V. Gybalo, A. A. Hrishov, and M. A. Zahovenko, 'Current issues and features of the training of military surgeons in the conditions of full-scale Russian aggression against Ukraine', *Ukrainian Journal of Military Medicine* 4/1 (2023): 5–12. https://doi.org/10.46847/ujmm.2023.1(4)-005.

[48] *Deployed Medicine*. https://deployedmedicine.com/.

[49] George Smolinski and Derek Licina, 'After the trauma: The role of rehabilitation medicine in U.S. DoD global health engagement', *Military Medicine* 188/1–2 (2023): 3–5. doi:10.1093/milmed/usac302.

[50] Nicola Petrosilloa, Eskild Petersen, and Sergii Antoniake, 'Ukraine war and antimicrobial resistance', *The Lancet Infectious Diseases* 23/6 (2023): 653–4.

these lessons need to be implemented by NATO countries. The paper by Hodgetts et al. highlights many of the observations from this chapter with a call for substantial reinvestment in UK military medical capacity in order to ensure that health service support to the UK armed forces can match the intensity and volume of casualties experienced by the AFU.[51] Schmidt makes similar recommendations for the German Bundeswehr medical service with an additional observation that Germany is likely to be a major transit hub for the medical evacuation of casualties from Western NATO nations with implications for NATO, EU, and German civil-military coordination.[52]

Perhaps the biggest change since the NATO experience of the campaigns in Iraq and Afghanistan has been the magnitude of direct and indirect threat of attacks by Russian forces on healthcare facilities both military and civilian. This has always presented a dilemma, especially in the use of the Red Cross to aid distinction of healthcare facilities.[53] Protecting healthcare will require a much higher level of tactical proficiency by military healthcare personnel and units in order to compensate for the relative lack of protection from the Geneva Conventions and other sources of international humanitarian law (IHL). It will also be necessary to ensure that healthcare services are protected against digital attack, both against denial of service and also for intelligence gathering.

The volume and severity of combat casualties are likely to cause a transition to MASCAL triage on multiple occasions. The concepts surrounding triage and MASCAL management are based on a single event that generates sufficient casualties to disrupt the normal functioning of a health system. The COVID crisis forced health systems to consider the management of a sustained health crisis with consequent reorganization of services in the medium term. This places triage policies at the operational and strategic level, rather than as a tactical response to a single MASCAL event.

The Ukrainian health system has been operating at a high level of stress for several years and thus has fundamentally changed the flow of patients within its health system. Horne et al. suggest six key strategies as an approach to redesigning a health system in response to a sustained crisis: distribute; decompress; delay; delegate (to the lowest capable provider); deliver faster (increased efficiency), and deliver better (increased efficacy).[54] Medical planners will need to consider how to use these mechanisms to adjust the distribution and size of the military medical system to adjust to fluctuations in casualty numbers. Even with these measures, some casualties will be

[51] Timothy J. Hodgetts, David N. Naumann, and Doug M. Bowley, 'Transferable military medical lessons from the Russo-Ukraine war', *BMJ Military Health* 2023: e002435. doi: 10.1136/military-2023-002435.

[52] Kai Schmidt, 'Lessons learned from the war in Ukraine for the Bundeswehr medical service', *European Military Medical Services* 2024: 10–14. https://military-medicine.com/article/4270-lessons-learned-from-the-war-in-ukraine-for-the-bundeswehr-medical-service.html.

[53] Martin Bricknell, Chu-Yi Lin, and Zac Bailey, 'Non-combatant status of military medicine and contemporary warfare: Old issues or new problems? *BMJ Military Health* 170/2 (2024): 97098. doi: 10.1136/military-2022-002161.

[54] Simon Horne, Philip Hunt, B. Hall, Steve Jefferys, J. Vassallo, and Ian Gurney. 'War and disaster are forcing a major rethink around mass casualty management', *BMJ Military Health* 170 (2024): 457–60. Doi:10.1136/military-2023-002489.

classified as T4, expectant, and will not be prioritized to receive resuscitation, surgery, or intensive care. Many clinical interventions for this group might be regarded as 'futile' (e.g. massive blood transfusion, time-consuming surgery). However, military medical services must ensure that they can still provide compassionate care to such patients who are doomed to die: similar to patients admitted to the 'moribund' wards of First World War military hospitals. These clinical decisions will cause significant emotional distress and will add to the mental consequences of the experience of war for military healthcare personnel. It has been argued that military medical services should place stronger emphasis on education for the ethical and emotional challenges of triage during war.[55]

Conclusions

In times of existential threat, it is clear that a country will need to mobilize its entire health sector in support of its population as a unified service. This was demonstrated in the COVID crisis and was an explicit policy decision for Ukraine. Thus military health services will need to focus on their 'role specialization' and entrust clinical care of military casualties to the civilian health system from the 'rear of the combat zone' until 'transition to rehabilitation' as these skills can be taught to any health professional. Military medical services in the combat zone will need to provide garrison healthcare to keep the armed forces fit to fight, and trauma-based medical support at the front line. Military medical services at the 'rear' should focus on treatment and recovery of military patients with a high likelihood of recovery for re-employment in the armed forces. Thus the narrative should move from civil-military cooperation to civil-military integration for health services during war.

The Ukrainian experience has reinforced the necessity of clinical standardization across the pathway of care for military patients from point of injury to rehabilitation. The large library of NATO standardization agreements for medical services needs to be maintained and broadened as an authoritative library of reference for the organizational and clinical procedures for managing military casualties. The use of these, and the US training material for deployed medicine, has shown the value of explicit knowledge for training and education. It has also proved useful for teaching civilian medical personnel how to care for war casualties. It may be appropriate to reintroduce education in disaster and conflict into the mandatory professional curriculum for all healthcare workers, civilian and military.

Clinical practice in the care of war casualties should be directed by a clinical governance system that monitors the performance of the whole system and conducts active research to improve clinical outcomes. This should be underpinned by a clinical registry which captures key clinical information on all patients which can be

[55] Frédérique Gignoux-Froment, Thibault Martinez, Jean Derely, Jean Turc, Simon-Pierre Corcostegui, Clément Derkenne, Vincent Peigne, Marion Trousselard, Angèle Bazin, and Antoine Lamblin, 'Ethical dilemmas faced by military health personnel during high-intensity conflict: The crucial issue of triage', *Military Medicine* 189/9–10 (2024): 234–8. https://doi.org/10.1093/milmed/usae165.

used to measure important outcomes such as how many died of wounds, case-fatality rate, and survival measures. The Ukrainian experience has shown the importance of a military medical academic institution as the focal point for clinical innovation, organizational learning, and professional education.

This chapter has been framed in the context of the land environment. There has been limited combat in the air environment because of the dominance of the Russian air force, limited Ukrainian aircraft, and very effective ground-to-air weapons on both sides. Combat in the maritime environment has been limited to littoral operations with no deep-water combat between adversaries. However, most of the observations and lessons apply to a complete military health system and are relevant to the organization of health services in support of air and maritime forces as well. Effective health service support to the armed forces is a vital element of the moral component of fighting power.

7
Maritime Lessons from the War in Ukraine

Deborah Sanders

Introduction

The Russian invasion of Ukraine in February 2022 has resulted in a prolonged and brutal conventional war in the heart of Europe. Now entering its fifth year, the conflict has been a wake-up call for all militaries and in particular all navies. Ukraine's success in limiting the Russian navy's ability to dominate the Black Sea has been little short of a miracle. At the start of the war, there was a consensus among military analysts and the intelligence community that Ukrainian forces would be quickly defeated and that they were particularly vulnerable in the maritime domain.[1] Despite these pessimistic forecasts, Ukraine quickly developed an innovative and adaptive maritime strategy that produced strategic effects with unconventional and limited maritime means. A state with little to no navy has developed an effective asymmetrical sea denial strategy that has heavily degraded the Russian Black Sea Fleet (BSF), its bases, and its maritime infrastructure, and constrained Moscow's freedom of manoeuvre in the Black Sea. Ukraine has also taken back control of key maritime assets and infrastructure and has opened a vital sea lane for its exports. Ukraine's success has raised important questions regarding what naval lessons could be effectively taken away from this conflict.

While lessons, particularly for the land and to a lesser extent the air campaign, have emerged, much less, however, has been written about the maritime, or more specifically naval, lessons from this conflict.[2] This is perhaps unsurprising given that the naval campaign has been largely secondary to developments in the land domain, as the emphasis has been on capturing and holding land in the east of Ukraine. This chapter addresses this lack of focus on the maritime domain and highlights the lessons from this conflict for contemporary navies.

[1] Eliot A. Cohen and Phillips O'Brien, 'The Russian-Ukraine war: A study in analytic failure', *Centre for Strategic and International Studies*, 24 September 2024.

[2] Frederik Mertens, Björn de Heer, Tim Sweijs, Paul van Hooft, and Frank Bekkers, 'Lessons from land warfare: One year of war in Ukraine', The Hague Centre for Strategic Studies, February 2023; Michael Stefanovic, Robert 'Chuck' Norris, Christophe Piubeni, and Dave Blair, 'The Somme in the sky: Lessons from the Russo-Ukrainian air war', *War on the Rocks*, 9 February 2023

Deborah Sanders, *Maritime Lessons from the War in Ukraine*. In: *Russia's War in Ukraine and Modern Warfare*.
Edited by: Mikael Weissmann and Niklas Nilsson, Oxford University Press. © Deborah Sanders (2026).
DOI: 10.1093/9780198970903.003.0007

There are, however, a number of caveats that need to be considered.[3] The first is that this conflict appears, unfortunately, to be far from over. Drawing lessons is generally best done after the bullets stop flying and the dust settles so that we can understand not only what happened but why, and assess its impact.[4] Second, important contextual factors may limit the applicability to other conflicts of lessons drawn from the war in Ukraine. Many analysts, for instance, have urged restraint when drawing maritime lessons, pointing out that the geographical and political context heavily shapes the nature of the naval battle in the Black Sea limiting the extent to which generalizations can be made. The confined nature of the enclosed Black Sea amplifies the importance of the land domain and lends itself to the adoption of a strategy of sea denial rather than maritime power projection, thus limiting the degree to which the lessons are applicable to larger blue water navies or to other maritime theatres.[5]

An additional contextual factor that also limits the extent to which lessons can be extracted is that the Black Sea is an enclosed sea, accessed through a choke point controlled by a littoral state. The decision to close the Bosporus by the Turkish government to the two 'warring parties' has exacerbated the confined nature of the maritime theatre with consequent effects on the maritime battle. For example, it has denied both parties the ability to reinforce their navies and it has also limited direct third-party engagement including naval support to Ukraine from allies. An additional contextual factor limiting our ability to draw out lessons has been that the Russian navy has not confronted a peer maritime competitor. This has very much been an asymmetrical naval battle where the assets and capabilities of the Russian navy have dwarfed that of the very small Ukrainian navy.[6]

A third point to bear in mind when drawing out maritime lessons from this conflict, is that strategy is, of course, interactive. Ukraine's success in challenging a dominant naval power cannot be understood in isolation from the strategy and failures of the Russian navy. As will be discussed later, qualitative factors such as poor training and lack of combat effectiveness have affected the performance of the Russian navy in the Black Sea. Ukraine's success therefore can be partially explained by Russian failures. The interactive nature of strategy also means that where Ukraine has innovated in the maritime domain, Russia has responded to counter these advances and vice versa. This can make drawing out lessons particularly challenging as success or failure is shaped by the behaviour and responses of the other in a conflict and will therefore be unique and not necessarily generalizable. For all these reasons, therefore, drawing

[3] Joseph Stieb, 'History has no lessons for you: A warning for policymakers', *War on the Rocks*, 6 February 2024.

[4] Frank G. Hoffman and George P. Garrett, 'A break in the clouds: Learning lessons from the sea', *Texas National Security Review*, 7/3 (Summer 2024): 100–112

[5] Ibid., p.108; Joshua C. Huminski, 'Learning the right lessons from Ukraine's naval war', *Engelsberg Idea*, 6 June 2024.

[6] Deborah Sanders, 'Rebuilding the Ukrainian navy', *Naval War College Review* 70/4 (Autumn 2017): 61–77.

maritime lessons from this conflict, in particular for larger blue water navies, has to be done carefully and with a degree of humility.[7]

In light of these caveats, this chapter will look at three key maritime lessons from the war in Ukraine and highlight the conditionality and complexity of each lesson. These three lessons are: the continuing importance of maritime power even if it should be understood as a much broader concept than just naval power; qualitative aspects of maritime power can also be decisive in naval warfare; and while amphibious operations remain challenging, they have demonstrated their utility at both the strategic and tactical levels. Having discussed these lessons, the last part of this chapter explores whether the naval battle in the Black Sea constitutes a change in the character of naval warfare.

The Continuing Relevance of Maritime Power

The first and most important lesson, applicable to large and smaller navies, is the continuing relevance of maritime power. This conflict demonstrates that maritime power, the ability to protect and advance a state's interests in the maritime domain offers policy makers a flexible and important tool to support a conflict, but its flexibility is not unconditional; it can be limited and effectively challenged.[8] A state's ability to exercise a degree of sea control, or at least deny sea control to an enemy, allows states to use the maritime domain to prosecute and support a military campaign and protect maritime trade and infrastructure. This means that navies matter even in campaigns that are heavily land oriented. This is because, as we have seen in the Black Sea, navies can play an important supporting role in land operations and in addition can also have strategic effect by damaging an enemy's economy by blockading and attacking its ports.

Even though the recently augmented BSF was not at the front and centre of the invasion, it did play an important role in the war. It protected the rear and southern flank of the land campaign, deterred navies friendly to Ukraine and in the first portion of the war it successfully blockaded Ukrainian ports strangling its economy.[9] The BSF provided local sea control which proved important in defending the southern maritime flank of Russia's initial advance and in helping to isolate defending Ukrainian forces from support and resupply. In the early part of the campaign, enhancing the capability and size of its navy, and building up its maritime infrastructure in the Black Sea, after the annexation of Crimea, significantly enhanced Russia's maritime power giving it the ability to use the sea unchallenged. The deployment of Bastion and Kalibr

[7] Steib, 'History has no lessons for you.'

[8] Deborah Sanders, *Maritime power in the Black Sea* (London: Ashgate, 2014).

[9] Deborah Sanders, 'What insights can be gained for navies from the conflict in Ukraine?', Korean Institute for Maritime Strategy, Periscope: 299. https://en.kims.or.kr/issubrief/kims-periscope/peri299/.

missiles in occupied Crimea and aboard the BSF provided Russia with a formidable strike capability in support of land operations.[10]

In addition, the use of naval exercises and deployments in the run-up to the invasion of Ukraine also created the enabling conditions for the use of Russian naval power in the early part of the invasion. Of note here was the exercise announced on 20 January 2022 of a sizeable number of ships, support platforms, and submarines in the waters around Russia and in the North Atlantic, the Mediterranean, the Sea of Okhotsk, and the Pacific.[11] These exercises served as cover for the deployment of significant naval forces to both the Mediterranean and the Black Sea allowing Russia to augment its maritime power in both theatres. The range of maritime assets available to Russia in the Black Sea has therefore allowed Moscow to perform a variety of missions and has played a role in deception with the pinning of Ukrainian forces by the latent threat of amphibious landings as discussed later.

While Russia's use of its navy demonstrates that conventional maritime power continues to matter and can be exploited in a variety of ways, this conflict also shows that naval power is but one, and not always the most important, way in which states can exercise maritime power. Ukraine has generated maritime power through an asymmetrical, evolutionary, and primarily land-based strategy that has been augmented with small attack crafts and sea mines. While the success of Ukraine's approach has been highly dependent on context, in particular the unique geography of the enclosed Black Sea, it does reinforce the importance of taking a more inclusive approach on how strategic effect can be generated and utilized in the maritime domain. Ukraine has effectively exercised maritime power without a navy, or at least without a conventional surface fleet, and adopted a successful sea denial strategy aimed at ensuring continued trade at its major ports and denying Russia the ability to attack mainland Ukraine from the sea.[12]

Ukraine's maritime strategy has evolved in three phases adapting to the initially limited capabilities and assets that were available. In the first phase of its sea denial strategy, Ukraine focused on degrading Russia's naval power and deterring Russian amphibious landing in Odesa and along Ukraine's adjacent southern coast.[13] Using its own domestically produced anti-ship missiles to attack the Russian navy, Ukraine demonstrated how it is possible to utilize what might be seen as non-traditional land-based means to achieve maritime goals.

The most important early success was the sinking of the Russian BSF flagship, the *Moskva*. The attack on the *Moskva* was more than a symbolic victory for the Ukraini-

[10] Ben Hodges, Steven Horrell, and Ivanna Kuz, 'Russia's militarization of the Black Sea: Implications for the United States and NATO', *Centre for European Policy Analysis*, 22 September 2022.

[11] 'Russian navy adds to pressure on NATO', *Navy Outlook*, 31 January 2022. https://www.navylookout.com/russian-navy-adds-to-pressure-on-ukraine-and-nato/.

[12] David Kirichenko, 'Ukraine's strategy for winning the battle of the Black Sea', Centre for Maritime Strategy, 2 November 2023.

[13] M. T. Harbib and S. M. Shamrir Al Af, 'Maritime asymmetric warfare strategy for smaller states: Lessons from Ukraine', *Small Wars & Insurgencies*, 21 September 2022.

ans, as this cruiser provided the air defence and command and control for the Russian fleet. Its destruction created an important window of opportunity for Ukraine's limited forces: it allowed Ukrainian drones to be more independent and less vulnerable.[14] The loss of its cruiser also forced the Russian navy to operate further off Ukraine's coast making it harder for its assets to employ maritime fires.

This success as well as the earlier sinking of the Alligator-class landing ship tank *Saratov* while pier side in the Ukrainian port of Berdyansk and the damage to two Russian Ropucha-class landing ships demonstrates the effectiveness of land-based missile strikes in achieving maritime objectives. Ukraine's successful attacks on Russian naval platforms essentially limited the freedom of action of the BSF forcing it to operate further away from Ukraine's littorals. Ukraine's use of land-based assets is a powerful reminder that the threats facing modern navies are as likely to be land as sea based.

Lacking surface combatants, Ukraine also reinforced its sea denial strategy in this first phase of its maritime strategy using more traditional and readily available means, in particular sea mines. Sea mines are a relatively cheap and effective weapon and Ukraine has utilized its large stock of Soviet era mines to good effect to protect its ports.[15] Ukraine hasn't been alone in using these capabilities.

Given their utility, Russia has also reportedly deployed mines in the waters off Crimea and more recently around the Kerch Strait Bridge. Estimates suggest that between 400 and 500 mines were deployed in the Black Sea at the start of the war by both sides posing a real threat to shipping as many have become untethered.[16] Regardless of whether Russia is responsible for setting these mines adrift in a false flag operation or whether the explanation is due to storms and weather conditions in the Black Sea, sea mines have proven a useful complementary weapon for both sides in augmenting their maritime power.[17] For the Russians, sea mines are a useful way of deterring Ukraine's Western partners from entering the Black Sea and also serve to push up the costs to Ukraine of exporting its grain. For Ukraine, they represent an effective means of defending itself against a potential Russian amphibious assault and limit Russia's ability to operate in the littorals.

Ukraine's early success in using primarily land-based missiles and mines for sea denial was followed by phase two of its maritime strategy again demonstrating the importance of thinking more broadly about maritime power. The aim in this second phase was to create a safe maritime zone that would allow for the resumption of Ukrainian trade.[18] The Russian seizure of key Ukrainian ports in the Sea of Azov early in the campaign and its bombardment of ports in the Black Sea, as well as its economic blockade essentially turned Ukraine into a landlocked state.

[14] Tayfun Ozberk, 'Opinion: Why did Russia give up Snake Island', *Naval News*, 3 July 2022.

[15] Mia Jankowicz, 'The Black Sea has a deadly naval mine problem that will long outlast the Ukraine war', *Military & Defense*, 26 September 2024.

[16] Tony Wesolowsky and Georgi A. Angelov, 'The Battle to clear the Black Sea of Mines', *RFE/RL*, 14 January 2024.

[17] Ben Rothchild, 'Ukraine symposium: "Dam the torpedoes!": Naval mines in the Black Sea', *Articles of War*, 15 March 2023.

[18] Harbib and Shamrir Al Af, 'Maritime asymmetric warfare strategy', 10.

As a result, the Ukrainian economy, in particular the agricultural sector, has been very badly hit. Ukraine produces as much as half of the world's sunflower seeds, a tenth of its wheat, and up to a fifth of barley and rapeseed, and prior to the United Nations (UN) and Turkey brokered Black Sea Grain Initiative[19] was unable to ship its products to international markets.[20] Creating a safe zone for the export of Ukrainian goods was therefore vital to the survival of the Ukrainian economy and its to ability to sustain a protracted war. Ukraine's liberation of the small, but strategically important, Snake Island, which had been occupied by the Russians, was a major step towards achieving this goal and again demonstrated how land-based assets, combined with drones, could be used to achieve important maritime objectives.

Snake Island had been captured by the Russians in the first week of the war and had become a symbol of Ukrainian resistance. For the Russians, it was an important forward logistical base for a possible amphibious operation, and it was ideally positioned to allow Russia to maintain its blockade of Ukrainian shipping.[21] Control of the island allowed Russia to block the movement of civilian vessels in all directions to the south of Ukraine at any time.[22] Taking back the island would allow Ukraine to create a vital humanitarian corridor for the export of its agricultural products. The inability of the Russian navy to hold Snake Island in the face of concerted missile and drone attacks by the Ukrainians made it impossible for Russia to reinforce and control the island. Ukraine's ability to target individual Russian air defence systems with long-range barrel artillery fire, assisted by Western-supplied ammunition and multiple launch rocket systems meant that Russia could not place the quantity of air defences on the island necessary to secure the location or to counter Ukrainian attacks on Russian landing craft.[23]

The 'voluntary' surrender of Snake Island by the Russian Federation is a reminder of the importance of maritime infrastructure and that even powerful navies can find it difficult to control ground, even a tiny island. Ukraine followed this success by demonstrating that it could further challenge Russian control of maritime infrastructure by attacking and then later capturing oil and gas rigs previously seized by Russia.[24] Taking back these important maritime economic assets effectively limited Russia's ability to use them for military purposes including situational

[19] The Black Sea Grain Initiative (or the Initiative on the Safe Transportation of Grain and Foodstuffs from Ukrainian ports, commonly called the grain deal in the media) was an agreement between Russia, Ukraine, Turkey, and the UN during the Russian invasion of Ukraine. The initiative specifically allowed for commercial food and fertilizer (including ammonia) exports from three key Ukrainian ports in the Black Sea.

[20] 'Importance of Ukraine and the Russian Federation of Global Agricultural Markets and the risks associated with the current conflict', *Food and Agriculture Organization of the United Nations*, 25 March 2022.

[21] Ozberk, 'Opinion'.

[22] Roman Petrenko, '"We will fight for the island as long as necessary": head of Ukrainian Chief Intelligence Directorate explains the importance of Zmiinyi (Snake) Island', *Ukrainska Pravda*, 13 May 2022.

[23] Seth Cropsey, 'Naval consideration in the Russo-Ukrainian war', *Naval War College Review* 75/4 (Autumn 2022): 1–26.

[24] Howard Altman, 'Ukraine attacks Russian-occupied oil platforms', *The War Zone*, 20 June 2022; Veronika Melkozerova, 'Ukraine reclaims Black Sea oil and gas platforms from Russia, Kyiv says', *Politico*, 11 September 2023.

awareness and reconnaissance, bases for forward deployment of assets, helicopter landing sites, and to position long-range missile systems.[25] Ukraine's success in achieving the key maritime goal of creating a permissive maritime environment for commercial vessels to operate, and denying Russia access to key maritime infrastructure, demonstrates how states can exert maritime power using non-traditional assets and capabilities and in circumstances where they may lack conventional naval platforms.

The third phase of Ukraine's maritime strategy has been to limit Russia's ability to use Crimea as a maritime base and push the BSF further away from Ukraine's coastline. This objective was facilitated by the development of a maritime drone fleet and the use of Western-supplied cruise missiles able to hit Crimea.[26] This phase again clearly demonstrates the evolution of Ukraine's asymmetrical maritime strategy and the ongoing utility of land-based assets, which combined with newer technology and traditional weapons can be used by the underdog to achieve sea denial. To achieve its maritime aims in this more recent phase, Ukraine has adopted a dynamic and experimental approach using new technology, in particular unmanned aerial vehicles (UAVs) and uncrewed surface vessels (USVs), to complement and enhance the effectiveness of its land-based assets.

Ukraine has demonstrated the utility of UAVs to not only successfully attack the Russian navy in Sevastopol and Russian ports hundreds of miles from Ukrainian-controlled waters, but also to hit infrastructure including the Kerch Strait Bridge which connects Russia to occupied Crimea. Using both USVs and UAVs, Ukraine attacked Sevastopol in October 2022, damaging the Russian frigate *Admiral Makarov* and minesweeper *Ivan Golubets*, and demonstrating how these weapons could be used to great effect at a low cost.[27] In light of the utility of USVs, the Ukrainian navy further developed its programme by building the Magura V5 drone, able to carry a 300-kilogram warhead, and the Sea Baby drone, able to carry about 800 kilograms of explosives.[28] Ukraine debuted the Sea Baby USV in July 2023 in an attack on the Kerch Strait Bridge that caused significant damage to the road lanes.[29] Ukrainian specialists have sought to improve the navy's drone capabilities, making them smaller, faster, and more manoeuvrable with increased range and payload.[30] These improvements have allowed Ukraine to successfully target high-value Russian military assets located in Crimea. Of particular note was the recent drone attack by Ukraine's military

[25] 'Latest defence intelligence update on the situation in Ukraine—27 August 2023', Statement released by the UK Ministry of Defence. https://x.com/DefenceHQ/status/1695669934699724819?t=VyjILXmPm9pHtCy_Rh7dLw&s=19.

[26] John C. K. Daly, 'Innovative naval tactics largely nullify Russia's Black Sea superiority', *Eurasia Daily Monitor*, 13 August 2024.

[27] Jon Hemler and Derek Bisaccio, 'Ukraine's USVs write new rules for asymmetric naval warfare', *Defense and Security Monitor*, 13 June 2024.

[28] Bryan Clark, 'Sea Drones in the Russia-Ukraine War Inspire New Tactics', *Hudson Institute*, 10 July 2024.

[29] Tayfun Ozberk, 'New video confirms Ukraine's latest bridge attack was done with kamikaze USV', *Naval News*, 16 August 2023.

[30] David Kirichenko, 'Ukraine's marauding sea drones bewilder Russia', *CEPA*, 30 January 2025. https://cepa.org/article/ukraines-marauding-sea-drones-bewilder-russia/.

intelligence which destroyed two Russian Be-12 Chayka amphibious aircraft and a Mi-8 helicopter.[31]

Through a process of learning and adaptation Ukraine has therefore extended the range of its maritime power. These new weapon systems have successfully driven the BSF out of Sevastopol, escalated the economic and military costs for Russia of maritime operations, and denied them sea control.[32] By arming UAVs and USVs with cruise missiles, Ukraine has extended the range at which it conducts naval attacks well beyond traditional coastal boundaries.[33] While the increasing utility and versatility of USVs is an important development and should be seen as a warning for all navies, it has also led some analysts to suggest that today's struggles in the Black Sea may herald a change in the future of naval warfare in which warships are increasingly vulnerable to nations that lack substantial navies.[34] The claim that rapid technological innovation and developments are dramatically changing the character of war in all domains, including the maritime environment, is an important ongoing debate that has intensified since the Russian invasion of Ukraine and is addressed in the final section of this chapter.[35]

What Does Ukraine's Innovative Use of Maritime Power Mean for Other Navies?

While there are important lessons that fall out of Ukraine's adaptive and innovative asymmetrical approach to maritime power there are questions as to the extent to which this approach applies to other navies. With the notable exception of the sinking of the *Moskva*, the majority of Ukrainian attacks have been against vessels at anchor and in ports.[36] While impressive, this suggests that Ukraine's ability to sink so much of the Russian fleet is contextual, and is as much about Russian failures as Ukrainian success. For example, analysts have argued that the *Moskva* was sunk because she was operating inside known enemy missile range, on a predictable patrol line, and with no layered anti-ship missile defence.[37]

There are also important questions about the durability of Ukraine's maritime success. Russia has adapted and developed counter-strategies and tactics to mitigate Ukraine's innovation in the maritime domain, suggesting that there are important

[31] Ronny Reyes, 'Ukraine's "Ghost" unit takes out Russian amphibious war planes for the first time', *New York Post*, 22 September 2025.

[32] H. I. Sutton, 'Evolution of Ukraine's maritime drone', *Covert Shores*, 28 July 2023. http://www.hisutton.com/Ukraine-Maritime-Drones-Evolution.html.

[33] Jennifer Parker, 'The Black Sea battle: Learning the right maritime lessons from Ukraine', *The Interpreter*, 11 October 2024.

[34] Scott Savitz and William Courtney, 'The Black Sea and the changing face of naval warfare', *RAND Commentary*, 31 October 2023.

[35] T. X. Hammes, 'Technologies converge and power diffuses', *Policy Analysis*, CATO Institute, 27 January 2016.

[36] Huminski, 'Learning the right lessons'.

[37] Tom Sharpe, 'Warship vulnerability: Lessons from the Moskva sinking', *The Naval Review*, 17 May 2022.

limits to the extent to which these lessons are generalizable for other navies. Facing the loss of its maritime assets, the Russian Federation has increasingly shifted towards the use of less traditional means to control the Black Sea such as using mines, missiles, and drone strikes from their shores rather than using sea-based assets. And while this does perhaps demonstrate the utility of adopting a more holistic approach to maritime power it also suggests that these measures can also be easily mirrored or countered. For example, Russia has installed nets and barges to protect critical infrastructure and potential targets and to limit the effects of Ukraine's USVs. It has also targeted suspected drone storage, assembly, and production sites and used patrol boats as advanced radar assets to detect drones and destroy them before they reach their targets.[38] In an attempt to counter Ukrainian USVs, Russian warships have also been equipped with additional firepower, including large calibre machine guns, to enable them to strike enemy drones.[39]

Russia is also developing its own USVs to detect and neutralize the Ukrainian naval drones.[40] To limit its vulnerability at sea from Ukrainian attacks, Russia has also looked at developing land-based solutions to the disruption of crucial supply lines transversing the Black Sea. In an interesting, if expensive, work, Russia is constructing new railways which link Rostov in southern Russia to the Russian-occupied cities in Donetsk and further south reducing the transit time for freight travelling from southern Russia to southern Ukraine.[41]

In addition, while Ukraine's successful asymmetrical strategy does offer states with limited traditional maritime power some important insights into how to challenge a larger and more powerful navy, this approach is also heavily dependent on key enablers such as reliable intelligence over large maritime areas, which itself requires advanced intelligence gathering techniques and real-time data analysis.[42] Ukraine's maritime successes have also been facilitated by the support of Western allies, a circumstance that might not apply to other actors.[43]

Ukraine's success in sinking so much of their enemy's fleet can also be explained by wider systematic problems in the Russian fleet. While the Russian navy is recognized as a powerful force, an aged fleet and a lack of investment in its larger blue water capabilities have caused endemic problems. The majority of Russia's larger blue water platforms, including the *Moskva*, are almost entirely Soviet era vessels that may look impressive but 'have serious flaws when it comes to modern sea combat'.[44] Originally

[38] Igor Delanoe, 'Russia's Black Sea Fellet in the 'special military operation' in Ukraine, *Eurasia Programme*, 7 February 2024.

[39] 'Shoigu orders to add firepower to Russia's Black Sea Fleet for downing enemy drones', *TASS*, 17 March 2023.

[40] H. I. Sutton, 'New Russian USV to counter Ukraine's', *Covert Shores*, 14 December 2023. http://www.hisutton.com/Russia-KMZ-BBKN-Dandelion-USV.html.

[41] Leon Hartwell, 'Ukraine's strategic resurgence: Redefining power dynamics in the Black Sea', *New Eastern Europe*, 9 April 2024.

[42] Eyal Pinko, 'The unique intelligence challenges of countering naval asymmetric warfare', *Centre for International maritime Security*, 1 April 2021.

[43] Taras Kuzio, 'The West reaps multiple benefits from backing Ukraine against Russia', *UkraineAlert: The Atlantic Council*, 12 January 2023.

[44] Nick Childs, 'The Moskva incident and its wider implications', *Military Balance Blog*, 29 April 2023.

commissioned into the Soviet navy in 1982 the *Moskva* was almost forty years old when it sank. Not only were its sensors and combat systems ageing, but there were critical malfunctions highlighted by inspections prior to the deployment. The publication of scans of the certificate of the technical conditions aboard the *Moskva* reveal a catalogue of problems with ship readiness. The certificate listed a number of faults in the missile and artillery units of the ship's air defence systems that highlighted the unsatisfactory condition of many critical systems and units. Taken together these certificates highlight the ship's low combat readiness.[45]

The Importance of Qualitative Factors When Thinking about Maritime Power

A second lesson emerging from this conflict is the importance of qualitative factors in shaping success in the maritime domain. The sinking of the *Moskva* highlights not only the vulnerability of ships in an enclosed sea but also how important qualitative factors are when thinking about maritime power. The *Moskva* sank when two of Ukraine's modern sea-skimming missiles evaded detection and bypassed the ship's defensive systems, striking the cruiser and causing a large fire and explosions. The sinking reveals the interactive nature of warfare, in that Ukraine's success cannot be entirely separated from Russian weaknesses and as such the importance of qualitative factors when thinking about maritime power.[46]

Qualitative factors are in many ways the key enablers that allow for optimum performance at sea particularly in naval combat. Ultimately, a state's maritime power is affected by several intangible, but important, factors including personnel training and combat readiness, morale and discipline, as well as naval leadership.[47]

The loss of the *Moskva* reveals how having a well-trained crew able to work effectively together as a team is a crucial enabler in naval warfighting. The rate at which the fire spread on the *Moskva* after it was hit has raised serious doubts about how combat ready and well trained the crew were. Many analysts have argued that the survivability of the *Moskva* was compromised by poor crew training.[48] After the sinking, there were reports that the *Moskva* was crewed by a significant number of conscripts with limited operational experience calling into question its operational effectiveness and survivability. Out of a crew of 534, almost 257 were conscripts and their average length of service had dropped to just one year, leading to claims that almost half of the crew were not properly prepared for combat.[49]

In addition, the failure to effectively address the fire suggests that crew members lacked the training, skills, and confidence to initiate solutions and act quickly

[45] 'How the Cruiser Moskva sunk', *Militarnyi*, 15 April 2023.
[46] Sharpe, 'Warship vulnerability'.
[47] Sanders, *Maritime power in the Black Sea*, 30–4.
[48] Tayfun Ozberk, 'Analysis: Chain of negligence caused the loss of the Moskva cruiser', *Naval News*, 17 April 2022.
[49] Christopher Carlson, 'The sinking of Moskva', Admiralty Trilogy Seminar, Histroicon, 2022.

and decisively to save the ship from sinking. The defence of a warship, 'requires constant vigilance by a trained and motivated crew, as well as capable equipment that is constantly maintained'.[50] This is particularly true when operating in a hostile warfighting environment. The effectiveness of a naval vessel during conflict hinges significantly on robust shipboard damage-control systems which are essential for minimizing casualties and preserving operational capabilities in the face of unforeseen emergencies.[51] Effective damage control requires close coordination among all members of the crew and the importance of clear communication, teamwork, and initiative; in many ways the ability to quickly assess the situation and take action to manage the situation.[52]

As well as having modern damage-control equipment, the survivability of a fleet is also dependent on continual training including daily damage-control drills to reduce the rate of vessel losses in wartime.[53] In addition, due to the lack of modern weapon systems on board, in particular the low level of automation of the *Moskva's* anti-aircraft missile system, the S-300F, it was even more important that the crew should be competent and well trained.[54] Reports that the crew abandoned their posts rather than fighting to keep the ship afloat also raise important issues about teamwork and the adaptability of crew members.

Crew complacency has also been identified as an additional explanation for the sinking of the Russian flagship. This complacency stemmed from the effect of fatigue, which itself was caused by the prolonged period in the early part of the conflict over which the crew had to maintain itself at high alert.[55] This reinforces the ways in which extended deployments during conventional war create physical and psychological effects that can undermine operational effectiveness and negatively impact the utility of maritime power.

The loss of the *Moskva* does therefore suggest that an important lesson is that even in an era of accelerating technological change, there is a need for an increased focus on the qualitative aspects of crew training and resilience, including damage-control training that stimulates real combat scenarios. Ultimately an agile crew trained for high-intensity operations allows ships to be able to respond quickly and effectively to threats, enhancing their survivability and their utility.[56]

[50] 'Russian cruiser Moskva sunk in the Black Sea—Assessing the implications', *News Commentaries*, 15 April 2022.

[51] Editorial, 'Effective strategies for shipboard damage control in warfare', *Combat Mastery*, 23 May 2024. https://combatmastery.com/shipboard-damage-control/.

[52] 'Why damage control training is essential for maritime Professionals', *Maritime Skills Academy*. 8 November 2024. https://www.maritimeskillsacademy.com/why-damage-control-training-is-essential-for-maritime-professionals/.

[53] Lyle Goldstein and Nathan Waechter, '2 years on, Ukraine's sinking of the Moskva intrigues China's naval strategists', *The Diplomat*, 15 April 2024.

[54] 'How the Cruiser Moskva sunk'.

[55] Captain (Ret.) Kamlesh K. Agnihorti, 'Sinking of the Russian cruiser Moskva: A cause-effect analysis', *National Maritime Foundation*, 19 May 2022.

[56] Parker, 'The Black Sea battle'.

Amphibious Operations Are Difficult, but Useful in the Maritime Domain

The third key maritime lesson that comes of out of the Russo-Ukraine conflict is that while amphibious operations remain difficult, nevertheless they remain useful at both the operational and tactical levels. In the first year of the war, there was much debate about the potential use by Russia of amphibious operations, in particular the likelihood of a large-scale landing. Having significantly augmented its naval forces in the Black Sea prior to the invasion of Ukraine with an additional six amphibious landing ships, Moscow had the ability to conduct major landings using two amphibious task forces. In February 2022 there were reports that Russia had successfully landed thousands of naval infantry ashore in Mariupol raising concerns that this marked the start of a broader campaign of amphibious activities.[57] However, despite sustaining the ongoing threat of an amphibious assault, Russia has been unable to repeat this early success on Ukraine's west coast.[58]

While the challenges Russia faces in conducting further large-scale amphibious landings are not new, they are an important reminder of just how difficult these operations are even for powerful states. One of the biggest challenges that all amphibious landings face is that of terrain. For Russia this problem is particularly acute in the Black Sea. The Ukrainian coast is ringed by steep cliffs and high ground with limited beach exits and few suitable beaches for a mechanized amphibious assault.[59] In addition, those coastal areas to the west not bounded by cliffs are heavily urbanized. After landing on these beaches, Russian forces would have to transition from beach assault to urban operations very quickly and while they could land forces sufficient to sustain a beachhead, they would struggle to land a large enough force to take a Ukrainian city. These challenges would be exacerbated by the fact that Ukraine has deployed counter-measures by positioning tanks and artillery in urban areas and taking up defensive positions along the coast.[60]

For the Russians, the loss of the *Moskva* and the geographically important Snake Island discussed earlier also significantly increased the costs of a further amphibious landing. Given the distance between Sevastopol and the Ukrainian coast, Snake Island would have acted as an important forward logistical base for the Russians essentially cutting in half the amount of time required to move new troops and assets. Without Snake Island, Russian amphibious forces would have to fight their way ashore and sustain the beachhead for more than twenty-four hours before any reinforcements arrived from Sevastopol by sea.[61] Getting sufficient Russian boots

[57] 'Russian navy carries out amphibious assault near Mariupol', *The Maritime Executive*, February 2022.
[58] 'Ukraine navy: Russia can't launch Black Sea landing due to lack of capability', *The New Voice of Ukraine*, 13 August 2024.
[59] Colonel Philip G. Wasielewski, 'Russia-Ukraine amphibious options are limited', *Proceedings*, 148/1, January 2022.
[60] Tayfun Ozberk, 'Russia's amphibious operation dilemma', *Naval News*, 20 March 2022.
[61] Ozberk, 'Opinion'.

on the ground during an amphibious landing would also be dependent on many additional factors including favourable sea and weather conditions, maintaining air superiority, and effectively coordinating and synchronizing all three domains. It would also be predicated on ensuring that there were no mechanical failures in key capabilities and assets such as landing craft and no sustained challenge from Ukrainian forces in the littorals.[62] Russian success would also depend upon having enough amphibious landing ships to conduct operations. Ukraine's concerted and successful attacks on Russian ships in the Black Sea, including attacks on amphibious assault ships, have degraded Russia's capabilities and assets and made landing much more difficult and costly.[63]

Despite these enduring challenges, the naval battle in the Black Sea has also perhaps paradoxically validated the importance of amphibious operations at the operational level and the tactical level. This is again an important lesson. While amphibious landings remain challenging, the conflict has shown that the ability to generate the credible threat of putting land forces ashore does have important benefits.[64] In the Black Sea, amphibious forces have been used as a strategic reserve to reinforce land operations, as well as important logistic connectors to make up for a lack of trucks and rail lines necessary to sustain Russian land operations.[65] Moreover, the very presence of an amphibious force off the coast has compelled the Ukrainian military to devote a portion of its limited resources to preparing to meet a potential assault from the sea.[66]

In the Black Sea we have also seen the utility of amphibious capabilities at the tactical level through the Russian employment of smaller raiding activities. This suggests that an important lesson is to think much more broadly and imaginatively about how amphibious forces can be used in a non-permissive environment. While landing amphibious forces depends on strong naval and air support to ensure air and sea superiority, amphibious raiding does not.[67] Lacking traditional amphibious capabilities such as landing craft and support vessels, as well as the ability to control the air, Ukraine has effectively utilized small-scale amphibious attack to bring the fight to the enemy, demonstrating perhaps one possible future for amphibiosity. An example of this occurred in August 2023 when Ukrainian special forces launched a successful amphibious raid into Crimea, attacking Russian personnel and equipment including an S-400 anti-aircraft system and damaging a Bastion anti-ship missile system.[68]

[62] Ibid.

[63] Kateryna Chursian, 'Ukraine says it hit Russian amphibious assault ship in Crimea', *Politics*, 26 March 2024.

[64] Walker Mills and Timonthy Mills, 'What can we learn about amphibious warfare from a conflict that has had very little of it? A lot', *Modern War Institute*, 22 April 2022.

[65] Ibid.

[66] Sidharth Kaushal and Mark Totten, 'Amphibious futures: The Royal Marines in contested new operating environments', *RUSI Occasional Paper*, January 2024, 9.

[67] Brett Friedman, 'Blood and water: The present and future of amphibious operations', *Military Operations* 1/2 (Fall 2012): 12–15.

[68] David Brennan, 'Ukraine's amphibious assault on Crimea sours Putin's big moment', *Newsweek*, 24 August 2023.

Later that year, Ukrainian forces also launched an amphibious raid across the Dnipro river securing a bridgehead to facilitate the launch of a counter-offensive south across Kherson's fields and marshes towards Russian-occupied Crimea.[69]

In light of the success of these operations, Ukraine has recently acquired high-speed armoured Swedish vessels, CB-90s, specifically designed for amphibious assaults and patrols. Operated by a specialist unit within its military intelligence these boats offer Ukraine the ability to defend coastal areas and to provide direct fire support to Ukrainian forces.[70] Their logistical capabilities, including the ability to transport supplies, personnel, and equipment give Ukrainian forces increased mobility and logistics sustainment.[71] More recently, Ukrainian forces have conducted a number of raids on the Tendra Spit, which is a narrow island in the northern Black Sea southwest of the occupied Ukrainian mainland in Kherson Oblast. The aim of these amphibious raids has been to undermine Russian logistics and to restrict enemy manoeuvrability in coastal zones. By landing a small groups of special forces on this target, Ukraine has successfully destroyed Russian armoured vehicles, a radar station, as well as electronic warfare systems and fortifications.[72]

While the military gains from these raids are small, they do demonstrate the ability of Ukraine to hit back at Russia using the maritime domain and the utility of amphibious raids. Moreover, Ukraine's amphibious success provides an important lesson for modern navies. While amphibious operations remain useful, if challenging, the form that they take might need to change depending on the context. Investing in smaller, lighter, more technologically advanced and agile units that operate independently could well increase the survivability and utility of future amphibious forces.[73]

Do These Three Lessons Herald a Change in the Character of Naval Warfare?

An important question that emerges from this discussion of the maritime lessons of the Russo-Ukraine War and the magnitude of Ukraine's success using non-traditional means is whether the naval battle in the Black Sea constitutes a change in the character of naval warfare. Ukraine's success in sinking Russian naval platforms has led to speculation that navies are becoming increasingly obsolete and easily replaced by

[69] David Axe, 'Amphibious tractors with tank engines: The secret weapons of Ukraine's river-crossing offensive', *Forbes*, 8 November 2023.

[70] Sofia Syngaivska, 'High-speed combat boats strengthen Ukraine's special forces at the sea', *Defense Express*, 11 October 2024.

[71] 'Ukrainian intelligence deploys CB90-class fast assault craft for amphibious operations and patrols', *Naval News*, 14 October 2024.

[72] Abbey Fenbert, '"Taking back what is theirs"—Ukrainian special forces successfully raid Russian-occupied island in Black Sea', *The Kyiv Independent*, 30 September 2025; Sofia Syngaivska, 'Ukrainian raiders trap and destroy Russian Vityaz system on Tendra Spit', *Defense Express*, 20 September 2025.

[73] RADM (Ret.) Mark Montgomery and Jerry Hendrix, 'Marines need to move beyond their amphibious assault past', *National Review*, 15 June 2023.

USVs and shore-based anti-ship missiles. This debate about the continuing utility of navies in many ways mirrors the debates about the utility of the tank in future land battles.[74]

Ukraine's innovative and adaptive use of both old and new technology has no doubt demonstrated the vulnerability of navies, but the conclusion that they are obsolete and that this heralds a new way of carrying out naval warfare is too simplistic and premature. There is no doubt that Ukraine's USV campaign in the Black Sea has in many ways had an 'outsized impact on public narratives about the war' threatening to skew lessons for contemporary navies.[75] USVs are not only considerably cheaper than ships, but they are also far less vulnerable than most warships to anti-ship missiles and torpedoes and attacks from shore-based threats.[76] Ukrainian use of USVs demonstrates that tools and tactics that have been viewed as niche capabilities in previous conflicts are increasingly becoming primary weapons systems that have escalated the risks for all navies.[77]

USVs do indeed have many advantages over crewed platforms, but they will not replace them, at least in the short to medium term. With some notable exceptions, most of Ukraine's accomplishments, including those conducted using USVs, have been against vessels that are either very old or have very limited capabilities.[78] USVs are not yet therefore a game changer even if they will continue to be a fixture of naval combat in the future.[79] While Ukraine's success in the maritime domain was not predicted, it is unhelpful therefore to suggest that this surprise heralds a change in the character of naval warfare. What the conflict in the Black Sea does offer is an important challenge to naval complacency.

Maritime doctrine has traditionally made bold, unsubstantiated claims about the utility of maritime power and navies.[80] Contemporary maritime doctrine claims that maritime power gives you access, poise, mobility, persistence, and versatility without really considering how these attributes are highly variable and dependent on context.[81] While we have not seen a fundamental change in the character of naval warfare, the conflict in the Black Sea is nonetheless a call to arms for all navies, both smaller and blue water. It offers an important opportunity to study a dynamic maritime conflict and assess elements of continuity and change and to make important adaptions

[74] Rob Lee, 'The tank is not obsolete, and other observations about the future of combat', *War on the Rocks*, 6 September 2022; Harrison Kass, 'Are tanks now obsolete?', *National Interest*, 13 December 2023.

[75] Michale B. Petersen, '"Russian navy and naval platforms": Assessing Russian plans for military regeneration, modernization and reconstitution challenges for Moscow's war machine', Research paper, Chatham House, 9 July 2024.

[76] H. I. Sutton, 'Uncrewed platforms have been critical to Ukraine's success in the Black Sea', *RUSI Occasional Paper*, 20 August 2024.

[77] Katie Crombe and John A. Nagl, 'A call to action: Lessons from Ukraine for the future force', *The US Army War College Quarterly: Parameters* 3/53 (Autumn 2023): 19–28.

[78] Sutton, 'Uncrewed platforms have been critical'.

[79] Petersen, 'Russian navy and naval platforms'.

[80] For an example, see 'UK Maritime Power', Joint Doctrine Publication 0-10, UK Ministry of Defence, 2017.

[81] Ibid., 36–9.

to our maritime strategy to enhance flexibility and adaptability. Navies and maritime stakeholders therefore have much to learn from the ongoing conflict in Ukraine.[82]

Conclusion

While there are many maritime lessons that might be taken from the war in the Black Sea, this chapter has focused on three. The first of these is the continued relevance of maritime power and that we need to have a much wider understanding of the ways in which this can be effectively used in the twenty-first century. Looking at the diminishing relevance of the maritime contribution to the Russian military campaign and arguing that this means that maritime power is less relevant in contemporary conflict misses the point.

As discussed earlier there were a variety of ways early on in the campaign that Russia was able to leverage maritime power in important ways, but the reason that the impact of Russian maritime power has diminished is because the Ukrainians found novel and innovative ways to generate their own maritime power. Ukraine's actions, its attempt to generate a form of maritime power to advance and protect its interests, clearly demonstrate the enduring relevance of maritime power. The war hasn't demonstrated that maritime power is irrelevant therefore, despite the scale of Russian losses, instead it demonstrates what happens when an inferior maritime power is able to generate ways to neutralize the advantages of its adversary. A focus solely on the Russian navy demonstrates the failures of maritime power; a focus on Ukraine demonstrates how important generating maritime power actually is in modern conflict. Maritime power is the only way of securing important strategic and operational-level objectives. In essence, just because the Russian navy appears to be a wasting asset does not mean that maritime power is unimportant.

In addition, the adoption by Ukraine of an adaptive maritime strategy making best use of its limited resources also demonstrates that maritime power needs to be understood more broadly as much more than naval power. Instead, as we have seen, Ukraine has utilized very traditional maritime means, such as sea mines, as well as land-based and land-controlled assets to achieve limited sea denial. By embracing new and novel technology Ukraine has further augmented its limited maritime power. USVs have played an important role in deterring the Russian navy from operating in the littorals and effectively made its key ports in Sevastopol untenable.

While Ukraine's strategy demonstrates that there are many ways of exercising maritime power, it does not, however, sound the death knell for warships. The particular context in which this battle has taken place in the Black Sea, including the unique geography and political situation, has enabled Ukraine to adopt a more

[82] Harrison Kass, 'Big U.S. navy warships: Did drones and missiles just make them obsolete?', *The National Interest*, 1 April 2024.

non-traditional approach in the maritime domain. And while elements of its success do indeed need to be studied by contemporary navies, Ukraine's strategy is unlikely to be successfully replicated with the same degree of success in other maritime theatres and by other navies.

A second lesson identified in this chapter is the importance of qualitative factors when thinking about modern navies. In many ways qualitative factors are the hidden, but in many ways decisive, factors when thinking about maritime power. As this chapter has argued, the sinking of the *Moskva* illustrates how the failure to invest in the qualitative aspects of maritime power can undermine a navy's ability to achieve an effect in the maritime domain.

The last lesson identified in this chapter has been that while amphibious operations remain important, if difficult, this conflict has demonstrated how they can also be used by the underdog to achieve effects at both the operational and tactical levels. We are therefore unlikely to see the end of amphibious operations in the twenty-first century. Instead, they will continue to be a flexible and adaptable tool for states in contemporary conflicts.

8
Air War over Ukraine

Viktoriya Fedorchak

'Прорвемось!'—We will break through!

Introduction

The war in Ukraine illustrated numerous trends and developments in modern warfare. Some of them were predicted, while others came across as a surprise on more than one occasion. While the revival of interstate warfare *per se* was a surprise for many, its land-centric nature was less so. However, in terms of the aerial domain and air war over Ukraine, many predictions and expectations were proven to be false regarding the Ukrainian air force (UkrAF), ground-based air defence (GBAD) units, and the Russian air force (Aerospace Forces, VKS). As the title suggests, this chapter explores various aspects of the air war over Ukraine. It is structured as follows. The first section outlines the main trends of the air war; this is followed by developments in piloted aviation for both Ukrainian and Russian forces. GBADs and the rise of drone warfare are then explored. The chapter ends with points for consideration regarding future wars.

The Main Developments of the Air War

Following the modern tradition of establishing air superiority over the opponent's airspace, during the initial attack on the night of the full-scale invasion, the Russians used simultaneous attacks on Ukrainian airports, airfields, military bases, command and control (C2) centres, and points of allocation of the military personnel, while jamming Ukrainian air defences. The purpose was to destroy Ukrainian military aviation on the ground prior to it becoming a threat to the Russians in the air. However, the Russians were not successful in achieving their objectives since various mobile elements of the Ukrainian air defences were relocated, and the UkrAF units withdrew from the areas under attack, dispersing to the tactical operating airfields deep into Ukrainian territory. While in some locations aircrew could disperse aerial assets

Viktoriya Fedorchak, *Air War over Ukraine*. In: *Russia's War in Ukraine and Modern Warfare*. Edited by: Mikael Weissmann and Niklas Nilsson, Oxford University Press. DOI: 10.1093/9780198970903.003.0008

and mobile infrastructure prior to enduring enemy attacks, in others this was done after the Russian strikes. This means that, sometimes, aircrew had to take off from partially damaged runways, under ongoing bombardment, and at night.[1]

Having survived the first contact with the enemy, the UkrAF had to engage in air battles over Kyiv to protect it from constant assaults while initially jammed GBADS could be restored. Many prominent Ukrainian pilots gave their lives in defending Kyiv, including the fighter pilot Oleksandr Oksanchenko, call sign 'the Grey Wolf'.[2] The Russians also tried to take over the international civilian Antonov Airport in Hostomel, aiming to use it as a forward-operating base, which could allow them to airlift and accumulate their forces on the doorstep of Kyiv. They managed to capture the airport, however it was then damaged by the Ukrainian forces, thus denying it to Russian exploitation.

During the three years of the Russian full-scale invasion, neither of the sides managed to establish air superiority. This was conditioned by a few reasons. First, during the eight interim years of war, the primary focus of the Ukrainian armed forces (UAF) was on strengthening and increasing the number of GBADs, including the Soviet era S300, the 9K330 Tor, the 2K12 Kub and the S-125 Neva. Accordingly, increasing the number of air fleet was not prioritized, but rather the modernization of and repairs to the existing assets. Second, from the perspective of the partner nations assisting Ukraine with military aid, defensive weapons were easier to justify under the conditions of the contested and constantly fluid political environments of the Western countries. Nevertheless, partner nations' assistance with air defences reinforced Ukrainian air denial. Third, a more significant accumulation of air defences across Ukraine led to an increase in the risks for Russian aviation to enter Ukrainian airspace as freely as they had initially intended. Fourth, consolidation of the front line in the late spring / early summer of 2022, allowed Russians to move in more of their air defences and hence to establish mutual air denial across various segments of the front line. Fifth, although various analysts expected Russian aviation to perform more effectively and to conduct air operations building on modern operational experience, this was not the case. In its turn, this occurrence can be explained by the land-centric Soviet and Russian tradition of thinking about air power only as artillery from the air rather than as a self-sufficient service capable of conducting independent operations. As will be mentioned in the chapter on the Ukrainian counter-offensive 2023, the Russians began using air cover and close air support (CAS) for their troops more after their artillery (read the traditional fire cover for their ground troops) was significantly degraded by the UkrAF. Finally, from the Ukrainian perspective, numeric and technological advantages were on the Russian side, and the battle for air dominance could cost the UkrAF even more in human lives and lost aircraft. In this regard, fleet reinforcement with Western aircraft like F-16s took a long time and was not brought into use until August 2024, with Mirages introduced only in January 2025.

[1] UAF (Ukrainian armed forces), 'Amazing video! How Ukrainian aviation works', YouTube, 30 July 2022. www.youtube.com/watch?v=gsMMnXjhlUc

[2] Mikhail Zhorohov, *Air war in Ukraine, February–May 2022* (Chernihiv: AirHub, 2022), 22.

Since the beginning of the full-scale invasion, the Russians have practised long-range mass attacks against various Ukrainian targets, employing their signature indiscriminate targeting. Initially, this campaign included ballistic and cruise missiles, and later on, from August 2022, Iranian Shahed drones, and most recently, glide bombs. This campaign also observed significant changes in the choice of targets: in spring 2022, the focus was placed on military targets, aiming to degrade Ukraine's ability to fight. Once the engagement of these targets failed to produce the desired effect, the focus shifted towards bigger targets as an illustration of some achievements. As a result, logistics infrastructure and civilian targets became predominant. In this regard, the Russians followed their traditional punitive bombardments of the civilian population and infrastructure each time the Ukrainians had successes in the battlespace. Furthermore, the situation deteriorated in October 2022 when General Sergey Surovikin, the person responsible for the indiscriminate bombing of Syrian cities, was appointed in charge of the Russian operation in Ukraine. The next day, civilians were bombed in the early morning when they were going to work. The following years showed many Russian tactics adjusted to create the biggest destructive impact on the civilian population, including: the timing of bombings, locations, concentration of people, etc. One of the most gruesome examples is the Russian bombing of the Okhmadyt children's hospital in Kyiv on 8 July 2024.[3]

In the attacks of late 2023 to early 2024, Russian barrages became more sporadic but with increased intensity, combining different weapons, including mass drone semi-swarms, as a means of reducing the cost of attacks while still draining Ukrainian air defences. For instance, in spring 2024 they would attack in waves, with the first wave of glide bombs (Unified Multi-Purpose Glide Bombs (UMPBs), followed by a wave of ballistic missiles and drones.[4] Following their modernization, the Shahed drones could be pointed in one direction and then be redirected to another area mid-flight.

Another significant development of the war in general, and the air war more specifically, was the reconsideration of the balance between numbers and quality and a shift from artificial mass (fewer but more advanced technologies) to a critical mass (sufficient number of available technologies including both advanced and previous-generation weapons) against the given enemy. Although throughout the war Ukraine was adopting more of the Western approach to the employment of force—precision, surgical strikes, focusing on effects rather than numbers of hit targets—eventually, the protracted nature and intensity of warfighting raised the question of increasing the number of available technologies across domains.[5] The issue was not only about

[3] Olha Fokaf, 'Hospital bombing was latest act in Russia's war on Ukrainian healthcare', Atlantic Council, 11 July 2024. https://www.atlanticcouncil.org/blogs/ukrainealert/hospital-bombing-was-latest-act-in-russias-war-on-ukrainian-healthcare/.

[4] Maria Varenikova, 'Ukrainian civilians in once safe city fear growing menace of glide bombs', *The New York Times*, 10 October 2024. https://www.nytimes.com/2024/10/28/world/europe/russia-glide-bombs-ukraine.html.

[5] Viktoriya Fedorchak, 'The mass approach in the air war over Ukraine: Towards identifying a critical mass', *Handlingar och Tidskrift* 1 (2023): 110–26.

accumulating more equipment to correspond to the increased intensity of warfighting but also about compensating for the lack of capabilities in one domain with those available in another. Accordingly, having a lack of aerial assets to satisfy multiple needs across all levels of warfare in numerous locations, the introduction of different types of drones of dual use and military grade, produced both abroad and locally, allowed Ukraine to build critical mass (sufficient numbers) to cover various gaps across tasks and services. As a result, drones are being used for everything from tactical intelligence, surveillance, and reconnaissance (ISR) to attack and transportation tasks in support of the ground troops and long-range attacks deep into Russian territory, targeting Russian airfields and oil refineries.

Piloted Aviation

UkrAF

After having been the third largest air force in the world following the collapse of the Soviet Union in 1991, the UkrAF suffered significant losses in its numbers and capabilities during the process of post-Cold War disarmament, with strategic bombers, missiles, and air-to-air refuelling capabilities being transferred to Russia in exchange for writing off various debts.[6] Through the 2000s, constant budgetary cuts and downsizing of the armed forces personnel had a grave impact on the air force, with shortages even in aircraft fuel, resulting in pilots sometimes having only a few flight hours per year.[7] The annexation of Crimea had further impact on the air fleet, with the Russians keeping some aircraft to themselves and returning the rest with significant damage and missing parts.[8] Despite this and a few losses during the war in Donbas, the UkrAF used the eight interim years of warfighting (2015–22) to prepare for further war with Russia. Accordingly, pilot training was focused on close-to-war situations, with training in short take-off and landing on ordinary highways, flying away from air attacks and tactics in engaging various incoming threats.[9] For example, during the large Sky Shield exercises in 2016, overhauled Su-24Ms landed on Kyiv-Odessa highway—marking the first time in UkrAF practice of landing these bombers on a highway. Prior to this, a similar landing was conducted by MiG-29s in the 2000s.[10] Other exercises involving Su-24s and Su-25s focused on simulating attacks on the enemy airfields and surface-to-air missiles (SAM) capabilities. The

[6] Viktoriya Fedorchak, 'The development of the Ukrainian air force 1991 on the road to 2022: Phoenix rising from the ashes', *The Norwegian Air Power Journal—Luftled* 2 (2022): 18–21.

[7] Babak Taghvaee, *Guardians of Ukraine: The Ukrainian air force since 1992* (Manchester: Hikoki Publications, 2020), 135.

[8] Andrij Kharuk, *Flying in uniform: Military and paramilitary aviation of Ukraine (1991–2021)*, (Kyiv: Vihola, 2025), 180–200.

[9] Jennifer Piggott, 'A closer look at clear sky 2018—A committed relationship between the U.S. and Ukraine', USAFE, 19 October 2018. www.usafe.af.mil/News/Article-Display/Article/1667802/a-closer-look-at-clear-sky-2018-a-committed-relationship-between-the-us-and-ukr/.

[10] Taghvaee, *Guardians of Ukraine*, 132.

practice was also focused on 'using iron bombs, S-8 unguided rockets and precision-guided air-to-surface missiles'.[11] In those years, although the number of the Ukrainian fixed-wing aircraft was not increased, the focus was on modernization and improving effective maintenance of the existing capabilities.

The most common tactics used during various aerial engagements with the enemy aircraft included flying under the radar, taking advantage of the air littoral or lower altitudes. Furthermore, the advantage and knowledge of the terrain often allowed Ukrainian pilots to lure Russian aviation within the range of the GBAD systems. The existing military and media coverage of the first stages of the warfighting during 2022 indicate sporadic use of air power by the UAF. Indeed, the air force had to focus on restructuring its capabilities and repairing the existing fleet. At the same time, various partner nations were gathering spare parts from Soviet aircraft across the world to assist in further repairs. Nevertheless, Ukrainian aviation continued to be involved in various missions and tasks. In spring 2022, during the siege of the Azovstal steel plant, the UAF conducted a series of helicopter mobility flights to the plant to deliver supplies and transport injured individuals. For the helicopters to be less noticeable, a Su-24 Fencer bomber was used as bait to distract Russian attention. It was not meant to be a sitting duck, but rather to distract the enemy by engaging targets of opportunity.[12] In essence, these were suicide missions, since the estimated success rate of flights deep into enemy-controlled territory was very low. According to Brigadier General Kyrylo Budanov, the head of the Defence Intelligence of Ukraine (DIU), during two missions, two helicopters were shot down, and a third helicopter sent for the rescue was also destroyed.[13] Nevertheless, the besieged received necessary supplies of weapons, ammunition, food, and medicine. Furthermore, the reinforcement of seventy-two soldiers of the Azov regiment was delivered while rescuing the severely wounded. These flights were eventually discontinued due to an information leak. Accordingly, 'the Russians strengthened their air defences, focusing more on the helicopters flying low at night'.[14]

With regard to other roles, Ukrainian piloted aviation was used across a wide variety of tasks and missions. It included assets from both the UkrAF and the Ukrainian Army Aviation, which is an air wing supporting the Ukrainian ground forces. Piloted aviation would often be re-sent in support of the advancement of the Ukrainian ground troops, as was the case during the Kharkiv and Kherson counter-offensives in summer 2022. It was used for air interdiction, degrading the enemy logistics supplies and C2 centres to prepare the battlespace for the further advancement of the Ukrainian ground troops. Furthermore, CAS was provided.

[11] Ibid.

[12] 'Interview with Yevhen Bulatsik, commander of the 7th Tactical Aviation Brigade', *Details*, 11 May 2023. https://www.youtube.com/watch?v=PuCgwApOPgw&list=PLVinHmL52La1kae0Gm2BRLRD1Div3mhDP&index=4.

[13] Howard Altman, 'Inside Ukraine's daring helicopter missions into Russian-occupied Mariupol', *The Drive*, 15 May 2022. https://www.thedrive.com/the-war-zone/exclusive-details-of-ukraines-daring-helicopter-missions-into-russian-occupied-mariu pol.

[14] Viktoriya Fedorchak, *The Russia-Ukraine War: Towards resilient fighting power* (London: Routledge, 2024), 84.

Although aerial assets were scarce and sometimes CAS had to be provided with drones, air interdiction roles were very prominent, including during the Ukrainian counter-offensive of 2023, when Ukrainian aviation was hitting various strategic targets in Crimea. They also participated in the liberation of Snake Island together with other branches of the UAF. They fulfilled such tasks as attack, airlift, and rescue. Although Ukraine did not have its own Airborne Warning and Control System (AWACS), the reconnaissance role was fulfilled by the modified Su-24MRs.[15] Pilots' tactics were also adjusted in order to overcome Russian numeric and technological superiority. This was further shown in flying very low, almost touching the ground, by both helicopters and fixed-wing aircraft.

Furthermore, the transfer of Western missiles into Ukrainian arsenals and their mounting on Ukrainian aircraft created the best of two worlds: the survivability and manoeuvrability of the Ukrainian aircraft and the precision of the Western weapons such as the British Storm Shadow cruise missiles and the French SCALP-EGs. With an additional 200-mile range, these weapons could be / were used against a variety of targets, including Russian logistics hubs, bridges, air defences, C2 centres, and warships.[16] A prominent example of the effective use of these weapons was the bombing of the Russian missile carrier corvette *Askold* when it was pierside in Kerch in Crimea on 4 November 2023, resulting in its significant damage.[17] In essence, the warships remain just big targets for air power with the right precision and explosive capacity of the missile, especially when they are stationed in ports or dry docks.

One of the most anticipated transfers of Western weapons to the UAF was Western aircraft. Although in spring 2023, at various air power events, some analysts stated that Ukraine would never get Western aircraft in its fleet, they were proved wrong in 2024. While in the earlier stages, some helicopters and Soviet era aircraft were transferred to Ukraine (Slovakia delivered MiG-29s in 2023[18]), a lack of sufficient numbers of fixed-wing aircraft was constantly felt in the UAF, while the tempo and intensity of warfighting was only increasing with each year. Finally, when the Americans approved permission, Denmark and the Netherlands were the first among Allied nations (Belgium and Norway also promised) to donate F-16s to Ukraine, with the first batches being delivered in the summer of 2024. The primary relevance of Western jets is not about beginning a fight with the VKS to establish air superiority—much bigger numbers of aircraft with long-range missiles and modern advanced radar technologies would be needed for that. Instead, more pragmatic considerations were

[15] 'Interview with Yevhen Bulatsik', 11 May 2023.

[16] David Axe, 'It's official. Despite losses, Ukraine's sole bomber brigade is bigger now than it was before the war', *Forbes*, 22 February 2024. https://www.forbes.com/sites/davidaxe/2024/02/22/its-official-despite-losses-ukraines-sole-bomber-brigade-is-bigger-now-than-it-was-before-the-war/.

[17] David Axe, '"How was your weekend?" The Ukrainian air force asked after its bombers blew up another Russian warship', *Forbes*, 6 November 2023. https://www.forbes.com/sites/davidaxe/2023/11/06/how-was-your-weekend-the-ukrainian-air-force-asks-after-its-bombers-blew-up-another-russian-warship/?sh=140a544c61a4.

[18] 'Slovakia to donate 13 MiG-29 fighter jets to Ukraine', *The Defence Post*, 19 March 2023. https:/www.thedefensepost.com/2023/03/19/slovakia-mig29-fighter-jets-ukraine/ (accessed 15 June 2023).

driving the transfer. First, they provide additional numbers of flexible and manoeuvrable aircraft that can engage more of the Russian incoming aerial threats. Second, they can increase the precision of the previously donated Western weapons, since they were designed for and are interoperable with those weapons. Third, increasing the aerial fleet enabled the start of employing Ukrainian aviation in a more Western style of suppression of enemy air defence (SEAD) missions. Strictly from a technological side, the donated F-16s were equipped with AN/ALQ-131s electronic counter-measure pods, allowing for better protection from the Russian air defences. Furthermore, they are equipped with the pylon integrated dispensing system (PIDS) and the electronic combat integrated pylon system (ECIPS), that 'spoof incoming radar- and infrared-guided anti-aircraft missiles'.[19]

Except for the F-16, another Western aircraft that joined the UkrAF fleet was the French Mirages 2000. The aircraft were swiftly delivered in January 2025 and were employed to intercept Russian missiles during the massive attack on Ukraine on 7 March 2025.

The combination of the technological characteristics of Western aircraft and the sophisticated skills of Ukrainian pilots has already illustrated remarkable achievements in December 2024. On 13 December 2024, the Russians conducted another mass attack on Ukrainian territory, using more than 200 drones, Kinzhal ballistic missiles, and ninety-four cruise missiles. The incoming hostile cruise missiles are the primary targets for Ukrainian aviation. During that attack, one of the Ukrainian F-16 pilots shot down six cruise missiles within a single mission, which is a historic record.[20] In the interview, the pilot explained that although the Russian cruise missiles had an individual anti-electronic warfare defence system, he engaged the first two targets with medium-range air-to-air missiles and, once approaching closer, the other two with short-range weapons. Although the F-16 can carry four air-to-air missiles, it also has a rapid-fire aircraft cannon, which allowed the Ukrainian pilot to take down two more of the Russian cruise missiles once he approached even closer within the firing range of the cannon.[21] Commenting on this achievement in an interview with *Business Insider*, retired Colonel John Venable, a twenty-five-year veteran of the United States (US) air force and a former F-16 pilot, he stated that the pilot had illustrated incredible skills in avoiding fragging (when the aircraft absorbs some explosive debris from the hit target in front of him); and that he had to come at a certain angle to achieve that success.[22]

[19] David Axe, 'France to the rescue! French-made Mirage 2000 jets could become Ukraine's most important aerial radar jammers', *Forbes*, 7 March 2025. https://www.forbes.com/sites/davidaxe/2025/03/07/france-to-the-rescue-french-made-mirage-2000-jets-could-become-ukraines-most-important-aerial-radar-jammers/.

[20] 'Ukrainian F-16 shot down six cruise missiles', *Militarnyi*, 7 January 2025. https://mil.in.ua/en/news/ukrainian-f-16-shot-down-six-cruise-missiles/.

[21] Official Air Force Command of UA Armed Forces interview with the F-16 pilot, Facebook post, 7 January 2025. https://www.facebook.com/kpszsu/posts/pfbid0AFiGdZXnw5ZQSbsvdgCGZFJRX5JGfUdhjUuJkyGzPhr1SUJFd1N2RLn9g5pNpRRil.

[22] Sinéad Baker, 'Record-breaking Ukrainian F-16 pilot showed great skill gunning down Russian missiles "without fragging himself," veteran US fighter pilot says', *Business Insider*, 10 January 2025. https://www.businessinsider.com/ukraine-f-16-pilot-skill-gun-russian-missiles-ex-pilot-2025-1.

Air power today is not just about the aircraft but also about the pilot—it was that synergy that has made the impossible possible more than once, especially in the case of the air war over Ukraine and the impossible that the UkrAF pilots face daily. One day, air warfare might become completely unmanned, but not today ... Wars of today and the future require all assets across domains and the full spectrum of automation. When one technology or capability is lacking, another can compensate—that is what improves survivability and allows chosen objectives to be achieved across different levels of warfare.

Russian Air and Space Forces (VKS)

Regarding the VKS, numerous features of the Soviet tradition have remained, while some of the operational experiences they gained in prior operations have proved to be less useful in Ukraine. The Russians continue with the mass approach in structuring and employing their air power. Despite the objectives of modernizing their air fleet and refocusing on more cutting-edge technologies like the multi-role aircraft Su-34 and Su-35, the Russians have paid little attention to the improvement of pilot training, with a significant decrease in flying hours per pilot.[23] Hence, the Soviet tradition of focusing more on equipment than the personnel has prevailed.

From the first days of the full-scale invasion, despite the initial surprise of facing opposition in the Ukrainian skies and suffering losses on their side, Russians continued to amass intense attacks, with some 140 sorties flown per day in the earlier weeks of the invasion. While initially the Russians managed to fly deep into Ukrainian territory, thanks to the initial interception of the UkrAF and eventual unjamming of GBADs, the Russian Su-25, Su-30, and Su-34 had to readjust to flying at lower altitude and often solo instead of in larger formations. In their turn, they became more vulnerable to the Ukrainian man-portable air defence systems (MANPADS). Furthermore:

> Besides the danger, the low-altitude flying meant very little time for acquiring, manoeuvring and dropping weapons on the given target, not to mention [a] lack of up-to-date maps and the challenges of attacking well dug-in Ukrainian forces that used the same equipment as Russian troops. Hence, the Russians were flying sorties against preassigned targets rather than taking advantage of time-sensitive targets or targets of opportunity. Furthermore, daytime low-altitude flights were considered too dangerous by more experienced Russian pilots.[24]

Other adaptations in their tactics included shifting most of the raids to the night time, since Ukrainian troops lacked night-vision goggles during spring 2022. However, this significantly undermined the precision of their strikes. Similarly, while in

[23] Justin Bronk, Nick Reynolds, and Jack Watling, *The Russian air war and Ukrainian requirements for air defence* (London: RUSI, 2022), 18.

[24] Fedorchak, *The Russia-Ukraine War*, 104.

Syria their Tu-22M Backfire bombers conducted carpet bombing of Syrian cities from medium altitudes, they had to employ them at lower altitudes in Ukraine, undermining their effectiveness.[25]

With the Russian loss of the Battle of Kyiv and withdrawal from northern Ukraine, they could refocus on the east and south, hence the establishment of the front line and consequent relocation of the GBADs and jamming systems across the front line, which eventually contributed to the establishment of the mutual air denial.[26] From the Russian sorties' perspective, they began using more medium and low altitudes closer to the front line. Losses among the Su-34 fleet resulted in them being used at the medium altitude against fixed targets at a distance of 18–15 kilometres.[27] In the facilitation of Kharkiv and Kherson counter-offensives, UkrAF used AGM-88 HARM missiles to suppress Russian GBADs, allowing Ukrainian Su-25s and Su-24s to conduct air interdiction and prepare the battlespace for further movement of the Ukrainian troops.[28]

The performance of the VKS was also undermined by the lack of air-to-air refuelling capabilities in their toolkit, since these were scarce and reserved only for the Russian strategic bomber fleet. This meant that often control air patrols (CAPs) would be shorter, more sporadic, and closer to the front line. Eventually, the Russians had to refocus on more standoff attacks with long-range missiles, usually Kalibr, while remaining airborne in the Russian airspace. Moreover, in the last year, in order not to jeopardize their aircrafts, the Russians refocused on employing glide bombs, the interception of which presented significant challenges for the Ukrainian air defences. Not having sufficient hardware and ammunition to scrap the Russian aircraft launching glide bombs, the most recent (as of March 2025) Ukrainian counter-measure has been to use electronic warfare (EW) to jam them.[29]

Another significant development at the beginning of the air war was the undermining of the traditional advantage of the Russian assaults, which was a combination of light aviation and paratroopers. The Russian airborne forces (VDV) dropped into an area were known to provide them with the fastest and most mobile means of seizing territory, which had been effectively used during the Russian wars post-Cold War.[30] However, during the battle for Hostomel airport, they proved to be easy targets for the light Ukrainian infantry with MANPADs, especially when one strike could destroy both the helicopter and the VDV team on it, hence increasing the cost-effectiveness of each used munition. This raised a lot of questions concerning the very utility of helicopters and paratrooper drops in highly contested environments. Some later attempts deep into Ukrainian territory were also unsuccessful.

[25] Ibid.

[26] Ibid.

[27] Bronk et al., *The Russian air war*, 18.

[28] Fedorchak, *The Russia-Ukraine War*, 105.

[29] David Axe, 'Ukraine is jamming Russian glide bombs all along the front line, erasing one of Russia's main battlefield advantages', *Forbes*, 26 February 2025. https://www.forbes.com/sites/davidaxe/2025/02/26/ukraine-is-jamming-russian-glide-bombs-all-along-the-front-line-erasing-one-of-russias-main-battlefield-advantages/.

[30] Bettina Renz, *Russia's military revival* (Cambridge: Polity, 2018).

Russia's continuous army-centric use of the air force was evident in the challenging air-land integration and less effective CAS missions. One of the most prominent examples was the Russian use of aerial assets during the Ukrainian counter-offensives in the Kharkiv and Kherson areas. According to the Ukrainian pilot, Andrii Pilshchykov, call sign 'Juice', who took part in these aerial missions in Kharkiv, despite the establishment of a new Russian mantra of risk avoidance in the Ukrainian skies, the surprise of the counter-offensive made them drop multiple aerial assets against rapidly advancing Ukrainian troops: 'They started this chaos once more. They were sending bombers, Frogfoots, and even fighters over the front lines, with the usual unguided bombs. And of course, our guys shot down a lot of them right there'.[31]

It may seem a paradoxical challenge, but in fact, if air-land integration is expected by default but is not effectively trained for on a regular basis, it is unlikely to be performed in the battlespace. At various stages of this war, the side capable of taking better advantage of this integration was more likely to achieve the tactical advantage. Accordingly, during the first year, the Russians were suffering aircraft losses during the random CAS missions, while the UAF had better integration with various assets and effectively provided CAS to the extent that, in order to undermine the Ukrainian CAS missions and 'to reduce fleet losses in CAS missions, the VKS began using Su- 35 and Mig- 31 high-altitude interceptors to target Ukrainian aircraft during their CAS missions in the frontlines, taking advantage of safe distances and preventing them from receiving return fire'.[32]

Another important occurrence was the issue with deconfliction between ground defence units and air forces. While this issue more significantly prevailed in Russian practice, some instances of friendly fire also took place on the Ukrainian side, especially with regard to MANPADs and helicopter aviation.

The Russians began to refocus on more effective employment of CAS on the front line only after the UAF had significantly degraded their artillery, which had traditionally been its primary fire cover for the ground troops. Hence, they had to execute effective CAS as they were conducting the warfighting. They showed successful performance in supporting their troops with helicopters during the Ukrainian counter-offensive in 2023, and in support of the Russian advance and capture of Avdiivka in February 2024.[33]

Air Defences and Russian Mass Long-Range Attacks

As mentioned, during the entire three years of their full-scale invasion, the Russians continued their long-range attacks involving ballistic and cruise missiles and

[31] Thomas Newdick, 'A MiG-29 pilot's inside account of the changing air war over Ukraine', The War Zone, 15 December 2022. https://www.twz.com/a-mig-29-pilots-inside-account-of-the-changing-air-war-over-ukraine.

[32] Fedorchak, *The Russia-Ukraine War*, 107.

[33] Angela Giuffrida, 'Russia claims full control of Avdiivka after Ukrainian retreat', *The Guardian*, 15 February 2024. https://www.theguardian.com/world/2024/feb/18/russia-claims-full-control-avdiivka-ukrainian-retreat.

various types of drones. During this time, Russian tactics and approaches to attack and Ukrainian responses changed, illustrating the enduring feature of warfighting—it does not occur in a vacuum but depends on constant interaction with one's enemy.

Over this time, Ukrainian air defences had significantly evolved in terms of both their overall effectiveness and sophistication. As was previously mentioned during the interim years of warfighting, Ukraine focused on strengthening its GBADs. However, it also witnessed numerous sabotage activities in targeting storage facilities where relevant ammunition used for GBADs and other Soviet era equipment were stored.[34] Accordingly, in the spring months of the full-scale invasion, the effectiveness of the Ukrainian air defences was at around 20–30%. However, after regrouping and adjusting to the specific tactics of the Russian attacks, the interception rate had already increased to 50–60% by June 2022, and went on to reach 70–80% by November 2022.[35] At the time, the commander-in-chief of the armed forces of Ukraine, General Valerii Zaluzhnyi, placed the effectiveness of the Russian long-range attacks at a coefficient of 0.76, which meant that out of a hundred Russian missiles, twenty-four would still get through the Ukrainian air defences.[36] In a recent study by the Center for Strategic and International Studies (CSIS), the percentage of total interceptions between 28 September 2022 and 6 October 2024 was estimated at 84.1%.[37] This significant improvement was achieved thanks to the greater availability of Western air defence systems such as the Norwegian advanced surface-to-air missile system (NASAMS)—a short-range air defence system produced by Kongsberg Defence & Aerospace, and delivered and already operational in Ukraine by November 2022.[38] Their effectiveness and versatility were assessed as follows:

> Since their deployment, NASAMS systems have been used to intercept approximately 900 aerial threats, achieving a reported 94% hit rate. Around 60% of intercepted targets were cruise missiles, including the Kh-101, Kh-555, Kh-59, and Kh-69, as well as Kalibr and Iskander-K missiles. The remaining 40% of intercepted threats—approximately 360 targets—were unmanned aerial vehicles (UAVs), primarily Shahed loitering munitions.[39]

Like many other Nordic countries, Norway trained Ukrainian operators for NASAMS in Norway to make sure the most systematic and practical training was provided with all relevant questions answered and the most detailed techniques taught by Western developers and practitioners. According to International Institute for

[34] Fedorchak, *The Russia-Ukraine War*, 57.

[35] Ibid., 108.

[36] 'An interview with General Valery Zaluzhny, head of Ukraine's armed forces', *The Economist*, 15 December 2022. www.econom ist.com/zaluzhny-transcript.

[37] Yasir Atalan and Benjamin Jensen, 'Breaking down Russian missile salvos: What drives neutralization?', Center for Strategic and International Studies, 24 February, 2025. https://www.csis.org/analysis/breaking-down-russian-missile-salvos-what-drives-neutralization.

[38] 'NASAMS air defence system proves highly effective in Ukraine, Norwegian official confirms', *Defence Industry Europe*, 25 February 2025. https://defence-industry.eu/nasams-air-defence-system-proves-highly-effective-in-ukraine-norwegian-official-confirms/.

[39] Ibid.

Strategic Studies (IISS) estimates, overall, during the 2022–4 period, Ukraine received different Western short-, medium-, and long-range air defence systems from eleven countries.[40]

One of the challenges for the Ukrainian air defences was that they were already running out of munitions for the Soviet era systems in autumn 2022. One of the solutions was to modify these systems to use Western weapons. Accordingly, Buk launchers were modified to launch the RIM-7 Sea Sparrow, while the Come Back Alive Foundation modified R-73 missiles to be launched from the 9K33 OSA.[41] Furthermore, the UK modified 'the MBDA Advanced Short-Range Air-to-Air Missile for ground launch, using the Supacat HMT 600 as the launch vehicle'.[42]

Another challenge for the Ukrainian air defences was their cost-efficiency, especially under the constant scarcity of resources and the need for uninterrupted delivery of the Western air defences and munitions for them. Although the Russians have traditionally employed mass attacks in their wars and had significant Soviet legacy munition arsenals, continued long-range attacks on Ukraine illustrated the need for a cheaper solution to sustaining their mass in those attacks over a prolonged time. This cheap mass was already introduced in August 2022, by using the Iranian Shahed drones in combination with the ballistic and cruise missile segments of the attacks.[43] This created a problem regarding the cost-efficiency of the Ukrainian response since the Iranian drones were considerably cheaper than munitions used in Western air defences. As a result, the most cost-effective solutions against Iranian drones were the German-produced self-propelled air defence artillery systems, the Flakpanzer Gepard, in conjunction with Soviet era anti-aircraft artillery ZU-23-2.[44] Furthermore, Ukrainian troops were also using MANPADs such as Igla to shoot down Shahed drones[45] and even cruise missiles.[46]

As was mentioned earlier, in the last two years of the full-scale invasion, Russian long-range attacks became more sporadic but with greater intensity, aiming to inflict damage and to significantly test and drain Ukrainian air defences. For instance, some of the intense attacks in the punitive Surovikin raids against Ukrainian civilians during autumn 2022 were 'on 15 November (around 100 missiles) and 23 November (70 missiles launched), resulting in severe blackouts due to disruptions of electricity production, transportation and distribution of electricity'.[47] On the other hand, a year

[40] Giorgio Di Mizio and Michael Gjerstad, 'Ukraine's ground-based air defence: Evolution, resilience and pressure', International Institute for Strategic Studies, 24 February 2025. https://www.iiss.org/online-analysis/military-balance/2025/02/ukraines-ground-based-air-defence-evolution-resilience-and-pressure/.

[41] Ibid.

[42] Ibid.

[43] Fedorchak, 'The mass approach', 119–20.

[44] Di Mizio and Gjerstad, 'Ukraine's ground-based air defence'.

[45] Valentyna Romanenko, 'Ukrainian air force video shows Shahed drones being downed by Igla MANPADS with thermal imaging', *Ukrainska Pravda*, 13 December 2023. https://www.pravda.com.ua/eng/news/2023/12/13/7432897/.

[46] 'Ukrainian air defense soldier shoots down cruise missile with Igla MANPADS', *Ukrinform*, 13 December 2025. https://www.ukrinform.net/rubric-ato/3937747-ukrainian-air-defense-soldier-shoots-down-cruise-missile-with-igla-manpads.html.

[47] Fedorchak, *The Russia-Ukraine War*, 108.

later, on 25 November 2023, the biggest Russian drone attack was focused on Kyiv alone, with seventy-five Iranian-made Shahed drones.[48] During the recent massive attack on 7 March 2025, 261 incoming aerial threats were identified.[49]

According to research by the CSIS, over the three years of full-scale invasion, on average the Russians fired 24.3 missiles and drones per day. Except for the evolving tactics, these attacks were also characterized by a greater diversification of missile models:

In certain months, over twenty-five unique missile and drone models were launched, with a daily maximum of fourteen unique models. The monthly average of unique missiles launched is 16.08. This diversification includes ballistic missiles, cruise missiles, kamikaze drones, and reconnaissance unmanned aerial vehicles (UAVs).

Alongside this growth, there have been notable changes in the mix of weapons. Compared to the October 2022–September 2023 period, the overall intensity of launches has doubled, largely driven by the increased deployment of Shahed-131/136 drones. In fact, these drones, alongside Kh-101 and X-555 air-launched cruise missiles, account for 75% of all launched missiles. In contrast, Kalibr missile use has fallen dramatically, shrinking to just one-fifth of its previous level in the 2022–3 period.[50]

Hence, in order to sustain continuity and mass characteristics, and widen the effects of their attacks, the Russians have focused on diversifying weapons in these attacks. In turn, this has required more significant manufacturing of various weapons. In the case of the Shahed drones, the primary modifications were aimed at improving their navigation and redirectability during flight, which would undermine the readiness of Ukrainian air defences in different regions of Ukraine. Another attempt was to increase the speed of the Shaheds to imitate the cruise missile signatures on radars. However, this attempt was futile—they were still much slower.

According to Andrii Yusov, a representative of the DIU, as of December 2024 Russian stockpiles included 1,400 long-range missiles: 350 Kalibr cruise missiles; 500 Onyx medium-range anti-ship missiles; fifty Kinzhal ballistic missiles; 130 Iskander missiles; and more than 400 X-101, X-55, and X-36 missiles. He further confirmed that the Russians were becoming more careful with using these weapons due to their scarcity and cost: 'On many positions, this is an untouched stockpile. Therefore, they pay a lot of attention to planning, and reconnaissance, and do not act recklessly as they did during the early stages of the full-scale invasion'.[51]

Despite depleting stockpiles of Russian long-range precision weapons, Russia continues to manufacture many of them. According to the DIU, the manufacturing rates for specific weapons are as follows: 3M55/3M55M Oniks/Oniks-M cruise missiles

[48] James Waterhouse, 'Ukraine war: Kyiv hit by biggest drone attack since war began', BBC, 25 November 2023. https://www.bbc.com/news/world-europe-67529571.

[49] 'This has never happened in history. For the first time! Ukraine raised mirage into the sky', *24 News Channel*, 7 March 2025. https://www.youtube.com/watch?v=fXVv3qymkQI.

[50] Atalan and Jensen, 'Breaking down Russian missile salvos'.

[51] Kateryna Serohina, 'Ukrainian intelligence directorate shares data on Russia's stock of long-range missiles', *RBC Ukraine*, 28 December 2024. https://newsukraine.rbc.ua/news/ukrainian-intelligence-directorate-shares-1735347294.html.

are at ten units per month; Kalibr 3M14—at thirty to forty units per month; 45 Kh-69 at one to three units per month.[52]

On the Ukrainian Layered Air Defences and Advantages of Diversification

The distinctive features of the Ukrainian air defences include their cross-domain origin and versatility. Accordingly, the layered approach to air defences included various means of early warning, conventional aircraft, GBADS and people on the ground with MANPADS, and anti-aircraft artillery such as German Gepards. The early warning and early identification of threats and their direction would allow choosing specific tools against various incoming threats. For instance, these days F-16s are being employed against cruise missiles, while Gepards and MANPADS are effectively used against enemy drones.[53] While some analysts argued that operating diverse air defences—both Soviet and Western ones—with different munitions, operational ranges and logistics support chains would be challenging if not counterproductive, the Ukrainian experience has illustrated that the other side of the coin is actually versatility and survivability of various systems. For instance, if some systems might not function, be jammed, or be missing some spare parts or updates, others would remain functional. Hence, a diversity of platforms creates an additional challenge for the enemy to disrupt the normal functioning of the air defences. They cannot all be switched off remotely, since they operate in a more dispersed and mission command manner.[54]

After the scandal during Ukrainian President Zelensky's meeting with US President Trump in Washington on 28 February 2025, the US cut off its military and humanitarian assistance to Ukraine.[55] As a result, any previous arrangements for weapons delivery were halted, and even intelligence sharing was stopped.[56] This had a direct impact on the effectiveness of Ukraine's capabilities for early warning regarding incoming aerial threats. The Russians did not wait long and conducted one of the biggest combined amassed long-range attacks on the entire Ukrainian territory on the night of 7 March 2025, focusing on reaching the Ukrainian civilian infrastructure, such as multi-storeyed apartment buildings, gas supply and storage

[52] Valentyna Romanenko, 'Ukraine's defence intelligence reports on Russia's missile stock and production rate', *Ukrainska Pravda*, 1 May 2024. https://www.pravda.com.ua/eng/news/2024/05/1/7453805/.

[53] Fedorchak, *The Russia-Ukraine War*, 108–9.

[54] Viktoriya Fedorchak interview given to Par Benoi Toussaint for the article 'Patriot, Crotale, IRIS-T … Le patchwork de la défense antiaérienne en Ukraine', *Le Soleil Newspaper Canada*, 22 December 2022. https://www.lesoleil.com/2022/12/22/patriot-crotale-iris-t--le-patchwork-de-la-defense-antiaerienne-en-ukraine-63ef0d81a4c842181af5f0e9d8153e93/.

[55] Andrew Roth and Oliver Holmes, 'US suspends all military aid to Ukraine in wake of Trump-Zelenskyy row', *The Guardian*, 4 March 2025. https://www.theguardian.com/world/2025/mar/04/us-military-aid-ukraine-pause-trump-zelenskyy-updates.

[56] 'US cuts off intelligence sharing with Ukraine', *The Financial Times*, 5 March 2025. https://www.ft.com/content/c58fccea-00c4-4fad-bc0a-0185b7415579.

facilities, and the electricity grid from Kharkiv to Ternopil and Ivano-Frankivsk[57] According to Andriy Kovalev, spokesman for the General Staff of the armed forces of Ukraine, during those twenty-four hours the Russians conducted eighty-nine air strikes against Ukrainian territory, dropping 149 guided bombs.[58] Further estimates of the Ukrainian radar forces indicated that the overall attacks included 261 incoming aerial threats—sixty-seven missiles of various types and 194 attack UAVs and different simulator drones.[59]

Of these incoming threats, Ukrainian aviation and air defences managed to down twenty-five Kh-101 cruise missiles, eight Kalibr cruise missiles, one Kh59 missile, and a hundred drones.[60] Although early warning capabilities were undermined, during this attack, for the first time the UkrAF, used Mirage 2000 aircraft to intercept the incoming threats. This was a remarkable and speedy development, considering these aircraft had been transferred to Ukraine only a month earlier.[61] Furthermore, on 14 March 2025, the French air force flew its E-3F AWACS escorted by two Rafale aircraft in a long reconnaissance mission over the Black Sea and near Crimea. The gathered intelligence was then shared with the UAF. This was a national French mission and not a North Atlantic Treaty Organization (NATO) one.[62] As the saying goes—where there is a will, there is a way.

On Drones

One of the most widely discussed and analysed aspects of the war in Ukraine is the proliferation of drone warfare in different shapes and forms on both sides. The initial lack of sufficient manned aviation assets across various roles and tasks of air power, from ISR to more kinetic tasks, inevitably created a significant gap in capabilities, undermining the UAF's airborne intelligence, consequent situational awareness, and firepower, both on tactical and operational levels. In this regard, drones became the means of covering these gaps, with the gradual source of establishing critical mass in aerial assets through foreign assistance with military drones, the customization of civilian drones and national manufacturing and repairing capacities.

During the earlier stages of the full-scale invasion, more expensive Turkish TB-2 Bayraktar drones were used to gather relevant ISR and attack Russian convoys, C2, and the areas of general force concentration. They were also used to lure Russian aviation to medium altitude and within the reach of the Ukrainian SAM systems. However, once the Russians had taken measures to defend their logistics supplies,

[57] 'Critical and gas infrastructure under attack! Russians carried out a massive bombing. News for March 7, 2025, *Podrobyci News Ukraine*, 7 March 2025. https://www.youtube.com/watch?v=sgLb9KzdNyc&t=265s.
[58] Ibid.
[59] 'This has never happened in history'.
[60] Ibid.
[61] Ibid.
[62] Visegrád 24, 'Yesterday, a French E-3F AWACS flew a long mission', *Twitter post*, 15 March 2025. https://x.com/visegrad24/status/1900978042027069485.

Bayraktars were used in more complex operations in conjunction with other aerial assets.[63] Situational awareness on both sides was achieved using such drones as the Ukrainian SKIF and Russian Orlan-10—both flying at medium altitudes and hence able to provide relevant imagery to coordinate rapid fires. For the tactical ISR purposes of ground troops and special operations forces (SOF), Russians used 'Granat' drones and different versions of the Orlan-10. For the SOF ISR purposes, the UAF used the Norway-produced Black Hornet microdrone, 'which is the smallest ISR system in the world and is often used by the SOF for immediate tactical intelligence'.[64]

One of the new developments in drone warfare in Ukraine was the widespread customization of commercial drones for various military purposes. The obvious starting point was the provision of the basic tactical ISR and the dropping of grenades on the enemy positions. However, as time passed these drones became more sophisticated and their use more advanced and systematic. Drones were used against moving targets such as tanks and armoured vehicles, resulting in the development of new tank defences further employed in conflicts outside of Ukraine.[65] A combination of aerial and naval drones was used in attacking Russian seaports and ships in Crimea, as well as the concentration of Russian troops on various segments of the front line and in the trenches. Furthermore, Ukraine manufactured long-range drones made it possible to conduct deep strikes into Russian territory and undermine their economic and production capacity by targeting oil refineries, which are the core of their economic and consequently military potential.

Another trend was the manufacture of drones for customized orders from specific units of the UAF. From students of Kyiv Polytechnical University to experienced engineers, drones were made according to the individual requirements of units on the front line, with various grassroots initiatives being sponsored through crowdfunding in and outside of Ukraine.[66]

Tactical bottom-up solutions in drone developments eventually resulted in the formation of specialized units focusing specifically on drone development and innovation in their employment in the battlespace. The most prominent units in the UAF are the 'Birds of Madyar' and K2. Both of these units were developing in parallel, with further collaboration and sharing of knowledge and innovative solutions.[67] The 'Birds of Madyar', led by Robert Brovdi, is a division of aerial reconnaissance in the UAF. They were not only the first most advanced drone unit providing timely tactical intelligence on the front line, and consequent attacks on the Russian positions, but they also became the innovation hub for using drones in modern warfare. The initial

[63] Fedorchak, *The Russia-Ukraine War*, 110.

[64] Ibid.

[65] Joseph Trevithick, 'Israeli Merkava tanks get improved anti-drone, magnet bomb defenses', *The War Zone*, 22 January 2024. https://www.twz.com/israeli-merkava-tanks-get-improved-anti-drone-magnet-bomb-defenses.

[66] Alina Matviychuk, 'E-cigarettes are disassembled to make combat drones for the armed forces: How KPI students help destroy the occupiers', *Vikna*, 31 July 2022. https://vikna.tv/istorii/rozpovidi/rozibraly-elektroczygarku-shhob-zrobyty-dron-dlya-zsu-istoriya-studentiv-kpi-yaki-dopomagayut-nyshhyty-okupantiv/.

[67] '"This will be the year of fiber optics"—Col Kyrylo Veres on drones, innovations in the armed forces of Ukraine and the truce', *DW Ukrainian*, 19 March 2025. https://www.youtube.com/watch?v=iGyt7PfeV9A.

stages of the invasion illustrated a high level of drone losses, while the demand for them in the high-tempo warfighting only increased. The division was innovative not only in terms of drone tactics but also in the optimization of the logistics and supply chains. In order to reduce dependency on long and costly delivery of spare parts, they purchased three-dimensional (3-D) printers and printed the required spare parts for their 'birds'. Furthermore, attention was paid to refocusing on self-sufficiency in servicing and repairing drones close to the front line, which meant that subunits would have engineers and tech support as part of their teams. Furthermore, they had already begun using fibre-optic first-person view (FPV) drones by December 2024.[68]

The same approach to self-sufficiency was adopted by the 20th Regiment of Unmanned Drones K-2 of the ground forces of Ukraine, established on the basis of the K-2 Mechanized Battalion from the 54th Mechanized Brigade. K-2 is led by its founder Lieutenant Colonel Kyrylo Veres. In the words of Lieutenant Colonel Veres, the self-sufficiency of the unit is illustrated in its constant 24\7 performance and the ability to fix any damage to a drone immediately after its flight, often overnight and returning it into deployment the next morning. This entails a necessity for qualified engineers and drone operators. Just like the Birds of Madyar, K-2 have their preparation centres across all roles and functions of their personnel. They also have their own recruitment mechanisms. In essence, this allows them to select the best cadres and enables them to train personnel in the right skills required for warfighting as it is at a given time.[69] Furthermore, increased self-sufficiency of both units is also conditioned by a greater focus on drones manufactured locally, thereby reducing dependency on supply chains, especially those coming from abroad.

Another significant development in aerial warfare is the manufacturing of heavier drones by the Ukrainian military industry. These drones are used for long-range attacks beyond the front line deep into the Russian territory, targeting oil refineries and gas infrastructure, airfields and air bases, radar sights, etc. The proliferation of these drones and smaller versions was possible thanks to the still relatively preserved military industry in Ukraine (*Ukroboronprom*) and continuous work on long-range drone prototypes during the eight interim years of war.[70] One-way attack drones (OWA UAVs) include propeller and jet-engine-powered models, and recently even civilian agricultural light aircraft converted into OWA UAVs.[71] Some of the most prominent examples include the multi-purpose UJ-22 Airborne produced by UkrJet, which can conduct both ISR and attack missions, having a payload of 20 kilograms and a range of 800 kilometres.[72] Since 2023, UJ-26 Beaver OWA drones have been used in mass attacks against enemy targets deep into Russian territory. The Beaver has a range of 1,000 kilometres and a payload of up to 20 kilograms. Another long-range

[68] '414th regiment Madyar's Birds to use fiber optic FPV drones', *Militarnyi*, 21 December 2024. https://mil.in.ua/en/news/414th-regiment-madyar-s-birds-to-use-fiber-optic-fpv-drones/.

[69] '"This will be the year of fiber optics"'.

[70] Fedorchak, *The Russia-Ukraine War*, 111.

[71] Joseph Trevithick, 'Ukrainian long-range attack drones to be tested by U.S. military', *The War Zone*, 14 March 2025. https://www.twz.com/air/ukrainian-long-range-one-way-attack-drones-to-be-tested-by-u-s-military.

[72] '"UJ-22 Airborne" UAS', *SPE UkrJet*. 2025. https://ukrjet.ua/eng.

drone, the AN-196 Lyutyi (Fierce), is often programmed with numerous waypoints in order to avoid Russian air defences. The mass attacks with Lyutyi drones deep into Russian territory are conducted in conjunction with smaller Rubaka kamikaze drones. Their role is to suppress enemy air defences and divert fire from the main attack drones.[73] On 29 September 2024, Ukraine conducted an attack on Russian military objects deep into Russia with a hundred drones, of which 30% were used for diversion, aimed to fly a range of 280 to 340 kilometres. Furthermore, 'to "spoof" Russian radar, strips of metal foil are attached to the wings of drones.'[74] Unlike the Russian poor battle-damage assessment (BDA), the success of the Ukrainian drone strikes is assessed through multiple channels of data collection: ground intelligence; Russian telegram channels; and, finally, satellite images.[75]

The importance of drones in the war in Ukraine is hard to exaggerate. They have covered significant gaps, not only in aerial assets but also in firepower and reach. Similarly, it is impossible to downplay the importance of Ukrainian long-range strikes deep into the Russian territory. First, the immediate necessity was to degrade Russian airfields and airbases closer to the border with Ukraine, and hence to reduce their range and time in airborne attacks on Ukrainian civilians. Second, targeting logistics and fuel storage facilities once again undermined the Russian ability to conduct operations effectively. Furthermore, the increasing threat of the glide bombs could be effectively handled primarily by targeting Russian aviation on the ground or undermining its logistics in the form of take-off and landing capacities and availability of air fuel or functional aircraft *per se*. Third, the primary core of the Russian economy and consequent ability to finance its war machine is its oil and gas industry. Oil refineries, storage facilities, and gas infrastructure were and remain legitimate military targets, aimed at undermining the Russian ability to wage war against Ukraine and its population. Furthermore, these strikes have dispelled the myth of Russian invincibility and the impenetrability of their anti-access/area denial (A2AD) bubble.

On the Russian side, prior to the full-scale invasion, drones were not prioritized in the manufacturing or modernization of their aerial assets. However, having observed the Ukrainian innovations with drones during the first year of the full-scale invasion, and experienced the need to use Iranian Shahed drones to sustain their long-range attacks, Russians began to copy Ukrainian tactics with drones and refocused their manufacturing capacities on the centralized mass production of drones and their diversification across functionality and range of use. According to their official strategy, they intend to manufacture one million drones by 2026 and 1.5 million by 2035. Hence, the intended average production rate is more than 13,000 drones per year for 2023–6.[76] In this regard, the Russians most certainly have and will continue to possess numeric superiority in the number of available

[73] 'Inside the secret world of Ukraine's long-range "Lyutyi" drone strikes on Russia', *Kyiv Post*, 18 October 2024. https://www.kyivpost.com/post/40691.

[74] Ibid.

[75] Ibid.

[76] Pavel Luzin, 'Russian military drones', FPRI (2023): 23. https://www.fpri.org/wp-content/uploads/2023/11/russian-military-drones-.pdf.

drones, since they have turned on their centralized mass production. On the other hand, Ukraine's advantage is in the diversity of manufacturers and customization of drones for different purposes and roles, which have a qualitative advantage over the Russian ones.[77]

Another development characteristic of both sides is the incorporation of artificial intelligence (AI) into drone technologies. The Ukrainian approach focused on employing AI in processing a significant amount of data collected from the battlespace. In this regard, it was used in a similar way to the Western practice of improving situational awareness and rapid decision making, reducing the kill chain to a few seconds.[78] This faster targeting cycle has allowed them to gain an additional advantage over the Russians with their more rigid decision-making processes. On the contrary, the Russian approach to AI and unmanned platforms has focused more on removing humans from the loop of decision making by having a potential full automation of the kill chain. From the perspective of innovation, Russia is adapting to the vulnerability of the operators of FPV drones, who are being specifically targeted on both sides. Second, Russians are known for giving little value to human life, so they do not have any ethical dilemmas about facilitating full AI control of the kill chain.[79]

Together with innovation in drone production and employment, various developments have also taken place in the area of anti-drone or EW measures which gave rise to the development of various drone-jamming weapons, from anti-drone guns to more complex systems. However, the high tempo of warfighting necessitates fast adaptations and innovations. The traditional radio-controlled drones would be vulnerable to electronic jamming. The newly developed and employed fibre-optic drones would not have this type of vulnerability. Fibre-optic drones are flying at lower altitudes and are linked to the ground with fibre-optic cables that unravel away from the drone controller. Besides their invincibility to jamming, they also send high-quality imagery of the target right before contact and elimination. Their speed is often 60 kilometres per hour within the range of 10 kilometres, before the fibre cable breaks. Furthermore, they have an additional advantage of penetrating areas previously more contested for drones. They can fly between buildings and into forest areas without losing connection with controllers like radio-controlled FPV drones would. Thus, Russian drones could penetrate deeper into Ukrainian positions, undermining the coverage the forest terrain would normally provide. On the other hand, the obvious limitation of the fibre-optic drones is their range, smaller payload, and the fact that 'the lines can become tangled or catch on objects, while the unwinding mechanism adds weight and reduces the loads the drones can carry'.[80]

In the context of recent developments in the battlespace, one of the primary Russian targets is the Ukrainian Bohdana artillery. This focus has strategic significance.

[77] '"This will be the year of fiber optics"'.
[78] Viktoriya Fedorchak, 'AI and unmanned platforms in Ukraine', *Luftled*, 1 (2024): 32–5.
[79] Ibid., 34.
[80] Roman Pahulych, 'Fiber-optic drones: The new must-have in Ukraine war', *Radio Free Europe*, 12 March 2025. https://www.rferl.org/a/ukraine-fiber-optic-drones-russia/33344310.html.

Ukraine boosted its manufacture of modernized Bohdana artillery to reduce its dependency on US-produced weapons.[81] Accordingly, Russian drones have focused on the identification and destruction of these specific artillery pieces. The proliferation of Russian drones and further increased visibility of the battlespace have resulted in the change from traditional artillery tactics of 'shoot and scoot' towards hiding artillery pieces in dugouts and buildings.[82] Another example is the Russian employment of fibre-optic drones in the Kursk region. Being surrounded by Russian troops from various sides of the Russian territory, it has allowed a wider use of these new drones. Furthermore, Russians changed tactics—instead of targeting Ukrainian personnel *per se*, they focused strikes on logistics supplies, which were essential in sustaining Ukrainian firepower in the territory of the Kursk region.

Overall, drone warfare in Ukraine illustrates that the age of drones has only just begun, and their further proliferation in different types of conflicts and warfighting is inevitable. Lessons learned from Ukraine have already spread to other ongoing conflicts like in Israel or Houthi attacks. As with any technology, the side that is the most adaptable and innovative can overcome posed obstacles and thrive.

Points for Consideration

The air war over Ukraine illustrated some aspects to consider for modern and future wars. Although the traditional gaining of air superiority has not been achieved by either side and mutual air denial has been established instead, control of the air remains a significant necessity in gaining freedom to manoeuvre across various domains. The extent of air superiority would vary from conflict to conflict and from one tactical situation to another. However, the undeniable necessity of establishing air superiority at least for a short period over a given segment of the battlespace remains a paramount function of air power. At the same time, this war has illustrated the tactical significance of air littoral in the battlespace and its impact on the extent of one's advancement on the front line. Manned aviation is still the primary means of establishing full control of the airspace across all altitudes, while SAMs and drones are beginning to take over in medium and lower altitudes. This trend illustrates that effective and systematic control of the battlespace across altitudes and ranges requires multiple assets across various domains. The lack of sufficient aerial assets in the UkrAF fleet resulted in compensations from other domains and often greater cost in Ukrainian military lives.

Air power has always been the most vulnerable on the ground or during take-off and landing. This inherent vulnerability will continue, illustrating the enduring necessity of using various elements of dispersion in basing, training, and employing air forces in times of both peace and war. Furthermore, as the battlespace becomes

[81] David Axe, 'Russia's latest fiber optic drones peek inside barns for Ukrainian artillery', *Forbes*, 10 March 2025. https://www.forbes.com/sites/davidaxe/2025/03/10/a-russian-fiber-optic-drone-slipped-into-a-camouflaged-dugout-and-discovered-a-valuable-ukrainian-howitzer/.

[82] Ibid.

more transparent, attention should be paid to limiting the logistical footprint and introducing more diverse means of passive air defences, such as camouflaged or underground hangars. Mobility and adaptability are the key elements of improving an air force's survivability and increased performance during warfighting.

While the unification of assets can have a force multiplication effect in any domain, the diversification of assets can also strengthen one's survivability. As was illustrated with Ukrainian air defences and recently with its artillery, overreliance on a single source of firepower or a type of ammunition can make one dependent on a single supply chain or a single manufacturer. The strength of the Ukrainian military assets lies, in fact, in their diversity. It does create additional logistical and maintenance challenges without a doubt but it reduces dependency on a single source that an enemy can target. Hence, the enemy constantly needs to target diverse routes and supply chains to each asset. This is applicable to Ukrainian fixed-wing aircraft and drones alike. Furthermore, although Russia can count on the centralized mass production of drones and various other assets, Ukrainian diversification of drone production allows for better quality and a more niche customization of drones for the requirements of the UAF.

This war shows that while the cutting-edge technologies are incredible across all domains in the battlespace, in large-scale interstate warfare having sufficient numbers is paramount for both gaining advantages and keeping up with various developments in warfighting. In the fight against the Russians' numeric superiority, Ukraine has had to innovate and come up with various new tactics to overcome and undermine that superiority. The obvious cost would be the wearing out of material and personnel and consequent human losses. In order to sustain one's fighting power, and in fact build a resilient fighting power in the air or any other domain, one must consider preparing for war during peacetime and have a functional mechanism to switch from peace to wartime production of various munitions and military material within a shorter time. In high-intensity war, time costs not money but blood and lives.

Finally, as much as analysts are focusing on the technical characteristics of military material, air power consists of both people and technologies, and the synergy between the two is what makes the impossible possible even if computer simulations predict otherwise. People are the ones to compensate for the lack of available military assets, they are the ones piloting fixed- and rotary-wing aviation and remotely piloted drones. People come up with new tactics and solutions according to situational specifications. Investment in people, improving their skills and readiness for a wide variety of tasks and challenges is what strengthens the resilience of the fighting power and, in fact, contributes to the survivability of personnel and military assets. The side that invests the least in its people is shooting itself in the foot no matter how high the numbers of its expendables or cannon fodder might be. I remain convinced that the primary advantage of Ukraine has always been its people, no matter which not needed role they play in the warfighting.

Conclusion

Overall, from all that is mentioned above, it can be concluded that although the air war over Ukraine did not illustrate the traditional fight for air superiority, control of the air remains a significant role of air forces. Lack of control of airspace has a significant undermining effect on freedom of manoeuvre across domains. The establishment of mutual air denial has not prevented either side from utilizing their aerial assets to a greater extent in support of various developments on the ground during the three years of full-scale invasion. Ukrainian manned aviation has shown that, despite being outnumbered, it has survived and continues to fight the Russians. The addition of F-16s and Mirages has only strengthened the strong will of the Ukrainian pilots to protect their country. The Ukrainian experience with air defences has illustrated the significant advantage of a layered approach and diversification combining both Soviet era and modern Western technologies. The proliferation of drones shows that current and future wars will be characterized by greater automation and diversification of UAVs, with the most adaptable and innovative side gaining the advantage at a given time. Any future interstate warfare is unlikely to be fought in a single domain. Hence, at least the effective deconfliction and routine integration between assets originating from different domains and services is required for one's effective performance.

PART II

ORGANIZATION, TECHNOLOGY, AND TRAINING

9

Force Generation and Training of the Armed Forces of Ukraine

Jack Watling

The opening phase of a conflict can be decisive. Many wars last a matter of days or weeks as the professional cadre of opposing militaries, expending a high rate of ammunition, on a battlefield that is comparatively sparsely occupied by combat units, are able to manoeuvre aggressively to achieve their objectives. Most wars are initiated under expectations of a short period of fighting. These expectations, however, are also often incorrect.[1] If the attacking party fails and the defending party is unable to mount a rapid counter-attack then conflicts tend to protract for years. This is because munitions become depleted and casualties impose a requirement for force generation on both sides. There is a strong tendency for reserves to be used to extend the period during which the first echelon can maintain a high tempo of operations, because the idea of a protracted conflict is anathema to policy makers who consequently often overextend their forces. This burning of the reserves, however, often leads to a period where the formation of new units sets the conditions for subsequent rounds of offensive operations, their tempo being determined by the speed and quality of the force generation process.[2]

The Russo-Ukrainian war fits this latter trend. Russia expended its professional force trying to achieve its objectives in 2022,[3] while Ukraine did so through the attrition imposed by defence and then in counter-attacks that autumn. The 2023 Russian offensive was conducted using rapidly mobilized personnel, while Ukraine built new units for its summer offensive. In 2024 the conflict was heavily attritional and positional, with both sides trying to sustain force generation.

The challenges of force generation in Ukraine bear careful scrutiny. This is the subject of this chapter. The chapter will discuss the pre-war Ukrainian training process, the impact of the onset of conflict on the training system, the force generation process inside and outside Ukraine, and the interaction with international partners.

[1] Lawrence Freedman, *The future of war: A history* (New York: Public Affairs, 2017).

[2] Hence the mantra that while professional soldiers start wars, citizen soldiers end them, see David French, *Raising Churchill's army: The British army and the war against Germany 1919–1945* (Oxford: Oxford University Press, 2001); Jonathan Fennell, *Fighting the people's war: British and Commonwealth armies and the Second World War* (Cambridge: Cambridge University Press, 2019).

[3] Michael Kofman and Rob Lee, 'Not built for purpose: The Russian military's ill-fated force design', *War on the Rocks*, 2 June 2022.

Jack Watling, *Force Generation and Training of the Armed Forces of Ukraine*. In: *Russia's War in Ukraine and Modern Warfare*. Edited by: Mikael Weissmann and Niklas Nilsson, Oxford University Press. © Jack Watling (2026).
DOI: 10.1093/9780198970903.003.0009

The chapter is based on direct observation of all elements of this process before and throughout the conflict. This chapter is exclusively concerned with the training of Ukrainian ground forces, air assault forces, and territorial defence forces. For reasons that shall be discussed there is considerable variation within and between these parts of the force, let alone examinations of the national guard, state border guards, UKROSOF, units of the special services, and marines. This chapter also excludes examination of the Ukrainian navy and air force. These other elements of the force deserve study, but the core ground component represents the centre of gravity of the armed forces of Ukraine (AFU) and is therefore the subject of this chapter.

Ukraine's Pre-2022 Training System

The Ukrainian training system between 2015 and 2022 had several elements. First, Ukraine maintained a series of training centres aligned with the different branches: infantry, artillery, radio-electronic troops, air assault forces, air defenders, and so on. The primary responsibility of these training centres was to receive recruits and to teach them how to handle their weapons and operate their equipment. The training centres would also introduce recruits to military culture and provide battle inoculation comprising various psychologically stressful tasks: running gauntlets, operating in close proximity to armoured vehicles, and killing animals, for example.[4]

Once a soldier completed their time at the training centre they would be assigned to a battalion (or division in the case of artillery and air defence (AD) units) and brigade. The battalion commander would have a deputy responsible for the preparation of troops, their integration into the unit, and to make them tactically proficient in carrying out the tasks of the unit. At the same time the deputy brigade commander for training would organize both collective training and exercises, cohering the units within the brigade and teaching them to work together and under command. Battalion training would usually be done on the brigade training area near the town where the brigade was based. Larger exercises would be carried out at national training centres and ranges. The brigade commander, with their staff, would arrange for certification of the unit's readiness through observed combined arms exercises. The final kind of training within the brigade would be led by the brigade commander and comprise mission rehearsals for specific combat tasks arising from brigade plans. This could be done either before or during a deployment.

Troops who had spent time with a unit and had demonstrated aptitude could be sent by the unit to the appropriate training centre to receive specialist training. Thus, a soldier might be qualified as a tank gunner or commander, and in doing so would also be trained to maintain and carry out basic repairs on the vehicle. Owing to the relatively short contracts soldiers signed—usually two years at a time for continuation of service—this upskilling often preceded personnel moving into the civilian sector.

[4] Author observations of Ukrainian initial training bases and interviews with instructors. Also a review of pre-war training literature.

In some cases these personnel would remain part of their brigade's reserve battalion, stationed near the town where it was based. Alternatively, they would transition from the military but retain reserve liability or become part of the brigade commanders' specialist pool of civilian volunteers, a phenomenon particularly prevalent among signals troops.[5]

Officer training in Ukraine was multi-layered. Many students at university took military courses that qualified them to join as officers and meant that they had reserve liability, but did not include significant tactical training or exercises. Professional officers would enrol in one of Ukraine's military academies, with different regions having different military specialist schools. Sumy, for example, hosted the artillery academy; Poltava the radio-electronic academy; and so on.

From 2015 to 2022 the cycle of force generation was largely geared around the need to sustain the line of contact in the Joint Forces Operations Area in Donbas. With around ten brigades on the line at any given time, Ukraine would cycle units so that they would carry out their tour and then return to their base. This would usually then see a cohort of contracts expire and the unit would receive new personnel from the training bases, before embarking on a collective training and exercise cycle to build up towards their next deployment in Donbas. There were some expeditionary rotations, such as the provision of a company of engineers to Kosovo Force (KFOR).[6]

Partner force capacity building was provided to the AFU prior to the full-scale invasion by the United Kingdom (UK), United States (US), and Canada, with some specialist support from other nations. The UK's Operation ORBITAL focused on tactical proficiency of the platoon and its specialist attachments like snipers.[7] Canada supported training to the AFU, focused on company and battalion through Operation UNIFIER.[8] The US delivered a broad set of training with a particular focus on battalion and brigade staffs.[9]

The Transition to War

The full-scale Russian invasion of Ukraine took the Ukrainian military by surprise. As a result, units were not positioned on some key axes. The professional force was still primarily deployed in Donbas. The axis north of Kyiv, and the southern axis from Crimea were both sparsely defended. Other areas like Kharkiv, Sumy, and Chernihiv

[5] Author interviews with Ukrainian brigade commanders prior to the full-scale invasion.

[6] Mission of Ukraine to NATO, 'Ukraine's participation in Alliance-led operations, missions, NATO response force and exercises', 16 June 2021. https://nato.mfa.gov.ua/en/ukraine-and-nato/nato-ukraine-cooperation-peace-support-operations (accessed 21 October 2025).

[7] Embassy of Ukraine to the United Kingdom: 'Operation ORBITAL', 1 May 2020: https://uk.mfa.gov.ua/en/partnership/881-ukrajina-velika-britanija/operaciya-orbital, accessed 21 October 2025

[8] Department of National Defence of Canada, 'Operation UNIFIER'. https://www.canada.ca/en/department-national-defence/services/operations/military-operations/current-operations/operation-unifier.html (accessed 21 October 2025).

[9] Clare Mills, 'Military assistance to Ukraine, 2014–2021', House of Commons Library, 4 March 2022. https://researchbriefings.files.parliament.uk/documents/SN07135/SN07135.pdf (accessed 21 October 2025).

had the advantage of having large training areas and associated units garrisoned in these communities. This meant that there were professional forces available, with their equipment to confront the immediate Russian invasion. On axes where forces were sparse, the AFU mobilized the reserve battalions of their brigades who operated for a time as separate battalions. They also took in large numbers of volunteers but needed officers to control these units. This was largely done by mobilizing instructors from the various training establishments, the early graduation of officer cadets, the mobilization of officers with reserve liability and the reassignment of officers in staff positions to leadership positions. This process allowed the AFU to rapidly expand and given the small size of the initial Russian invasion force, it proved sufficient to blunt the separate Russian columns advancing on separate axes.[10]

A key problem for Ukraine at this stage, however, was that once it became apparent that Russia would transition to a protracted war, it was also clear that Ukraine would need to continue to generate troops.[11] Its training system had been thoroughly disrupted, however. First, the deputy battalion and brigade commanders normally responsible for receiving recruits and training and exercising them were largely in the combat zone. Second, equipment that would normally be used for training was being sent to combat units. Third, many of the training areas had been occupied, were being actively struck,[12] or were heavily damaged. Fourth, instructors had been deployed to combat units. In the summer of 2022 the challenge for the AFU was basic training of battlefield replacements. In an attempt to expand and improve basic training, the UK launched Operation INTERFLEX, bringing Ukrainian troops to the UK for five weeks of basic individual training before returning them to Ukraine.[13] This rapidly became a multi-national effort. At the time it was evident that if Ukraine was to sustain the war it would also need to re-establish an effective means of training and cohering brigade staffs and providing extended tactical training to recruits. The UK explored these options but additional training in these areas was rejected by then UK Defence Secretary Ben Wallace on the basis that the burden should be shared more broadly across Europe. Europe failed to step up and so this training took a long time to be started and was piecemeal and incoherent. This should be marked as one of the significant strategic failings of the effort to support Ukraine by its international partners, and this is not a failure identified in hindsight but one for which the consequences were briefed to decision makers at the time.

[10] For a detailed description of this period, see Mykhaylo Zabrodskyi, Jack Watling, Oleksandr V. Danylyuk, and Nick Reynolds, 'Preliminary lessons in conventional warfighting from Russia's invasion of Ukraine: February–July 2022', Royal United Services Institute, 30 November 2022. https://www.rusi.org/explore-our-research/publications/special-resources/preliminary-lessons-conventional-warfighting-russias-invasion-ukraine-february-july-2022 (accessed 21 October 2025).

[11] This was something the Ukrainian commander-in-chief voiced publicly at the end of 2023 but had been emphasized privately throughout 2022, see Valentyna Romanenko, 'Zaluzhnyi on recruiting and mobilisation: Dismissed military enlistment officers were professionals', *Ukrainska Pravda*, 18 December 2023.

[12] One training area visited by the author for example had been struck by eighteen cruise missiles.

[13] UK Ministry of Defence, 'Operation Interflex reaches three-year milestone', 26 June 2025. https://www.gov.uk/government/news/operation-interflex-reaches-three-year-milestone (accessed 21 October 2025).

One reason for the failure to put a robust training system in place at a time when there was the capacity to do so was that the Russian withdrawal from Kyiv, the culmination of the Russian Donbas offensive, and the clear shortage of troops on the Russian side allowed Ukraine's pre-war training mechanisms to be partially revived for the purposes of taking new recruits and cohering them into brigades. Combined with the consolidation of brigades with their reserve units, this allowed Ukraine to mass the reserves necessary for the offensives to recover Kherson and Kharkiv, which proved successful, but also triggered Russian partial mobilization.

The Emergent Ukrainian Training System

From late July 2022 planning had started to conduct a deliberate offensive operation to recover the occupied territories. By the time the plan had been presented to Ukraine's international partners, and equipment pledged to the operation, however, Russia had begun to implement its mobilization. The result was that experienced Ukrainian brigades became fixed on the front. In this context the decision was made to form new brigades who would receive equipment sets from different international partners, receive mentoring for their staffs, and specialist training to operate the equipment provided. The idea was that these units would be better cohered. In principle this was a reasonable proposition, except that there was insufficient time to properly train the units. Equipment arrived just weeks before the planned operation and training was short and fragmented. Without a backbone of experienced troops in the units, and with inexperienced brigade staffs, these formations were not highly capable. Attrition, meanwhile, among Ukraine's experienced brigades began to climb during the fighting for Bakhmut.[14]

The Ukrainian summer offensive of 2023 failed for a range of reasons. Insufficient concentration of forces, an inadequate supply of materiel relative to the adversary build-up, and the commitment of underprepared forces all saw Ukraine take heavy losses in equipment and personnel for very modest gains. After four months on the attack Ukraine culminated with its units needing replacements to bring themselves up to strength.[15] Russia, meanwhile, building on the initial expansion of its forces through partial mobilization had been on a recruiting drive that saw the size of its forces grow significantly.[16]

[14] For a detailed discussion of these pressures, see Jack Watling, Oleksandr V. Danylyuk, and Nick Reynolds, 'Preliminary lessons from Ukraine's offensive operations, 2022–23', Royal United Services Institute, 18 July 2024. https://www.rusi.org/explore-our-research/publications/special-resources/preliminary-lessons-ukraines-offensive-operations-2022-23 (accessed 21 October 2025).

[15] Yuri Clavilier and Michael Gjerstad, 'Combat losses and manpower challenges underscore the importance of "mass" in Ukraine', IISS, 10 February 2025. https://www.iiss.org/online-analysis/military-balance/2025/02/combat-losses-and-manpower-challenges-underscore-the-importance-of-mass-in-ukraine/ (accessed 21 October 2025).

[16] Dara Massicott, 'Russian military wartime personnel recruiting and retention 2022–2023', RAND, 16 July 2024. https://www.rand.org/pubs/research_reports/RRA2061-4.html (accessed 21 October 2025).

From the autumn of 2023 Russian forces began to field units with line companies and assault detachments. These units would take up positions on a sector of front with the line units holding ground and the assault detachments attacking. The unit would continue to prosecute attacks until the assault troops were killed or wounded and then rotate out of the line to receive replacements. The result was that while Ukraine could inflict massive losses in aggregate on Russian units, it was primarily attriting an undertrained contact layer and never really brought the Russian units below 70% strength. In this way the Russians managed to maintain a near-continuous offensive over 2024. It was not an effective method for taking ground, but it did fix Ukrainian units on the line of contact and impose a steady rate of attrition on the AFU that demanded replacements.[17]

It was in this context that the Ukrainian training system began to stabilize. Following the 2023 offensive the AFU had a shortage of combat capable infantry while its brigades were fixed in defence. The training process that emerged endeavoured to adhere to the pre-war Ukrainian process in some important respects. First, troops would be mobilized and sent to their respective training areas. They were supposed to receive a minimum of five weeks of training. In practice, a lack of equipment and experienced instructors often saw the intensity of training decline so that it became closer to three weeks of training spread over five weeks.[18] This was also affected by the frequency of Russian ballistic missile strikes on training areas.[19] After this personnel would be assigned to a brigade.

Because combat capable staff were retained in the fighting units the centralized training areas largely lacked up to date knowledge of the relevant tactics. Most brigade commanders therefore retained a platoon of instructors who would take personnel from the training areas and put them through additional training to bring them up to date on how the brigade fought. The result was divergence in training as different brigades focused on different problem sets. Quality of training also varied considerably depending on the quality of the brigade staff and the tactical situation in the brigade's area of responsibility.[20]

Ukraine has had no higher tactical echelon to manage the rotation of brigades until the formation of the corps echelon in 2025.[21] Brigades therefore have tended to rotate their battalions on and off the line such that tactical training is provided for the battalion out of contact. Personnel are therefore either given training with this battalion and then rotated onto the line or are assigned to a combat battalion on the line and, if they survive, do collective training when the battalion is rotated

[17] Author engagements with Ukrainian brigades on the line of contact throughout 2024.

[18] Author visits to Ukrainian training centres, 2023–5.

[19] Roman Kohanets, 'Russia launches ballistic missile strike on Ukrainian training base, casualties reported', United24, 16 October 2025. https://united24media.com/latest-news/russia-launches-ballistic-missile-strike-on-ukrainian-training-base-casualties-reported-12563 (accessed 21 October 2025).

[20] Author visits to brigades, 2023–5.

[21] Taras Safronov, 'Already carrying out tasks: Ukraine completes transition of armed forces to corps structure', Militarnyi, 1 October 2025: https://militarnyi.com/en/news/already-carrying-out-tasks-ukraine-completes-transition-of-armed-forces-to-corps-structure/ (accessed 21 October 2025).

back. For brigades under heavy pressure the scale of training off the line is often at a much smaller scale. Within the brigade the commander is responsible for determining requirements for specialists, whether that be communicators, UAV operators, or vehicle crews.

The Ukrainian system of training in the unit is manifestly inefficient when the units in question cannot be rotated off the front. It leads to undertrained personnel being committed to combat and a higher rate of casualties, and it prevents brigades conducting training on a larger scale, limiting the size at which they can plan and execute operations. The shortage of training has seen a steady decline in the combat performance of Ukrainian infantry. Conversely, as shall be discussed later, the much better training of specialists has seen a progressive improvement in UAV operations.

The system of training in the unit has made it especially difficult for Ukraine to leverage the training offered by its partners outside of Ukraine. First, experienced brigades cannot be spared and so their personnel are not generally sent abroad. Second, even if some personnel are sent to training areas, the AFU has insufficient equipment to make it available to move abroad. As a result, troops end up without relevant equipment, or must use equipment by proxy. Moreover, many of the tactics—especially as regards unmanned aerial vehicle (UAV) and electronic warfare (EW) operations, critical to fighting in Ukraine cannot be integrated into training in foreign training areas because of regulatory constraints.

To try and circumvent this problem the AFU continued to generate fresh brigades which would be equipped by one of Ukraine's partners and theoretically train collectively abroad, before being brought to Ukraine. This proved very problematic. Desertion rates were high. The officers for these units lacked experience. Moreover, training struggled to be made relevant to Ukrainian conditions. In many cases this was because the foreign directing staff were not familiar with the tactical drills relevant today at the front in Ukraine. They were teaching an outdated syllabus. Even this syllabus, however, could not be easily taught because of the compressed training timeline. Because Ukrainian troops are undertrained and there is insufficient time to do successive training packages that progressively make them safe to practice complex manoeuvres, personnel could not be certified as safe to practice fire and manoeuvre at scale. As a result Ukrainian officers became frustrated that they could not accurately replicate the combat environment in training areas, while partners became frustrated at the perception that they did not know how to fight. The result was that the creation of new brigades and collective training outside Ukraine was discontinued.[22]

The combination of these factors meant that when Ukrainian units returned from France, Germany, and Poland they either performed poorly on the front, or were

[22] Volodymyr Ivanyshyn, 'Ukraine will no longer create new brigades, official says', *Kyiv Independent*, 12 February 2025. https://kyivindependent.com/ukraine-will-no-longer-create-new-brigades-official-says/ (accessed 21 October 2025).

broken up and used as replacements for more experienced brigades.[23] The latter approach also created problems because it meant that the brigade staffs did not know the capabilities or drills used by their newly attached battalions. The overall result has been that Ukrainian brigades have often taken on more and more battalions, while these battalions—whether veteran units or newly raised ones—have been attrited such that they probably each contain between one and two companies of combat capable infantry, and maybe only a platoon considered capable of offensive action.

The Fragmentation of Ukraine's Training System

In 2023 the burgeoning specializations of assault infantry and UAV operations saw the creation of a dedicated UAV school and assault infantry courses. The intent was to have an efficient centralized system. These schools tended to offer thirty-day training packages on reconnaissance, first-person view (FPV), or bomber operations.[24] In 2024, however, Ukraine sought to massively expand the use of UAVs through the 'Wall of Drones', and in doing so needed to train more UAV operators than the centralized system was funded or staffed to manage. Moreover, UAV operations became highly dependent upon the technology being used, and because each brigade or emergent UAV unit used different platforms, so too did the necessary tactics and skills differ. As money and resources began to flow to the new UAV units, while larger brigades with private financial backing like 3rd Assault Brigade, Khartiia, and others, established internal UAV detachments, training began to spread out, along with resources to sustain it. Expertise in UAV operations then became concentrated in an independent UAV branch.[25] The result has been growing proficiency in the employment of these tools but also a decline in the capacity of the force to use them with a combined arms approach.

A similar trajectory has affected assault infantry. In 2023 brigades would select personnel to form assault platoons which were usually held at the brigade echelon to spearhead offensive action.[26] In 2024 Ukraine began to establish independent assault battalions. The training of these units involved what would be considered basic light infantry skills in a North Atlantic Treaty Organization (NATO) military. These units, because they were not committed to hold ground, could spend more time training and therefore became centres of excellence for dismounted close combat in the AFU.

[23] Tim Zadorozhnyy and Emmanuel Chaze, '"Negative lesson"—Ground forces chief acknowledges desertions, mismanagement in French-trained brigade', *Kyiv Independent*, 6 January 2025. https://kyivindependent.com/negative-lesson-ukraines-ground-forces-chief-acknowledges-desertions-mismanagement-in-french-trained-brigade/ (accessed 21 October 2025).

[24] Author visits to the UAV school and assault infantry training courses in 2023.

[25] Ministry of Defence of Ukraine, 'About us', Uncrewed Systems Force. https://mod.gov.ua/en/about-us/unmanned-systems-forces (accessed 21 October 2025).

[26] Author observation of Ukrainian offensive operations in 2023.

These units were subsequently turned into regiments and, then, their own branch.[27] Each unit has its own approach to training. As they have time, they reach a level of proficiency. But there is a limited opportunity to bridge the gap between the training they receive and the capabilities of the rest of the force.

AD has evolved in a similar direction. In 2023 most AD assets were moved to be coordinated by the air force. The air force, however, had an AD doctrine and a command and control structure that was not optimized for managing Ukraine's large numbers of mobile fire teams, critical for defeating Russian UAVs, or their integration with point defence. Consequently a dedicated AD task force for Kyiv was established, which pioneered novel approaches to AD. In 2025 a comparable task force was established for Kharkiv, and thereafter orders were given to create one for each oblast. This sudden scaling, however, did not have a mechanism for training AD operators on how to build and run these task forces. The task forces themselves had innovated substantially from established practices but while they had trained their own personnel, they were not staffed to train others.[28]

The new forces—uncrewed systems forces, AD task forces, and assault forces—each managing training within the unit, have produced new and innovative ways of fighting. The imposition of a 15–30 kilometre attrition belt along the front has essentially eliminated the concept of a forward line of own troops (FLOT). The assault infantry, meanwhile, integrates new tools into traditional light infantry tactics that offer a promising approach to combined arms manoeuvre. The AD task forces are leading the world in understanding how to fight a complex AD battle against a layered attack. But the training system has also fragmented and without higher headquarters the ability to practice integrating these capabilities is severely limited. Mapping how to laterally move knowledge within the force to ensure all units are adhering to best practice and building the corps' echelon to enable combined arms training is a priority for Ukraine in 2026.

Lessons for the Future

There are important lessons from the evolution of Ukraine's training structure that should inform Ukraine's partners and the AFU after a ceasefire as it endeavours to build a force capable of deterring future Russian aggression.

First, it must be emphasized that many of the force generation decisions made by Ukraine made sense in the immediate context in which they were made and had significant longer term downsides. When a state mobilizes for war it is likely to have more people come forward than it can equip and manage, even where there is a low proportion of the population eager to volunteer. The limitation will be the state's

[27] Andrzej Wilk and Peotr Zochowski, 'President of Ukraine announces the creation of assault forces. Day 1308 of the war', OSW, 23 September 2025. https://www.osw.waw.pl/en/publikacje/analyses/2025-09-23/president-ukraine-announces-creation-assault-forces-day-1308-war (accessed 21 October 2025).

[28] Extensive author engagement and observation of AD-TF Kyiv and Kharkiv.

capacity to absorb recruits. If at all possible, this should be used to ensure that a proportion of the force is being trained thoroughly. Mass should enable longer training. The rewards of continuing to train officers and capable infantry will take time to be felt but will prevent a nation from suffering from a degradation in force quality over time.

Second, the Ukrainian approach of training in the unit, without a higher echelon able to rotate units off the line, is a bad one. The approach should be to significantly increase the duration of training in the training centres and to have training battalions, staffed by experienced officers and wounded personnel from front-line units who can ensure that teaching reflects up to date tactics and is realistic. These training battalions should bring personnel up to a level of reasonable tactical proficiency. They should then be passed to combat battalions that are rotated off the line so that the unit comprises a mix of veteran and green personnel, who can then go through collective training to be cohered into a force ready for deployment to the front.

Third, while progression of training to ensure a reasonable level of safety is sensible—and can be maintained if there is enough time to progress—regulation must be prepared to enable an effective transition to war. In wartime the need to practice with actual combat equipment in combination is critical to ensuring the readiness of combat units. It also prevents trainers from stagnating and teaching obsolete tactics by imposing stasis on the training environment. Ensuring that, in war, legislation allows the military to take risks in training is critical.

Fourth, Ukraine struggled to manage the process of dissimilar rearmament. Many brigades trained personnel to operate vehicles they never received, while other brigades received equipment for which they had a shortage of trained operators. The diversity of equipment used by the AFU is particularly dramatic given the number of countries donating gear. However, most Western militaries lack stocks of spare vehicles with which to equip newly mobilized personnel and their defence industries are not well suited to the rapid scaling of existing platforms. It is therefore inevitable that standardization will reduce in war and that equipment will change. This should be anticipated and training reflect this reality rather than a presumption that equipment will adhere to pre-war regulated standards. Pragmatism on qualifying personnel across multiple platforms will therefore have to prevail.

For Ukraine specifically, the establishment of an army corps echelon to function as a divisional layer, managing the tactical rotation of multiple brigades, is a welcome reform. It is too early for this to be felt at time of writing, in October 2025, but after the conclusion of a ceasefire, retaining this echelon and ensuring that there is a proper process for rotating combat brigades will be fundamental to ensuring the force can train and maintain force quality in the event of a future conflict, and thereby deterring Russia by demonstrating Ukraine's capacity to wage a protracted struggle.

Finally, Western training for the conduct of combined arms manoeuvre is often concentrated in time, such that commanders are rewarded for being aggressive and bold, even if it leads to a high rate of attrition. Once a conflict protracts, however, a force's ability to preserve experienced and hard to replace personnel becomes critical for its long-term competitiveness. A training system that emphasizes short-term gains over long-term sustainability is liable to create significant misalignment between higher headquarters and tactical commanders in an actual conflict. This is an area, therefore, where operational commanders should be trained so that they understand the trade-offs and constraints in how they employ a force relative to the force's force generation capacity. Playing longer exercises significantly alters the incentives.

10
Intelligence and Russia's War in Ukraine

Niklas Nilsson

Introduction

Russia's war in Ukraine illustrates or has functioned as a catalyst for a set of developments underlining important challenges and trends in wartime intelligence, acquiring increased centrality with the war. This chapter examines four significant categories of intelligence practices that have been actualized during the war.

Highlighted by divergent pre-war assessments, the *limitations of forecasting and prediction* illustrate the complexity of gauging adversarial intent and military capacity. In particular, the dominant perception that the war would conclude with a quick Russian victory proved false, underlining the exaggerated understanding of Russian capabilities and underestimation of Ukraine.

The utilization of *declassified strategic warning intelligence for strategic communication*, which the United States (US) and United Kingdom (UK) employed to call out Russian military preparations, represented a proactive adaptation to the contemporary information environment. While providing a largely successful prebunking of Russia's attempts to legitimize the war and facilitate a cohesive Western response to the invasion, controversial aspects raised in the aftermath include short-term risks to operational security as well as longer-term concerns regarding the potential politicization of intelligence.

The war also accelerated *intelligence sharing* between the US and allies, and Ukraine. Particularly in connection to decisions to supply Ukraine with long-range weaponry, requiring access to targeting intelligence, the US relaxed restrictions on intelligence sharing. The degree of the cooperation, its significance to the war effort, and the inherent vulnerabilities of intelligence dependencies were underlined by the Trump administration's temporary halt on intelligence sharing and its effects for Ukraine's targeting capacity.

Finally, the war has seen a rapid *technological evolution in intelligence* given the need to process and integrate the enormous amounts of data produced by the transparent battlefield. Commercial tech companies have flocked to Ukraine, eager to test and develop their products in authentic conditions, while Ukraine has made consistent and successful efforts to build its own defence tech sector.

Niklas Nilsson, *Intelligence and Russia's War in Ukraine*. In: *Russia's War in Ukraine and Modern Warfare*.
Edited by: Mikael Weissmann and Niklas Nilsson, Oxford University Press. © Niklas Nilsson (2026).
DOI: 10.1093/9780198970903.003.0010

Ultimately, Russia's war in Ukraine has been a driver of significant developments in intelligence, many of which are only beginning to be comprehended and analysed. This chapter has no pretentions of being exhaustive, yet provides an overview of some central developments in intelligence in relation to the war.[1]

The Limitations of Prediction and Forecasting

Assessments from Western intelligence organizations as well as the broader security and military analytic community regarding the probability that Russia would make the final decision to launch a full-scale invasion of Ukraine diverged considerably in the months preceding 24 February 2022. Whereas US and UK intelligence agencies forcefully made the case that the invasion was imminent, several European allies remained sceptical, referring to the risks, costs, and irrationality attached to such a decision. While the US and UK assertions proved correct, predictions regarding how the war would proceed once initiated proved predominantly wrong, across the Western intelligence and civilian analytic communities.

The majority of analysts and experts assessed that Russia would defeat a far smaller and militarily inferior Ukraine in a matter of weeks or months. This analysis also seems to have been the predominating one informing policy-making circles, not least given the reluctant and slow process of scaling up military assistance to Ukraine with advanced weaponry on the part of the Biden administration. This stemmed from a calculated averseness towards the risk of the war escalating beyond Ukraine, yet was also informed by a perceived futility of sending equipment that would likely end up in Russian hands.[2]

When it comes to the divergence in assessments regarding Russia's decision to invade in the first place, Driedger and Polianskii attribute the failure in the analytic community to predict it to an overreliance on rationalist, cost-benefit arguments, highlighting the likely economic, diplomatic, and security costs Russia would incur, and assuming Putin's regime would act in accordance with a logic of cost minimization.

Among the points made at the centre of this argument were that Russia's elites would not go ahead with a large invasion because crippling Western sanctions were nearly certain; the cost of seizing and holding Ukraine's vast territory and controlling its population of forty million seemed prohibitively high; and that diplomatic isolation, battered energy exports, and damage to Russian business interests in the West would undermine the Russian regime. Instead, many leading analysts and think

[1] In preparation of this work, the author utilized generative artificial intelligence (AI) and AI-assisted technologies to support the writing process, including ChatGPT, Atlas.ti, and Unriddle. The author carefully reviewed, edited, and validated the content to ensure its accuracy and integrity and takes full responsibility for the final published work.

[2] Eliot A. Cohen and Phillips O'Brien, *The Russia-Ukraine War: A Study in Analytical Failure* (Washington, DC: Center for Strategic and International Studies, 2024); Mike Eckel, 'How did everybody get the Ukraine invasion predictions so wrong?', *Radio Free Europe/Radio Liberty*, 17 February, 2023, https://www.rferl.org/a/russia-ukraine-invasion-predictions-wrong-intelligence/32275740.html

tanks asserted that the Kremlin's strategy aimed at brinkmanship or limited escalation to force concessions from Ukraine and North Atlantic Treaty Organization (NATO).[3] This utility-based focus overlooked the Kremlin's disregard for or underestimation of costs, and its acceptance of risk for ideological, domestic-political, or historical revisionist reasons. In this context, not least Putin's personal obsession with restoring Russia's sphere of influence in the former Soviet space and his denial of a legitimate Ukrainian national identity, were not appropriately factored into the assessment. Moreover, Russia's experiences of the limited fallout and consequences of its wars in Georgia 2008, Ukraine from 2014, and Syria from 2015 added to its risk appetite when it came to invading Ukraine. The decision was underpinned by the intelligence failure of Russia's Federal Security Service (FSB), whose obsequious pre-war analyses had painted an overoptimistic picture of the realism in the 'special military operation'.[4]

Analyses predicting the trajectory of the war consistently overestimated Russian, and underestimated Ukrainian capabilities, based on flawed assumptions about Russian military power, a dearth of corresponding knowledge about Ukraine, and the inherent difficulties of forecasting war. Russia's overhaul of its military between 2008 and 2014 in the 'New Look' reforms intended to professionalize large segments of its force, and invested heavily in modern equipment.[5]

In 2014, Russia's rapid capture of Crimea surprised and alarmed Western officials, whereas its expeditionary deployment to Syria conveyed to outside observers that the Kremlin could quickly project force abroad, gaining Russia the epithet of a 'near-peer' adversary of NATO. Western military assessments and policy debates amplified this image. Bettina Renz argues that the overestimation of Russian capabilities stemmed from an overcorrection as earlier neglect of developments in the Russian military spurred a sense of alarm in the US and parts of Europe, resulting in an inflated depiction of Russia's armed forces. Worst-case thinking, repeatedly portraying the Russian armed forces as advanced and globally agile, served to justify a stronger NATO posture. Moreover, the elevated notion of Russian 'hybrid warfare' attributed an unrealistic degree of strategic cunning to the Kremlin. Although numerous experts warned that Russia's success in Crimea was as much about local conditions and Ukraine's limited ability to respond as about Russian prowess, those caveats found little traction in a policy environment that overemphasized potential threats.[6]

The preoccupation with the quantities of Russian military's hardware also overshadowed subtler considerations about organization, command, intelligence, force

[3] Jonas J. Driedger and Mikhail Polianskii, 'Utility-based predictions of military escalation: Why experts forecasted Russia would not invade Ukraine', *Contemporary Security Policy* 44/4 (2023): 544-560. https://doi.org/10.1080/13523260.2023.2259153.

[4] Dara Massicot, 'What Russia got wrong: Can Moscow learn from its failures in Ukraine?', *Foreign Affairs* 102/2 (2023): 78–93; Shlomo Shapiro, 'Intelligence and the Ukraine war', *NSF* 24/1 (2023): 7–18. doi:10.37458/nstf.24.1.1.

[5] Cohen and O'Brien, *The Russia-Ukraine War*.

[6] Bettina Renz, 'Western estimates of Russian military capabilities and the invasion of Ukraine', *Problems of Post-Communism* 71 (2024): 219–31. https://doi.org/10.1080/10758216.2023.2253359.

employment, and logistics that proved to function quite differently in battle with a live opponent than had been showed off at various Russian large-scale scripted exercises in the preceding years. The human dimension has also featured prominently in post-fact analyses of the reasons for Russia's failed invasion, citing poor morale and incompetence, e.g., manifested in desertions, corruption, and counterproductive tactical behaviour.[7] Quantitative data on the number of tanks, aircraft, or precision-guided missiles overshadowed unquantifiable concerns that nevertheless had a detrimental impact on the Russian army's ability to sustain complex operations.[8] While acknowledged by many analysts, the coverage of intangible aspects rarely enjoyed the same visibility as more spectacular claims about hyper-sonic missiles or doctrinal breakthroughs.[9] In reality, intangible aspects of military power such as cohesion and the will to fight are both critical and difficult to measure. These elements can shift rapidly, are context dependent, and are nearly impossible to assign precise numeric values.[10]

Conversely, many forecasts tacitly assumed Ukraine would collapse quickly. This assumption stemmed partly from a perception of Ukraine as politically weak, divided, and corrupt, with an incohesive national identity unable to serve as a solid foundation for resistance.[11] While a substantial analytic community in the West monitored the Russian military reforms after 2008, expert knowledge on Ukraine was limited and, consequently, so was the understanding of Kyiv's post-2014 military build-up.[12] By 2022, the Ukrainian forces could not match Russia's overall quantity but had already fought a war against Russia for eight years and drawn on that experience to conduct a far-reaching reform of its military, with NATO support. This combination proved decisive: Ukrainian troops remained cohesive, integrated incoming Western arms more effectively than anticipated, and repeatedly adapted on the battlefield.

The erroneous assessments of the outcomes of the war speak to a generic intelligence problem: the challenge of accommodating qualitative factors into assessments of military capability. Assessments are traditionally dominated by observable and material quantitative measurements such as the availability of equipment and soldiers, as well as demonstrated capabilities during exercises. However, these types of factors do not provide a full picture of how a military force will perform in actual battle and may provide inaccurate assessments and in extension poorly informed and even detrimental decisions in war.

[7] Robert Dalsjö, Michael Jonsson, and Johan Norberg, 'A brutal examination: Russian military capability in light of the Ukraine war', *Survival* 64/3 (2022): 7–28. https://doi.org/10.1080/00396338.2022.2078044.

[8] Cohen and O'Brien, *The Russia-Ukraine War*.

[9] Renz, 'Western estimates'.

[10] Ben Connable, Michael J. McNerney, William Marcellino, Aaron Frank, Henry Hargrove, Marek N. Posard, S. Rebecca Zimmerman, Natasha Lander, Jasen J. Castillo, and James Sladden, *Will to fight: Analyzing, modeling, and simulating the will to fight of military units* (Santa Monica, CA: Rand Corporation, 2018).

[11] Cohen and O'Brien, *The Russia-Ukraine War*.

[12] Renz, 'Western estimates'.

Intelligence Disclosure and Strategic Communication

The months leading up to Russia's February 2022 invasion of Ukraine saw a surge in public intelligence disclosures by the US and UK. Officials released a steady stream of information regarding Russia's force build-up, the preparation of false-flag operations as pretexts to invade and a plot to oust the Ukrainian government and install a Kremlin-aligned regime.[13] From late 2021 onward, commercial satellite imagery and declassified intelligence assessments provided detailed snapshots of Russia's military positioning around Ukraine. Intelligence agencies and defence ministries published detailed accounts of Moscow's war planning and subsequent political agenda for subduing Ukraine.[14] Equally striking were the frequent post-invasion updates: for instance, the British Ministry of Defence began issuing daily briefings outlining battlefield developments and analysing Russia's tactical setbacks.[15] Intelligence releases have also drawn attention to Iranian and North Korean weapons transfers to Russia and were instrumentalized to deter Chinese arms shipments.[16]

This approach represented a revaluation of the relationship between secrecy and openness in the utilization of strategic warning intelligence. Although it failed to dissuade the Russian leadership from invading Ukraine, the disclosures played a crucial role in preparing Ukraine and European allies for the invasion and countering Moscow's narrative legitimizing the war.[17] The unprecedented amount of strategic intelligence disclosures demonstrated a concerted effort to deny Russia the ambiguity it exploited in earlier aggressions and to maintain a consistent flow of accessible information, to warn and convince Ukraine and European allies that war was imminent (although largely unsuccessfully), to consolidate perceptions of the war's causes, and to galvanize support for Ukraine.[18] Although the invasion still occurred, exposing the Kremlin's designs for Ukraine helped establishing a cohesive perception among allies of what was happening and the consequences thereof in the immediate aftermath of 24 February.[19]

[13] Mark Phythian and David Strachan-Morris, 'Intelligence & the Russo-Ukrainian war: Introduction to the special issue', *Intelligence and National Security* 39/3 (2024): 377–85. DOI: 10.1080/02684527.2024.2330132.

[14] Ruxandra Buluc, Rubén Arcos, and Cristina Ivan, 'When spies go public! Lessons learnt from the instrumentalization of intelligence for strategic communication in the run-up to the Russian-Ukrainian war', *Intelligence and National Security* 40/1 (2025): 42–57. https://doi.org/10.1080/02684527.2024.2405255.

[15] Karla Adam, 'Invasion prompts U.K. intelligence to lift the veil—a bit', *The Washington Post*, 23 April 2022.

[16] Vivian Salama, William Mauldin, and Nancy A. Youssef, 'U.S. considers release of intelligence on China's potential arms transfer to Russia: Western nations have intelligence that Beijing might end its self-imposed restraint on weapons supplies to Moscow', *Wall Street Journal*, 23 February 2023.

[17] Max Colchester, 'U.S., Allies fight information war with Russia to deter Ukraine invasion: Public airing of secret intelligence about Kremlin's plans aims to avert a Russian attack', *Wall Street Journal*, 9 February 2022.

[18] Huw Dylan and Thomas J. Maguire, 'Secret intelligence and public diplomacy in the Ukraine war', *Survival* 64/4 (2022): 33–74.

[19] Seto Takashi, *Weaponized disclosure of intelligence in the Russia-Ukraine War* (Tokyo: National Institute for Defense Studies, 2022).

The approach taken to disclosure and strategic communication of intelligence to influence domestic and foreign audiences was not unprecedented. US Ambassador Adlai Stevenson's display at the UN Security Council of aerial photos during the Cuban Missile Crisis in 1962 and the US leveraging of intelligence about Iraq's purported Weapons of Mass Destruction programme in 2002–3 are famous cases in point. However, the systematic approach with which intelligence was utilized for political messaging in the run-up to the Russian invasion arguably represented a different calculus regarding the balance between the immediate benefits of disclosure versus potential immediate and longer term operational risks.[20]

To a significant extent, the approach was permitted by the growing open-source analysis ecosystem of private and civil society actors.[21] The interaction between government and non-government produced information allowed various investigative networks to hone in on developments indicated by official intelligence disclosures and corroborate these in near-real time.[22] At the same time, the public availability of open-source material allowed the official disclosure of redacted intelligence while sustaining acceptable protection of sources and methods.[23] An important case in point was the utilization of commercial satellite imagery to replace secret footage from US military satellites.[24]

The intelligence disclosures failed to convince the leadership of most allies and Ukraine that an invasion was imminent in the run-up to the war.[25] Yet once the invasion started, they provided for a unified understanding of events around which Western allies could unite in the backing of Ukraine. The disclosure of secret intelligence on this scale, and the frequency, coordination, and highly public nature of it, was a novel practice in the utilization of strategic warning intelligence for strategic communication purposes. In turn, the communication strategies employed in relation to Russia's war in Ukraine were both necessitated and enabled by the contemporary contested information environment. Previous experiences, not least from 2014, had highlighted the need to quickly establish a dominant perception of events among an abundance of competing narratives. The exponential growth of available information and possible interpretations of it compelled the proactive dissemination of the authoritative evidence offered by intelligence assessments and placing these in

[20] Joshua C. Huminski, 'Russia, Ukraine, and the Future Use of Strategic Intelligence', *Prism* 10, no. 3 (2023): 9-25; Julian E Barnes, Adam Entous, 'How the U.S. Adopted a New Intelligence Playbook to Expose Russia's War Plans', *New York Times*, 23 February 2023.

[21] Niklas Nilsson, Mikael Weissmann, and Björn Palmertz, 'Hybrid Threats and the Intelligence Community: Priming for a Volatile Age', *International Journal of Intelligence and CounterIntelligence* online first (2025), https://www.tandfonline.com/doi/epdf/10.1080/08850607.2024.2435265?needAccess=true.

[22] Huminski, 'Russia, Ukraine, and the future'.

[23] Takashi, *Weaponized disclosure*.

[24] Shane Harris and Paul Sonne, 'Russia planning massive military offensive against Ukraine involving 175,000 troops, U.S. intelligence warns', *The Washington Post*, 3 December 2021, https://www.washingtonpost.com/national-security/russia-ukraine-invasion/2021/12/03/98a3760e-546b-11ec-8769-2f4ecdf7a2ad_story.html; Paul McLeary, 'Russian buildup near Ukraine gains steam, new satellite images show', *Politico*, 23 December 2021, https://www.politico.com/news/2021/12/23/russia-buildup-ukraine-satellite-images-526109.

[25] Shaun Walker, 'A war foretold: How the CIA and MI6 got hold of Putin's Ukraine plans and why nobody believed them', *The Guardian*, 20 February 2026. https://www.theguardian.com/world/ng-interactive/2026/feb/20/a-war-foretold-cia-mi6-putin-ukraine-plans-russia.

a frame of reference of one's own choosing, while the expanding availability of private and public capacity for collection and analysis enabled the combined disclosure of secretly derived conclusions with substantiation offered by open sources.[26]

The extent to which the measures employed in relation to the war is foreboding of a significant shift in the utilization of intelligence, and in the activities, practices and roles of intelligence agencies remains to be seen. Whereas US and UK decision makers apparently perceived the utility of intelligence disclosures to clearly outweigh the drawbacks in this case, the calculation may very well be different in another conflict, for example in the event of an imminent Chinese invasion of Taiwan. Intelligence disclosures certainly come with risks attached, including the 'disclosure dilemma' of balancing the utility of disclosure against the detriment of endangering sources and methods.[27] More fundamentally, if the perceived success of intelligence-based strategic communication in relation to the war in Ukraine raises political expectations that intelligence should to a larger extent be produced with this utility in mind, rather than in support of decision-making, these expectations come with risks that intelligence might become increasingly politicized in the process.[28]

Ukraine itself has developed a sophisticated approach to strategic communication since 2014, involving an extensive network of government and civil society entities, which has enabled the country to put its experience of fighting Russia in the information domain to good use throughout the war.[29] Intelligence disclosures is an integral component of Ukraine's broader information strategy, leveraged to shape both domestic and international perceptions. Ukraine's employment of intelligence in its strategic communication also represents hard-earned experience from 2014 and the occupation of Crimea, when Ukrainian authorities failed to mount an effective response. In recognition of the crucial importance of shaping the international understanding of events, the Security Service of Ukraine (SBU) rapidly released intercepted calls linked to the downing of Malaysia Airlines Flight MH17 in July 2014.[30]

During the war, SBU and Ukraine's military intelligence service (Main Directorate of Intelligence of the Ministry of Defence of Ukraine, HUR), have continuously provided selectively declassified intelligence for strategic communications purposes including intercepted phonecalls from Russian soldiers and tactical drone footage

[26] Niklas Nilsson, *Intelligence and Strategic Communication* (Riga: NATO Strategic Communications Centre of Excellence, 2025).

[27] Austin Carson Allison Carnegie, 'The Disclosure Dilemma: Nuclear Intelligence and International Organizations', *American Journal of Political Science* 63, no. 2 (2019): 269–285, https://doi.org/10.7910/DVN/JQEUBQ; Ofek Riemer, 'Politics is not everything: New perspectives on the public disclosure of intelligence by states', *Contemporary Security Policy* 42, no. 4 (2021): 554–583.

[28] Joshua C. Huminski, 'Russia, Ukraine, and the future use of strategic intelligence', *Prism* 10/3 (2023): 9–25.

[29] Ivar Ekman and Per-Erik Nilsson, *Ukraine's information front—Strategic communication during Russia's full-scale invasion of Ukraine* (Stockholm: Swedish Defence Research Agency, 2023). https://www.foi.se/en/foi/reports/report-summary.html?reportNo=FOI-R--5451--SE.

[30] Niklas Nilsson, 'De-hybridization and conflict narration: Ukraine's defence against Russian hybrid warfare', in Mikael Weissmann, Niklas Nilsson, Björn Palmertz, and Per Thunholm (eds), *Hybrid warfare: Security and asymmetric conflict in international relations* (London: I. B. Tauris, 2021), 195–213.

from the battlefield. Day-to-day updates in a variety of social media channels document Russian military inaptitude, low morale, and war crimes; front-line progress; acts of sabotage and arrests of collaborators, aligning with a broader unified national narrative of resistance. Ukraine's emphasis on real-time intelligence sharing, citizen engagement, and strategic storytelling demonstrates a comprehensive approach to wartime communication. The strategy extends beyond simple messaging to encourage public participation through interactive channels, such as online chatbots and apps for reporting Russian troop movements and suspected espionage.[31] Digital platforms have created a feedback loop in which public engagement informs operational decision making while reinforcing the notion that ordinary citizens are part of the national defence effort.[32]

The communication strategy is intended to underscore Ukrainian ingenuity by showcasing innovative tactics, ranging from targeted drone strikes to the continuous disruption of Russia's command and control capabilities.[33] The image projected is one that clearly contrasts Ukrainian resilience and military aptitude with Russian disarray and incompetence. By documenting destruction of Russian equipment, leadership disputes, and violations of humanitarian law, Ukraine's intelligence-driven communication has inspired confidence in Ukraine's armed forces as well organized, technologically adept, and poised to make effective use of external support.

US-Ukraine Intelligence Sharing and Cooperation

Another important aspect of the utilization of intelligence during the war, as well as Ukraine's defence, has been the extensive sharing of secret intelligence between Kyiv and its Western partners. Ukraine's intelligence cooperation with the US Central Intelligence Agency (CIA) and UK Secret Intelligence Service (SIS or MI6) had been growing considerably in the years following the 2014 annexation of Crimea, with the US and UK agencies increasingly committed to aiding, training, and providing high-tech equipment for Ukraine's SBU and HUR, as well as setting up a network of listening stations and forward-operating intelligence bases in the country.[34] In addition to US and UK intelligence, other European intelligence agencies have provided substantial intelligence support for Ukraine. Importantly, Baltic intelligence services have long experience and significant signals

[31] Peter Schrijver, 'Beyond counterintelligence: Understanding the SBU's social media outreach on Telegram during wartime', *Intelligence and National Security* 39/3 (2024): 525–38. DOI: 10.1080/02684527.2024.2321692.

[32] Peter Schrijver, 'Ukraine's fight on the front lines of the information environment', Modern War Institute, 12 September 2023. https://mwi.westpoint.edu/ukraines-fight-on-the-front-lines-of-the-information-environment/.

[33] Ibid.

[34] Adam Entous and Michael Schwirtz, 'The spy war: How the CIA secretly helps Ukraine fight Putin', *New York Times*, 25 February 2024; Patrick Reevell, 'How the CIA and Ukrainian intelligence secretly forged a deep partnership', *ABC News*, 17 January 2025. https://abcnews.go.com/International/cia-helped-rebuild-ukraine-intelligence-russia-invasion/story?id=116909361.

intelligence capabilities directed at monitoring the Russian military, and have provided especially important information particularly on developments, movements, and positions in the eastern part of Ukraine.[35] In return, the Ukrainian services have provided its Western partners with large quantities of valuable intelligence on Russia.

From the very beginning, the US approach to intelligence cooperation constituted a balancing act between assisting Ukraine's capabilities and ability to defend itself against Russia and avoiding incidents risking inciting an unnecessary deterioration in US-Russia relations. Before the full-scale invasion, US intelligence sharing was therefore subjected to restrictions barring information sharing with potential lethal consequences. Due to occasional Ukrainian operations viewed in Washington as violating this agreement, such as a series of assassinations of rebel leaders in Donbas, the Obama administration considered cancelling the cooperation.[36]

The full-scale invasion in February 2022 accelerated intelligence cooperation significantly, as the US downgraded restrictions and CIA personnel relayed data enabling the armed forces of Ukraine (AFU) to target Russian forces advancing on Ukraine's territory, assistance that Ukrainian sources have deemed crucial in repelling the initial attack on Kyiv and Hostomel Airport.[37] Restrictions were lifted on sharing intelligence on Russian positions in the south and east of Ukraine, which had not previously been provided out of concern that it could have been utilized for offensive operations against Russian and separatist positions on territory occupied since 2014.[38]

Yet the balancing act continued although the goal posts were moved. The Biden administration considered it crucial to avoid the war escalating into a direct confrontation with Russia and thus kept US military support to Ukraine at a level where the US could not be construed as directly partaking in the fighting. Therefore, the US retained restrictions against sharing information that Ukraine could utilize to hit Russian targets outside of Ukraine. Restrictions also applied to information that would enable Ukraine to target the Russian strategic leadership. According to US officials, it also did not provide information regarding the precise location of individual Russians. However, it did share the locations of mobile command posts, which combined with Ukraine's own signals intelligence would indicate the presence of senior command figures, resulting in Ukraine managing to target and kill several Russian generals.[39] This illustrates the fine line observed between sharing intelligence useful to Ukraine in its own targeting, and sharing 'targeting intelligence', a legal and policy definition that would have indicated a more active US role in the targeting process and

[35] Shane Harris, Paul Sonne, Dan Lamothe, and Michael Birnbaum, 'U.S. intelligence helped Kyiv sink Russian warship', *The Washington Post*, 6 May 2022.
[36] Entous and Schwirtz, 'The spy war'.
[37] Ibid.
[38] Harris et al., 'U.S. intelligence helped Kyiv'.
[39] Julian E. Barnes, Helene Cooper, and Eric Schmitt, 'U.S. intelligence is helping Ukraine kill Russian generals, officials say', *New York Times*, 4 May 2022.

thus heightened the risk of eliciting a response from Russia directed at the US.[40] The cooperation adopted the semantic solution of terming intelligence shared with the UAF 'points of interest' rather than targets.[41] This distinction enabled the US to share increasing amounts of intelligence with Ukraine throughout the war, yet maintain that this was in order to help Ukraine defend itself and that the US was not actively participating in the war.

This arrangement was put to the test several times during high-profile Ukrainian strikes against Russian targets. One highly controversial issue was the aforementioned successful targeting of Russian generals. Another, and perhaps the most significant, was Ukraine's success in sinking the Russian flagship of the Black Sea fleet, the *Moskva* on 14 April, 2022. The successful strike was a major embarrassment for Moscow and forced the Black Sea fleet to withdraw from the vicinity of Ukraine-controlled shores. The strike was preceded by information provided by the US on the identity of the ship and its location. US officials, however, denied prior knowledge of or involvement in the decision, or knowledge of Ukraine's possession of the Neptune missiles utilized to sink the ship, reiterating that while the US shared intelligence with Ukraine, it refrained from advising on the objects of targeting.[42]

Intelligence cooperation deepened as Ukrainian, US, and coalition personnel set up an intelligence fusion centre, Task Force Dragon, in Wiesbaden, Germany, in April 2022, originating in the need to assist the UAF in delivery, training, and efficient use of M777 howitzers. The establishment of Task Force Dragon constituted the starting point for a far more systematic and integrated intelligence sharing arrangement between the US and partners, and Ukraine.[43]

Steps towards sharing increasingly detailed and comprehensive information utilized for targeting were driven by decisions to provide Ukraine with progressively more advanced and long-range weapons systems, requiring comprehensive overview and prioritization regarding 'points of interest'.[44] The US decisions to provide long-range precision systems to Ukraine, in the form of HIMARS in 2022 and ATACMS in 2023, were considered highly risky by the Biden administration. Each strike with these systems required meticulous targeting processes that were closely coordinated with US personnel and came with restrictions against targets outside Ukraine. Ukraine repeatedly requested that restrictions be removed regarding intelligence on targets inside Russia.[45]

[40] Shane Harris and Dan Lamothe, 'Intelligence-sharing with Ukraine designed to prevent wider war', *The Washington Post*, 11 May 2022.

[41] Adam Entous, 'The partnership: The secret history of the war in Ukraine', *New York Times*, 30 March 2025.

[42] Ken Dilanian, Courtney Kube, and Carol E. Lee, 'U.S. intel helped Ukraine sink Russian flagship Moskva, officials say', *NBC News*, 5 May 2022. https://www.nbcnews.com/politics/national-security/us-intel-helped-ukraine-sink-russian-flagship-moskva-officials-say-rcna27559; Harris et al., 'U.S. intelligence helped Kyiv'.

[43] Entous, 'The partnership'.

[44] Ibid.

[45] Helene Cooper, Julian E. Barnes, Eric Schmitt, and Michael Schwirtz, 'Ukraine asks Americans to give more intelligence on targets inside Russia', *New York Times*, 18 May 2024.

During Russia's advance on Kharkiv in June 2024, Ukraine was for the first time allowed to use HIMARS against Russian targets across the border.[46] In November 2024, the Biden administration also lifted restrictions on Ukraine's use of ATACMS in the Kursk salient, in response to North Korea's provision of troops to the Russian military.[47] According to media reports, whereas the US has provided daily general intelligence for attacks with short-range weapons, it has provided 'strike packages' for long-range drone and missile attacks.[48]

While the US has maintained its official stance regarding the nature of its intelligence sharing with Ukraine, this enlarged assistance in targeting represented a considerable change in practice. Particularly the provision of ATACMS, with a reach of 300 kilometres, had long constituted a red line for the Biden administration but was eventually approved in preparations for a strike campaign against Russian targets in Crimea and the Black Sea Fleet. ATACMS complemented the UK/French provision of Storm Shadow / SCALP missiles, which were also prohibited from use on Russian territory and dependent on US intelligence for precision targeting.[49] Germany has hitherto declined providing Ukraine with Taurus missiles out of concern for escalating the confrontation with Russia, although the new government under Chancellor Merz has been open to the possibility of doing so.

The Trump administration's decision to halt intelligence sharing with Ukraine, other than information needed for force protection, on 4 March 2025, in order to pressurize the Ukrainian government to accept US terms for peace negotiations with Moscow, exposed Ukraine's dependency on US intelligence for battlefield situational awareness and precise targeting. Ukrainian sources described losing access to data allowing for precision long-range strikes into Russia with US-provided weaponry as well as Ukraine-made drones. Ukraine also lost access to commercial satellite imagery provided by Maxar and experienced reduced efficiency in its use of front-line drones.[50]

[46] Samya Kullab, 'As U.S.-supplied weapons show impact inside Russia, Ukrainian soldiers hope for deeper strikes', *AP News*, 22 June 2024. https://apnews.com/article/ukraine-war-us-weapons-long-range-missiles-7f061044c7b50a1946409eea9e9c399f.

[47] Ido Vock, 'Atacms: What we know about missile system Ukraine has used to strike Russia', *BBC*, 20 November 2024. https://www.bbc.com/news/articles/cx2nrlq1840o; Adam Entous, Eric Schmitt, and Julian E. Barnes, 'Biden allows Ukraine to strike Russia with long-range U.S. missiles', *New York Times*, 17 November 2024. https://www.nytimes.com/2024/11/17/us/politics/biden-ukraine-russia-atacms-missiles.html.

[48] Warren P. Strobel, Siobhán O'Grady, Ellen Nakashima, Missy Ryan, and Kostiantyn Khudov, 'U.S. pauses sharing of intelligence Ukraine uses for strikes on Russia', *The Washington Post*, 5 March 2025. https://www.washingtonpost.com/national-security/2025/03/05/us-ukraine-intelligence-sharing/.

[49] Entous, 'The partnership'; Arthur Parashar, 'British-supplied Storm Shadow missiles are useless in Ukraine now President Trump has blocked US intelligence from being shared with allies', *Daily Mail*, 6 March 2025. https://www.dailymail.co.uk/news/article-14467869/British-supplied-Storm-Shadow-missiles-useless-Ukraine-President-Trump-blocked-US-intelligence-sharing-allies.html.

[50] Strobel et al., 'U.S. pauses sharing of intelligence'; Andrew Roth and Luke Harding, 'Trump says Putin launching massive strike on Ukraine is "what anybody would do"', 7 March 2025. https://www.theguardian.com/us-news/2025/mar/07/trump-says-it-is-easier-to-deal-with-russia-and-putin-wants-to-end-the-war.

The pause allowed Russian forces to advance, accelerating assault operations particularly in the Kursk salient.[51] On 12 March 2025, US director of national intelligence Tulsi Gabbard affirmed the resumption of intelligence sharing after a preliminary understanding for the conditions for a ceasefire was reached between the US and Ukraine in Jeddah, emphasizing the utilization of intelligence sharing as a bargaining tool.[52] The US halt to intelligence sharing underscores Ukraine's vulnerability in this regard, and how the new US administration utilized intelligence sharing, as well as advanced weapons deliveries, as leverage to induce concessions from Ukraine.[53]

Technology Innovation and Adaptation

The war has seen a remarkable pace of adaptation and adoption of new military technologies for intelligence purposes. Ukraine has, as a consequence of its armed conflict with Russia since 2014 and especially since 2022, invested heavily in the development of defence technology. Tech developers from the country's private sector and civil society have become closely integrated with the war effort, developing a number of artificial intelligence (AI)-driven solutions for military use.[54] International tech companies have flocked to the war, offering their services in supplying and developing systems for information management, cloud services, and intelligence fusion for the AFU.

Media reporting describes how Ukraine has become an AI war laboratory, where large Western defence tech companies and startups alike have been competing for access to the combat environment to verify and enhance their products, and where Ukraine has become a testing ground for cutting-edge military AI and data analytics tools. This has given rise to a context in which a very large share of the development of critical technologies and technological innovation in intelligence, for which the war has functioned as a catalyst, comes from commercial, rather than state, actors. Palantir was an early mover in this regard by offering Ukraine AI services capable of gathering, integrating, and analysing massive amounts of battlefield intelligence from a multitude of open and classified sources. Tech giants Amazon Web Services and Microsoft have offered their cloud solutions to salvage and protect Ukraine's data. Ukraine, for its part, has seized the opportunity to attract investment and build its own tech sector.[55]

[51] 'Russian offensive campaign assessment', Institute for the Study of War, 2025. https://www.understandingwar.org/backgrounder/russian-offensive-campaign-assessment-march-9-2025.

[52] Martin Fornusek, 'US resumes sharing intelligence with Ukraine, Gabbard confirms', *The Kyiv Independent*, 13 May 2025. https://kyivindependent.com/us-resumes-sharing-intelligence-with-ukraine-gabbard-confirms/.

[53] Erin Banco and Jonathan Landay, 'US cuts intelligence sharing for Ukraine, adding pressure for Russia peace deal', Reuters, 6 March 2025. https://www.reuters.com/world/us-cuts-off-intelligence-sharing-with-ukraine-ft-reports-2025-03-05/.

[54] Kateryna Bondar, *Understanding the military AI ecosystem of Ukraine* (Washington, DC: Center for Strategic and International Studies, 2024).

[55] Vera Bergengruen, 'How tech giants turned Ukraine into an AI war lab', *Time*, 8 February 2024.

Indeed, the ability offered by AI tools to make quick use of the vast amounts of information and data generated by the war has accelerated the development and exploitation of these services. For example, Ukraine's 'Delta' situational-awareness platform merges data from a wide variety of sources and is utilized across all levels and units of the UAF in support of battlefield decision making.[56]

The technology dimension of the war in Ukraine has also demonstrated the extreme transparency of the modern battlefield, which is in turn enabled in large part by commercial solutions. Companies such as Planet Labs, Maxar, and Capella Space have provided near-real-time satellite imagery and earth data analytics that Ukraine has been able to integrate into command decision making and targeting, further enabled by AI-driven image analysis tools.[57] Unmanned aerial systems is an area that has seen a very rapid development in the collection and analysis of battlefield intelligence, from Ukraine's deployment of the large and expensive Bayraktar TB2 early in the war to the mass production and employment of cheap, commercial drones refitted for military purposes with appropriate optical equipment and munitions. Another major development has been the utilization of Clearview AI's effective but highly controversial facial recognition software, which Ukraine has utilized to identify Russian soldiers, saboteurs, and collaborators.[58]

Technological developments have also facilitated the active participation of society at large in the war effort, where civilian and public participation in Ukraine's defence has become an important element of intelligence production. The gathering and dissemination of information by Ukrainian and international volunteers and civil society organizations illustrates how smartphones and social media are now closely integrated with mobilization, intelligence, and direct engagement in conflict.[59]

Civilians have been empowered to participate through the development of dedicated apps and chatbots to transmit geotagged data of for example troop movements or the impact of long-range fires. This has allowed for crowdsourced intelligence where citizens with smartphones can document and upload information of significance to the AFU. With apps such as eVorog and ePPO, as well as mapping portals tied to the AFU, locals can submit pictures and short videos of vehicles, convoys, equipment storages, or overflight paths.[60] These uploads inform military targeting as the data is integrated with the intelligence fusion process. This was particularly significant in the initial stages of the full-scale invasion in 2022, when data gleaned from civilian footage helped verify Russian advances and prepare coordinated defensive actions. Open-source analysis naturally also plays a major role in wartime intelligence, where

[56] Kateryna Bondar, *Does Ukraine already have functional CJADC2 technology?* (Washington, DC: Center for Strategic and International Studies, 2024).

[57] Anna Mysyshyn, *Advanced technologies in the war in Ukraine: Risks for democracy and human rights* (Washington, DC: GMF, 2024); 'How Ukraine is using AI to fight Russia', *The Economist*, 8 April 2024.

[58] Mysyshyn, *Advanced technologies*.

[59] Matthew Ford, 'From innovation to participation: Connectivity and the conduct of contemporary warfare', *International Affairs* 100/4 (2024): 1531–49. doi: 10.1093/ia/iiae061; Joanna York, '"World's first TikTok war": Ukraine's social media campaign "a question of survival"', *France24*, 24 February 2023. https://www.france24.com/en/europe/20230224-world-s-first-tiktok-war-ukraine-s-social-media-campaign-a-question-of-survival.

[60] Ford, 'From innovation to participation'.

dedicated intelligence analysts as well as private actors, non-government organizations (NGOs) and volunteers have been able to generate large amounts of actionable intelligence from social media platforms. Geolocation software and text and image classifiers provide extensive processing capacity for the vast amount of generated content.

The employment of new technologies for collecting, analysing, and safeguarding intelligence has evolved rapidly during the war in Ukraine. Commercial and volunteer-driven innovations have been integrated with military intelligence collection and fusion to significantly enhance the capacity for information processing and support decision making in what has been declared the hitherto most extensively documented war in history. Among the results are a virtually transparent battlefield on which concealment and deception become exceedingly difficult, and a vastly accelerated kill chain and targeting process. Similarly, the Russian military pursues data fusion and AI-supported analysis in support of processing enormous volumes of information and supporting quick decisions.[61]

The impact of these developments is debated and inflated statements about the game-changing impact of technology on war should be met with healthy scepticism. However, it is clear that the frontiers of technological possibilities in battlefield intelligence have taken a serious leap as an effect of the war. Future wars will likely see progressively more advanced applications of these possibilities, with a range of attached ethical concerns including autonomous lethal actions, privacy, and civilian participation.

Conclusion

Russia's war in Ukraine has highlighted challenges as well as incentivized development and change in practices, functions, and innovations in intelligence that will remain salient beyond this conflict. One of the most revealing experiences stems from the difficulties of forecasting and prediction. Despite clear indications of a large-scale Russian assault, many assessments both overestimated Moscow's ability to conduct a swift, decisive campaign and underestimated Ukraine's capacity for resistance. These misplaced forecasts underscored how challenging it is to account for qualitative factors such as morale, leadership, and adaptability, which can dramatically affect outcomes in war.

Simultaneously, the innovative use of intelligence for strategic communication marked a reprioritization between the utility of openness and the imperative of secrecy. US and UK agencies disclosed highly sensitive insights to strengthen Ukraine's defence, enable a cohesive Western response, and debunk Russia's own strategic communication regarding the war and its causes. This considerable and coherent utilization of intelligence to substantiate strategic warning was enabled by

[61] Samuel Bendett, *The role of AI in Russia's confrontation with the West* (Washington, DC: Center for a New American Security, 2024).

an expanding open-source ecosystem and commercial satellite imagery. It nevertheless remains to be seen whether similar approaches will be employed in future conflicts, given the inherent risks involved.

The increasingly progressive US intelligence sharing with Ukraine raised complex questions over the limits of participation, including where intelligence support might be perceived as crossing into direct combat involvement. The Trump administration's temporary halt to intelligence sharing also underscored the perils of established intelligence dependencies and the extent to which it can be employed as political leverage.

Finally, the war has been a catalyst for technological advancements affecting all aspects of intelligence, from collection to analysis and dissemination. As new technologies enable the increasing transparency of the modern battlefield, they also produce enormous amounts of data that are beyond human capabilities to integrate and analyse. Commercial AI and cloud solutions are thus becoming increasingly significant, and the ability offered to tech companies to test and develop their products in authentic conditions in Ukraine has made a substantial mark on the defence tech sector.

This chapter has highlighted a few of the many lessons to be learned and conclusions to be drawn regarding how the war in Ukraine is affecting conceptualizations and practices of intelligence. Many of these will likely be re-evaluated with time and in relation to future conflicts. However, as is the case with many other aspects of modern warfare, Russia's war in Ukraine will in all likelihood be perceived as a highly instructive case of wartime utilization of intelligence for decades to come.

11
UAV/Drone Dimension of the War in Ukraine

Samuel Bendett

Introduction

As the war in Ukraine is clocking its fifth year, the use of unmanned aerial vehicles (UAVs) and aerial drones has become the main technological and military pillar of operations for both sides in the conflict. Both forces' development, testing and use of military-grade drones, combined with an unprecedented application of commercial drone technologies, have made this war one of the most unique in recent history. Alongside the legacy defence companies, numerous volunteer and private citizen-driven efforts on both sides have also enabled rapid prototyping and testing of different UAV classes and types in combat, sometimes bypassing or complementing official military institutions put in place to oversee UAV development and acquisition. Such complementing and overlapping efforts have resulted in a highly unique ecosystem aimed at the fastest utilization of available technologies for greater impact on the battlefield.

By early 2026, drone warfare in Ukraine has grown to include a range of tactical, close-support, mid-range, and long-range drones and UAVs, and, today, it is common to come across a news article or an in-depth analysis of the Ukraine war that mentions the impact of drones and UAVs in combat. Simply put, today, these technologies are a key factor that influence how both belligerents plan and execute their military operations. The sheer quantities of UAVs today represent a challenge to conventional combat thinking previously defined by the ability to mass forces for attack or conduct largely unobstructed logistical and support operations in the field. As the forces in Ukraine and around the world seek to recalibrate their approach to ground warfare, logistics, supply, evacuation, and related missions, the focus on tactical and mid-range UAVs as a key adversarial asset is emerging as an important element of formulating new strategies and tactics.

Over the past four years, the evolution of drone warfare in Ukraine has resulted in the nearly ubiquitous presence of different UAV classes operating at different altitudes that execute a range of missions—this environment has enabled round-the-clock

Samuel Bendett, *UAV/Drone Dimension of the War in Ukraine*. In: *Russia's War in Ukraine and Modern Warfare*.
Edited by: Mikael Weissmann and Niklas Nilsson, Oxford University Press.
DOI: 10.1093/9780198970903.003.0011

observation of adversary assets, positions, locations, movements, and manoeuvres. In short, aerial drones and UAVs hold a key part in our understanding of this particular war.

This chapter will trace key developments witnessed on the battlefield today, concentrating on the tactical elements shaped by such ubiquitous drone use. Specifically, long-range UAV applications will be excluded to focus more on the above-stated topic, given significant international coverage of Russian and Ukrainian long-range drone development and strikes.[1] Given a rapidly growing body of analysis and literature that is already devoted to analysing drone warfare in this war, this text will not restate some of the major themes and topics that are discussed in open sources—rather, this chapter will attempt to analyse how aerial drones and UAVs are impacting the tactical environment of Ukraine. Specifically, this chapter will concentrate on the Russian military and will emphasize the analysis and writing on that topic from both official and unofficial Russian sources. This is done to add an additional analytical element to the rapidly growing body of literature that highlights many Ukrainian technological developments in the Western and English-language media—such analysis tends to provide detailed coverage of Ukrainian developments and capabilities given a relatively free and unobstructed access to Ukrainian combatants, military decision makers, open-source analysts, government officials, and technology developers.[2]

Russian-language unofficial sources such as numerous Telegram channels covering the war are of special interest—while these are mostly pro-Kremlin and pro-Russian biased sources, they nonetheless provide a range of information and commentary on the war's progress overall, and on many specific technologies and combat tactics in particular. Occasionally, such Telegram sources are the only outlets covering specific sections of the front, or even specific military engagements, showing images and videos of drones and UAVs used, along with providing feedback and commentary on how certain military technologies are performing. This chapter will also utilize a range of English-, Russian-, and Ukrainian-language public sources such as news and analytical outlets, news sites, state media outlets, military blogs, public messaging and social media platforms, and official military journals, where appropriate.

[1] Examples include: Svitlana Vlasova, Victoria Butenko, Kosta Gak, and Kathleen Magramo, 'Massive Russian drone attack slams Kyiv and hits maternity hospital in Odesa, Ukrainian officials say', *CNN*, 10 June 2025. https://www.cnn.com/2025/06/09/europe/russia-ukraine-kyiv-odesa-drone-attack-intl-hnk; Iryna Blanchuk, 'Ukraine's general staff confirms strike on Russian facility producing special receivers for Shahed drones', *Pravda.com.ua*, 9 June 2025. https://www.pravda.com.ua/eng/news/2025/06/9/7516271/; Mary Ilyushina, 'Russia's deadly drone industry upgraded with Iran's help, report says', *Washington Post*, 29 May 2025. https://www.washingtonpost.com/world/2025/05/29/russia-iran-drone-cooperation-industry/.

[2] Examples include: Kateryna Stepanenko, 'The battlefield AI revolution is not here yet: The status of current Russian and Ukrainian AI drone Eefforts', *ISW*, 2 June 2025. https://understandingwar.org/backgrounder/battlefield-ai-revolution-not-here-yet-status-current-russian-and-ukrainian-ai-drone; Kateryna Bondar, 'Ukraine's future vision and current capabilities for waging AI-enabled autonomous warfare', *CSIS*, 6 March 2025. https://www.csis.org/analysis/ukraines-future-vision-and-current-capabilities-waging-ai-enabled-autonomous-warfare; Stacie Pettyjohn, 'Evolution not revolution: Drone warfare in Russia's 2022 invasion of Ukraine', *CNAS*, 8 February 2024. https://www.cnas.org/publications/reports/evolution-not-revolution; 'On the ground in Ukraine: Winter 2025—The Russia contingency with Michael Kofman', *War on the Rocks*, 24 February 2025. https://warontherocks.com/episode/therussiacontingency/33700/on-the-ground-in-ukraine-winter-2025/.

Discussion

Select Technology Developments

According to multiple analytical sources that track the adaptation of the Russian Ministry of Defence (MoD) and military-industrial complex to this war, by the end of 2024, the Russian government, industry, and military have revamped their defence industry to adapt to Western sanctions and launched import-substitution programmes to mitigate the lack of certain key components needed for both military and civilian economies.[3] While challenges remain, the Russian defence industry and the overall UAV development efforts are still capable of supplying the forces with necessary technologies, while the UAV fleet is continuing to grow and today includes loitering munitions and one-way attack drones; intelligence, surveillance, and reconnaissance (ISR) UAV platforms; and numerous tactical-level first-person view (FPV) drones.[4]

The above-mentioned UAV types are utilized the most in the tactical space that is the focus of this chapter. They include Lancet loitering munition variants used to target Ukrainian equipment and artillery at up to 70 kilometres; Kub loitering munition used alongside Lancets (in smaller numbers) that are being improved with greater ranges and better payload capacities; and tactical drones such as Molniya (Lighting) fixed-wing one-way attack drones that gained notoriety in 2024 as a cheap, easy-to-assemble single- and double-engine drone made from plywood and plastic; today, its improved variants feature an improved aerodynamic fuselage, two engines, and an increased payload.[5]

In 2025, Russia's ISR drone fleet has grown to include Orlan-10, Orlan-30, Zala, Supercam, Skat S350 and Albatros variants used for fire control, intelligence gathering, and reconnaissance. According to open sources, the Russian military is modernizing its reconnaissance drones with increased protection against electronic warfare (EW) counter-measures deployed by the Ukrainian defenders.[6] As an example, in March 2025, there were reports of Orlan drones that were equipped with radio reconnaissance systems allegedly capable of detecting radio emissions and signals from Ukrainian air defence radars, EW systems, and Starlink satellite terminals.[7] Supercam drones are reportedly receiving upgrades for long-duration operations that include aerodynamic improvements and new batteries.[8]

The Russian defence industry is improving ISR UAVs with the machine vision technology to reduce the likelihood of being targeted and shot down by Ukrainian

[3] 'Main trends in the development of the Russian military-industrial complex for the 2nd quarter of 2025' ('Основні тенденції розвитку російського ВПК за I квартал 2025 року'), *Drukarnia*, 24 April 2025. https://drukarnia.com.ua/articles/osnovni-tendenciyi-rozvitku-rosiiskogo-vpk-za-kvartal-2025-roku-AgPv3#heading-2-353.

[4] Ibid.

[5] Ibid.

[6] Ibid.

[7] Ibid.

[8] 'Kalashnikov upgrades Supercam scout drone', *Tass*, 26 January 2024. https://tass.com/defense/1737637.

interreceptor drones, while also experimenting with equipping its drones with portable EW stations. In March 2025, a downed Zala reconnaissance drone examined by Ukrainian forces was equipped with such technologies.[9] Other reconnaissance drones like the Supercam and Albatros are used widely across the front, with newer modifications featuring longer operational ranges and better optics. Russia's ISR drones remain among the strongest and most capable components of the Russian ongoing operations, prompting Ukrainian forces to come up with counter-measures like the interceptor UAVs to take out Russian ISR and combat UAV platforms.[10] In addition to technologies mentioned above, Russian ISR drones are now using active manoeuvres while in flight and climbing to different altitudes for the duration of a mission in an attempt to avoid various Ukrainian counter-measures.

People's VPK and Defence Industry Nexus

By early 2026, Russia continues to actively develop and utilize cheap and attritable FPV drones in larger numbers for strikes against targets that include any sized units, down to individual soldiers, as well as vehicles, tanks, dugouts, locations, buildings, and other targets. At this point in the war, these UAVs are assembled both on state orders and among numerous volunteers, startup and even individual initiatives from available commercial components and a growing number of domestically made parts. Some of these volunteer and startup groups exist outside of the official military-industrial complex ecosystem and are partially compensating for the slowness of the state-owned defence enterprises when it comes to rapid innovation, iteration and technical updates and improvements on the front-lines.[11] This is one of most defining features of this conflict, as Russian civil society efforts started to mobilize around calls to procure equipment, supplies, and systems that were found lacking across the force starting in the spring 2022.

Beginning with purchases of commercial, Chinese-made DJI quadcopters that are still supplied to the Russian units in large numbers, these efforts also started to purchase Chinese-made drone components and to assemble FPV-type and quadcopter/multirotor drones, modifying them based on front-line feedback, as well as training drone operators to pilot these drones.[12] From 2022 onwards, many of these initiatives took on the burden of delivering to the Russian military both ready-made drones and the experience needed to operate them.[13] Over time, such efforts have

[9] 'Main trends in the development'.

[10] Vlada Toropkova, 'Drone-on-drone war: How Ukraine's FPV interceptors are beating Russia in the sky', *United24*, 22 April 2025. https://united24media.com/war-in-ukraine/drone-on-drone-war-how-ukraines-fpv-interceptors-are-beating-russia-in-the-sky-7770.

[11] 'Main trends in the development'.

[12] 'Russians imported millions in Chinese drone parts, some with Western chips', *Kharon*, 6 August 2025. https://www.kharon.com/brief/russians-imported-millions-in-chinese-drone-parts-that-use-western-chips.

[13] 'Specialists in Crimea improve FPV drones' communication system, double range', *Tass*, 24 May 2024. https://tass.com/defense/1792717.

grow into nation-wide originations and have matured into fairly sizeable enterprises capable of manufacturing large quantities of quadcopters and FPVs for the front, such as those manufacturing Ovod, Upyr, Hortensiya, VT-40, and other FPV and quadcopter variants; such enterprises are now aided by regional governments and direct MoD contacts, while a few smaller efforts still rely on private donations from regular citizens and wealthy individuals.[14] In 2024, Russian Defence Minister Belousov remarked that his forces were receiving close to 4,000 drones daily, amounting to an overall delivery of roughly 120,000 drones per month, and up to 1.4 million for that year.[15] That amount likely includes tactical drones made by both defence industry and some of the country's startups, and has at least doubled throughout 2025.

The flip side of this overlapping use of drones manufactured by defence-industrial enterprises and a number of Russian volunteer efforts and startups is the diverse quality of such UAVs, which sometimes has a detrimental effect on actual combat.[16] Since there is still no overarching standard for testing and evaluation of such quadcopters and FPVs, some performed better than others, some did not make it out of the box, and some could allegedly fail mid-flight—not to mention the varying quality of drone-enabled combat from one section of the front to the next that itself depends on the UAV pilot experience, training and preparation.[17] The Russian MoD has understood the impact of such efforts in combat operations and is building closer cooperation with the relevant initiatives. The current growth of the country's Unmanned Systems Forces that were officially launched in November 2025 is likewise going to ensure that certain UAV development and use standards are going to be enforced across this and the larger Russian force structure.

One of the biggest reasons for such non-official VPK (or *voyenno-promyshlennyi* (military-industrial) complex) technology streaming to the front is the amount of paperwork required for Russian soldiers, officers, unit commanders, and drone operators in order to receive a tactical drone. In contrast between 2023–2025, a typical DJI Mavic quadcopter or a batch of FPV drones or quadcopters could still be potentially procured 'as is' from volunteers willing to help the Russian military, without filling out any official documentation. Given that over the past four years, many of these volunteer efforts have established their own delivery pipelines to specific units across the force, some Russian military units that have such direct links are supplied with the some of the latest UAV and counter-UAV technologies in adequate numbers; other units that do not have such relationships can still be found lacking certain tactical drone capabilities.[18] The Russian MoD has responded with the establishment of

[14] 'Tech firm delivers thermal-imaging drones to Russian troops in special military op area', *Tass*, 26 February 2024. https://tass.com/defense/1751365.

[15] 'Four rotors and a prayer: How Russia built a billion-ruble drone industry using Chinese parts', *TheInsider*, 16 May 2025. https://theins.ru/en/inv/281354; Olivia Savage, 'Ukraine conflict: Russia producing 4000 FPV UAVs per day, minister claims', *Janes*, 31 July 2024. https://www.janes.com/osint-insights/defence-news/air/ukraine-conflict-russia-producing-4000-fpv-uavs-per-day-minister-claims.

[16] Romanov Lite, Telegram post, 1 August 2024. https://t.me/romanov_92/44931.

[17] KCPN, Telegram post, 12 June 2024. https://t.me/kcpn2014/2355.

[18] Rusich Army, Telegram post, 8 June 2025. https://t.me/rusich_army/24028; Svarschiki, Telegram post, 21 May 2025. https://t.me/svarschiki/11706.

platforms that enable such startup technology to be tested for potential military application, such as establishing technology accelerators for technical startups to aid the country's defence-industrial complex, but there is not a lot of information available on these efforts in the public domain.[19]

Another avenue used in 2022–2025 for the 'Peoples' VPK' to test and deliver their technologies to the front was by donating their systems to front-line semi-volunteer units such as Bars-Sarmat (that has been folded into the country's Unmanned Systems Forces structure in 2026) and Espaniola Brigades that claim to test such UAVs and other unmanned technologies in actual combat in order to provide feedback to manufacturers.[20] The MoD has also stood its own drone, UAV, and robotic systems' testing and evaluation effort—the Rubicon Center has emerged as the focal point for testing and applying the latest tactical technologies in combat that also include Peoples' VPK products, as it did with using 'Prince Vandal Novgorodsky' (KVN in Russian) fibre-optic drones against the Ukrainian forces in the 2024–5 Kursk region counter-offensive that eventually expelled Ukrainian units from Russian territory.[21] The Rubicon Center's precision and lethality in going after Ukrainian targets has been recognized as one of the most dangerous developments out of the Russian MOD in this war.[22]

There is a certain level of distrust between large defence enterprises responsible for weapons production and the rapidly innovating volunteer and startup efforts that will be hard to resolve, as each sees the other as standing in the way of progress.[23] This attitude is amplified by the fact that throughout Russia's modern history (dating back to the early 17th century), all weapons developed has been a monopoly of the state, yet today, there are certain private-sector efforts developing what essentially amounts to lethal weapons and components for them, often on their own initiative and with private funding. This novel reality dictated by the Ukraine invasion is new and unfamiliar to many in the Russian defence-industrial sector and the government. Moreover, the defence-industrial enterprises that normally test and evaluate their technologies over a lengthy period of time to make sure they meet specific MOD standards are wary of small, nimble Russian enterprises that were sometimes set up in homes and garages and that can rapidly innovate on a specific technology in response to front-line requests, but do so without requisite testing prior to front-line delivery.[24] Such smaller grassroots efforts can indeed produce a desired solution but cannot scale

[19] Official Rivir Center. https://center-rivir.ru/ (accessed 5 June 2025).

[20] Espaniola, Telegram post, 2 June 2025. https://t.me/spainrus/2387; TSN 'Bars-Sarrmat', Telegram post, 9 June 2025. https://t.me/russian_shock_volunteer_brigade/2706.

[21] ITsPBP 'Rubicon', Telegram post, 9 June 2025. https://t.me/icpbtrubicon/241; David Axe, 'Lying in wait on the ground, Russia's best attack drones devastate Ukrainian convoys', *Forbes*, 16 March 2025. https://www.forbes.com/sites/davidaxe/2025/03/16/lying-in-wait-on-the-ground-russias-best-attack-drones-devastate-ukrainian-convoys/.

[22] Kollen Post, 'Deadly, elite, and scaling — inside Russia's Rubikon front-line drone system,' *Kyiv Independent*, 3 December, 2025, https://kyivindependent.com/russians-really-scale-how-moscow-built-a-drone-system-that-now-threatens-to-tilt-the-balance-on-the-battlefield-in-ukraine/.

[23] Chadaev, Telegram post, 13 Aug 2024. https://t.me/chadayevru/3145.

[24] 'Expert says merger of military-industrial complex with national industry will create healthy competition in Russia' ('Эксперт заявил, что объединение ВПК с народным создаст в РФ здоровую конкуренцию'), *Tass*, September 19, 2024, https://tass.ru/ekonomika/21911927.

up their efforts, given the general lack of funding and significant production capabilities stemming from small-scale investments in such efforts—while larger defence majors produce large quantities of pre-ordered weapons and systems built to certain specifications that often cannot be quickly modified as needed.

At the same time, there are key outliers, such as the KVN fibre-optic controlled drone that was designed and built by Ushkuynik, a Russian volunteer-driven technology accelerator effort set up in 2023 that became instrumental in Russia's counter-strikes against Ukrainian forces in Kursk in 2024, and that continues to be used in greater numbers today across the front.[25] By September 2025, the Russian state is backing increased manufacturing of this drone across the country, supporting Uhskuynik's original efforts with greater funding to manufacture 50,000 KVN drones each month.[26] Some defence majors such as Kalashnikov and its subsidiaries have also Gotten considerably better at iterating and improving their own military-grade UAVs based on frontline feedback.[27]

There are also indications by 2025 of an emergent middle ground between truly grassroots, smaller drones and UAV research and development efforts (R&D) and the larger domestic defence enterprises. Some startups like the manufacturers of Ovod, Hortensiya, and Upyr drones and light quadcopters have become mid-level defence enterprises with the assistance of regional governments, and the MoD itself.

An example of the speed of iteration at the front was demonstrated in February 2025, when the Russian FPV designers acknowledged that the technological race in Ukraine combat is pushing drone developers to innovate at an ever-increasing rate. As an example, the chief engineer of the Ovod FPV drone noted that the technology to operate this drone needs to be updated every two to three weeks to keep pace with battlefield changes. Therefore, Ovod manufacturers put a 'pryanik' (a famous gingerbread cookie baked in the Tula region where these drones are assembled) in every box with an Ovod inside.[28] According to the chief engineer, 'the gift serves as a barometer for technological updates'. If the cookie is stale by the time it reaches drone operators, then it took the UAV longer than three weeks to get to the front and, therefore, the equipment may already be outdated and should undergo extensive updates, or should no longer be used.[29]

The tension described above between the official VPK and 'Peoples' VPK' is one of the defining features of this conflict, as the Russian military and government came face to face with an unprecedented, nation-wide effort that amounts to a private military industry born in a country where, historically, all military and weapons development and procurement were normally conducted and controlled by the state.

25 'Novgorod-made FPV drone Prince Vandal causes $300 mln damage to NATO', *Tass*, 15 January 2025. https://tass.com/defense/1899767.

26 'Prince Vandal drone to have complete import substitution', *Tass*, 18 January 2025. https://tass.com/defense/1901261; 'Over 50,000 Prince Vandal drones are produced in Russia every month', *Tass*, 6 September 2025. https://tass.com/defense/2013567.

27 'Kalashnikov introduces major software update to its SKAT-350M UAV last year', *Tass*, 4 February 2026, https://tass.com/defense/2081813.

28 Tulaovod, Telegram post, 15 February 2025. https://t.me/tulaovod/482.

29 Ibid.

Many Russian volunteers and startups were wary of MoD hinting at greater cooperation with the Peoples' VPK, fearing a takeover of such small efforts by large defence manufacturers or by unscrupulous individuals with close personal ties to the MoD's procurement process, which for them would eventually amount to the end of rapid innovation and iteration that marked this unusual evolution of Peoples' VPK as a key supplier of tactical drones and UAVs to the front.

At the same time, the Russian government, at the highest levels, has recognized the significance of Peoples' VPK projects, with President Putin actually ordering both VPK ecosystems to cooperate closely for better and faster weapon development and procurement.[30] There are no benchmarks presently to evaluate how such cooperation is expected to unfold, given the complete state control of weapons development and manufacture in Russian and Soviet history. There are efforts underway by Kalashnikov to engage with certain People's VPK efforts, such as the 2025 agreements to cooperate with the Ushkuynik enterprise and Project Archangel volunteer efforts to scale up certain drone and counter-drone technologies from these organizations.[31] Therefore, the MoD's announcement that it is also launching its own unmanned systems forces (USF) was met with an understanding that such a military branch could also be responsible for establishing certain common and enforceable standards for UAV and drone technology development. This new military branch is also likely to unite the different UAV units across the force, with their disparate drone quality and acquisition, although the historical Russian preference for centralization and control may inhibit the flexibility and innovation that are still evident in many Russian grassroots drone development initiatives to date.[32]

In 2025, Russian military bloggers and commentators that provide frequent and continuous feedback on the war via the Telegram messaging app debated the development of the USF, arguing that there can be two possible pathways based on the current combat in Ukraine. The first option was to create special testing units dealing with uncrewed and robotic systems. Such units could concentrate on the research and application of combat robotic vehicles in the interests of a given military grouping.[33] The second possible USF option was to create separate combat units in which UAVs and robotic systems play key roles but are not the only type of weapon used. This option favours FPV-equipped units roaming along the front, along with the creation of combined units that are built around FPV-type drones as the main weapon that can also be equipped with a range of capabilities to support artillery and assault units.[34]

[30] 'Putin noted the contribution of the "people's defense industry" to Russia's overall success at the front' ('Путин отметил вклад "народного ОПК" в общий успех России на фронте'), *Tass*, 25 October 2024. https://tass.ru/armiya-i-opk/22226383; 'Expert says merger'.

[31] 'Kalashnikov gunmaker launches production of new Archangel UAVs for Russian troops', *Tass*, 27 August 2025, https://tass.com/defense/2008115; 'Kalashnikov head gives high marks to Knyaz Vandal of Novgorod fiber optic drone', *Tass*, 11 August 2025, https://tass.com/defense/2000615.

[32] 'Main trends in the development'.

[33] Zhivoff, Telegram post, 6 March 2025. https://t.me/zhivoff/19432.

[34] Ibid.

Several examples of current units structured in this way exist in the Russian military, such as the previously mentioned Rubicon Center and the Espanola Brigade. For many Russian commentators who follow the war, the key factor as the Russian military proceeds with the USF is not to 'spread' the FPV availability and expertise 'too thinly' across all troops and military institutions, given their estimation that there are not enough qualified drone specialists for all units.[35] This line of argument also supports additional options for structuring the Russian forces that use unmanned and robotic systems to allow competition to apply different methods, tactics, and technologies as appropriate to avoid the rigid command, control, and application structure normally associated with the Russian MoD.[36]

Another recent Russian commentary pointed out successful tactics used by Ukrainian forces, including their 'dominance in the sky' at certain sections of the front via round-the-clock monitoring by reconnaissance UAVs, in turn resulting in a large number of FPV drones striking Russian movements and forces.[37] As a result of such saturation, the Russian infantry often has no chance to reach its intended targets, resulting in massive numbers of soldiers killed and injured, while Russian military equipment used in such assaults 'generally lives only in one direction and is needed exclusively to quickly reduce the distance [to the Ukrainian positions]'.[38]

Russian military bloggers and commentators have also acknowledged that in addition to the losses to infantry, Russian logistics capabilities suffer as well, while the Ukrainian forces' constant hunt for Russian UAV crews takes a serious toll, given that such units must operate closer to the line of contact to be more effective.[39] The Russian commentary also notes that Ukrainian forces quickly adapt to Russian improvements in EW, such as quickly changing drone operating frequencies to render EW countersues ineffective, while many Russian EW systems can work on fewer than half of the frequencies Ukrainian forces use.[40]

Heavy Tactical Drones

Along with tactical small-scale drones, the Russian defence industry has been attempting to replicate Ukraine's success with the development of heavy multi-rotor drones dubbed 'Baba Yaga' that are frequently used by the Ukrainian forces at night. Many such drones originated as heavy Chinese-made agricultural UAVs, and some of these heavy drone variants are now domestically manufactured in Ukraine. Despite there being no discernible obstacles to the procurement of such heavy Chinese UAVs, the Russian defence industry is still struggling, trying to develop its own equivalents

[35] Ibid.
[36] Ibid.
[37] T3mny, Telegram post, 8 March 2025. https://t.me/t3mny/2243.
[38] Ibid.
[39] Ibid.
[40] Ibid.

for bombing and logistics missions. Certain domestic developments highlighted by state media are not manufactured in quantities large enough to make a difference on the front-lines. For example, the Russian state media highlighted developments like Kukushka (Cuckoo Bird) with a claimed 35 kilogram load capacity, designed for combat and logistics missions, although there is little available evidence of this drone's use in combat at this point.[41] Another typical domestic development highlighted by state media is the Koshchei hexacopter capable of carrying up to 15 kg for bombing and logistics missions; and Vorobey-15 octocopter that can also carry a maximum of 15kg for combat and support missions.[42]

Another Russian attempt to build an equivalent to Baba Yaga highlighted by the Russian media is the MiS-50 quadcopter, an improved version of its MiS-35 predecessor.[43] Russian developers claimed that it was obvious to them that a completely new UAV should be developed that would make maximum use of the technical solutions tested in Ukraine combat that had proven itself successfully on the MiS-35.[44] As a result, MiS-50 allegedly features high maintainability, modularity, and reliability, although it can only carry a maximum payload of 5 kilograms, with flight ranges up to 10 kilometres.[45]

Despite such numerous defence industry efforts, the Russian military still lacks significant numbers of heavy quadcopters that are true equivalents of Ukraine's vaunted heavy drones delivered in quantities needed by frontline troops. Instead of waiting for the industry to deliver heavy drones in large numbers, the Peoples' VPK stepped in in some cases, with soldiers and their supporting volunteers trying to assemble their own heavy multi-rotor drones from available components. Some of these drones were dubbed 'Vobla' by the soldiers (*vobla* is the name of a Russian dried fish normally eaten with beer, a decades-long culinary and cultural tradition), are built from large wooden or metallic X- or square-shaped frames, with FPV drones attached to the ends in lieu of heavy propellers and motors.[46] Given the large number of FPVs available to the Russian forces, this design likely came together out of a desperate need to deliver heavy munitions or conduct supply runs. Such designs have quickly progressed to being capable of delivering a heavy TM-62 anti-tank mine, along with ammo and food, to front-line troops.[47]

According to Vobla developers, the growing mass production and ease of assembly of these drones allows them to be constructed directly at the front line from available

[41] 'Expert Kondratyev: Russia has created a drone better than the Ukrainian "Baba Yaga"' ['Эксперт Кондратьев: в России создали дрон лучше украинской «Бабы Яги»'], AIF, 18 February 2025. https://aif.ru/society/ekspert-kondratev-v-rossii-sozdali-dron-luchshe-ukrainskiy-baby-yagi; KondratyevMaksim, Telegram post, 18 February 2025. https://t.me/KondratyevMaksim/1153.

[42] 'Russia's bomber drone Koshchei begins to be used in special operation zone', *Tass*, 16 September 2025, https://tass.com/defense/2017261; 'Russia begins mass production of Vorobey-15 bomber drone', *Tass*, 19 March 2026, https://tass.com/defense/2103855.

[43] Misdronekb, Telegram post, 16 February 2025. https://t.me/misdronekb/116.

[44] Ibid.

[45] Ibid.

[46] Eban_warrior, Telegram post, 18 February 2025. https://t.me/eban_warrior/2112.

[47] Ibid.

materials, and therefore enables their growing use.[48] Some Russian units are repairing captured Ukrainian "Baba Yaga" drones to put them into operation against previous owners. Russian builders have indicated that Ukraine is also using a makeshift design against Russian forces. Some Russian commentators acknowledged the usefulness of this temporary solution but also noted that this effort arose out of 'hopelessness' at the current lack of heavy Russian multi-rotors.[49] Video evidence of Vobla is available throughout 2025 on Russian social media and messaging apps, but it includes few examples of successful strikes—most of the content is related to the drones' assembly and aerial tests.

Artificial Intelligence in Tactical Combat

At the same time, despite some of the setbacks mentioned above, the Russian defence industry and, on occasion, Peoples' VPK as well continue to develop and utilize artificial intelligence (AI) mechanisms in aerial drone and UAV operations. Specifically, such development amounts to enabling the drones to fly to targets on their own once such targets have been confirmed by the operator. In such cases, drones are equipped with image recondition, target lock, and terminal guidance to go after intended targets. Russian media also alludes to these capabilities as 'autopilot' based on pre-loaded information housed on the drone itself.

Some examples of such developments were seen on military-grade loitering munitions such as Lancet and V2U, or on the afore-mentioned ISR drones to evade incoming Ukrainian interceptor drones.[50] A few Russian quadcopters made by the defence industry and by defence-industry-affiliated enterprises likewise claim to feature such technology, although the actual numbers of Russian drones enabled by such AI applications are not publicly known. For example, in a summer 2025 engagement described by the Russian state media, Russia's Ovod FPV 'enabled by machine vision struck Ukrainian targets in the Black Sea', although such statements normally do not feature detailed descriptions of technologies or tactics used.[51]

Multiple other FPV types such as Rusak-S are advertised by state media as having related capabilities—specifically, Rusak is described having an 'a

[48] David Hambling, 'Russians lash FPVs together to make Frankenstein heavy lift drones', *Forbes*, 25 February 2025. https://www.forbes.com/sites/davidhambling/2025/02/25/russians-lash-fpvs-together-to-make-frankenstein-heavy-lift-drones/.

[49] Eban_warrior, Telegram post, 18 February 2025. https://t.me/eban_warrior/2112.

[50] Spencer Faragasso, 'Russian Lancet-3 kamikaze drone filled with foreign parts: Western parts enable Russian Lancet-3 drone to have advanced targeting and anti-jamming capabilities', *ISIS-online*, 18 December 2023. https://isis-online.org/isis-reports/detail/russian-lancet-3-kamikaze-drone-filled-with-foreign-parts/; 'Obscure Russian V2U Drone Unraveled By Intelligence: Autonomous Loitering Munition Powered By Nvidia Chip', *Defence Express*, 9 June 2025, https://en.defence-ua.com/weapon_and_tech/obscure_russian_v2u_drone_unraveled_by_intelligence_autonomous_loitering_munition_powered_by_nvidia_chip-14798.html.

[51] 'Russian Black Sea fleet destroys Ukrainian armed forces equipment near Crimea using AI' ('Черноморский флот РФ с помощью ИИ уничтожил оборудование ВСУ у Крыма'), *Ura*, 9 June 2025. https://ura.news/news/1052945844.

single-board computer with a built-in neural network that [allegedly] recognizes artillery, armored and transportation vehicles, with the neural network trained on terabytes of video footage'.[52] By mid-2025, there are no official figures on how many AI-enabled Ovod or Rusak-S drones are actually fielded in combat.

Despite years of defence and government research into drone swarming applications, an actual AI-enabled UAV swarm, where fully autonomous swarm elements collectively carry out a mission, has not yet been officially fielded across the Russian force. However, starting in 2025, Ukrainian military experts noted the Russian military using a potential swarm of up to six V2U drones with different coloured wings for drone-to-drone recognition. According to these Ukrainian sources, this drone swarm reportedly has a range of up to 80 kilometres, has inertial and satellite navigation systems, and apparently heavily relies on foreign-made components such as a Western module for video recognition and processing, and a high-speed hard drive.[53] The same sources reported that one such drone model has a gas engine, which increases the drone's operational range to over 100 kilometres.[54] The same Ukrainian sources also reported that Russian units have been launching between thirty and fifty of these drones daily in multiple operational directions in test configurations of two to six drones. These UAVs reportedly rely on visual terrain navigation to fly towards a target; can allegedly autonomously detect, classify, and selects targets; and apparently do not require an operator to authorize the final strike decision.[55] No official confirmation of this drone exists in the Russian media or the MoD.

In June 2025, Ukrainian GUR (military intelligence) published additional information on the above-mentioned drone. According to Ukrainian sources, this drone can allegedly autonomously search and select targets, 'with its computing system is based on the Chinese Leetop A203 minicomputer, which uses a high-speed assembly based on United States (US)-made NVIDIA Jetson Orin as the central processor'.[56] GUR further indicated that this AI-enabled drone is equipped with only one global positioning system (GPS) module likely due to the effect of Ukrainian EW counter-measures. GUR also confirmed that 'computer-vision'-enabled navigation is utilized in this UAV, with the drone comparing the image from the camera with pre-loaded photos of a given area.[57]

[52] Maksim Kondratyev, Telegram post, 2 November 2024. https://t.me/KondratyevMaksim/826; 'Russia launches serial production of Rusak-S with AI' ('В России запустили серийное производство «Русака-С» с ИИ'), *Lenta*, 6 August 2024. https://lenta.ru/news/2024/08/06/v-rossii-zapustili-seriynoe-proizvodstvo-rusaka-s-s-ii/.

[53] Stepanenko, 'The battlefield AI revolution'.

[54] Ibid.

[55] Ibid.

[56] Ibid.; Evgenia Gubina, 'GUR revealed details of a new Russian AI drone that searches for targets for attack', *Mezha.media*, 9 June 2025. https://mezha.media/en/oboronka/gur-detali-novogo-rosiyskogo-drona-302543/.

[57] Ibid.

It's likely that as the war continues, the Russian industry will unveil more AI-enabled UAVs for both short-range quadcopter and FPV missions, and for longer ranged applications such as those described above. It's also likely that the MoD's own AI technologies directorate that was set up in 2022 is closely involved in the development of AI technologies for the forces, including AI applications, data analysis, and dataset development for UAVs and aerial drones.[58]

Impact of Tactical Drones on the Russian Ground Forces' Operations

The Ukraine combat is forcing both belligerents and Western militaries to consider how tactical drones impact conventional operations.[59] In the US, the military is observing the impact of ubiquitous drone surveillance of the battlefield and ever-present FPV threat that impact combat, logistics, manoeuvre, and supply operations, potentially undoing many years of US Department of Defense doctrines that were built on the ability to conduct such operations without significant tactical aerial threats.[60] Examples include the 2024 US army revamping of its 'Forge', the capstone event of basic training for over half a decade, based on lessons learned from the war in Ukraine, which now includes lessons on how soldiers can conceal themselves from tactical drones.[61] The US Army also launched the Unmanned Advanced Lethality Course at Fort Rucker that is designed to rapidly train soldiers on the lethal employment of small UAS that include FPV drone operations.[62]

In April 2025, US Marine Corp (USMC) set up a drone attack team—units equipped with FPV-type drones to study the lessons of Ukraine combat and to study such drones' impact on USMC tactical operations.[63] And in May 2025, international news outlets reported the UK military's alleged plans to make itself more flexible,

[58] 'Russian defense ministry introducing AI technologies in army', *Tass*, 25 August 2022. https://tass.com/defense/1497995.

[59] Examples: Joseph Trevithick, 'Growing drone threats lead to army modular shelter kit improvements', *TWZ*, 9 June 2025. https://www.twz.com/news-features/growing-drone-threats-lead-to-army-modular-shelter-kit-improvements; Benjamin Jensen, 'Fewer soldiers, more drones: What Ukraine's military will look like after the war', *CSIS*, 28 April 2025. https://www.csis.org/analysis/fewer-soldiers-more-drones-what-ukraines-military-will-look-after-war; Neil Hollenbeck, 'How to transform the army for drone warfare', *The War Room*, 9 January 2025. https://warroom.armywarcollege.edu/articles/transform-for-drones/.

[60] Michael T. Hackett and John A. Nagl, 'A long, hard year: Russia-Ukraine war lessons learned 2023', Army War College Publications, 29 April 2024. https://publications.armywarcollege.edu/News/Display/Article/3890256/a-long-hard-year-russia-ukraine-war-lessons-learned-2023/.

[61] Jeff Schogol, 'Army revamps basic training to simulate battlefield stalked by drone swarms', *Task and Purpose*, 17 October 2024. https://taskandpurpose.com/news/army-training-drones-conceal/.

[62] Leslie Herlick, 'U.S. Army Aviation Center of Excellence Launches Unmanned Advanced Lethality Course to Equip Soldiers for Future Warfare', *Army.mil*, 18 August 2025, https://www.army.mil/article/287907/u_s_army_aviation_center_of_excellence_launches_unmanned_advanced_lethality_course_to_equip_soldiers_for_future_warfare.

[63] Jeff Schogol, 'Marine Corps stands up "attack drone team" to take lessons from Ukraine and teach them to grunts', *TaskandPurpose*, 4 April 2025. https://taskandpurpose.com/news/marine-attack-drone-team/.

with greater numbers of tactical drone and loitering munitions, while moving away from decades-old conventional ground force structure as a response to Ukraine battlefield lessons.[64] The urgency to update Western ground forces' fighting concepts and tactics was demonstrated during the 'Hedgehog 2025' exercises held in May 2025, when a small group of Ukrainian drone operators defeated a combined combat group of several thousand NATO troops who were attempting to carry out a mechanized attack.[65]

For their part, multiple Russian military scholars, military academics, and military experts have weighed in on the impact of drones on modern combat in general, and on the war in Ukraine in particular. In February 2025, the director of Moscow-based Center for the Analysis of Strategies and Technologies (CAST[66]) described the impact of tactical UAVs by noting that the slow progress exhibited by Russian forces against the Ukrainian military in 2025 is still not overcoming the predominantly positional nature of military operations in Ukraine, in place since 2022, reflecting the established long-term trends in military affairs.[67]

This positionality in Ukraine is due to many factors, including the high dispersion and low density of forces and their battle formations as a result of sharply increased capabilities for reconnaissance, detection, target designation, and high-precision strikes. These capabilities that are often enabled by aerial drones and UAVs today result in the 'vulnerability of forces ranging from lower tactical units to operational formations to individual military equipment, combat vehicles, and individual personnel on the battlefield'.[68]

This near-'total' transparency on the battlefield is explained by a significant number of constantly deployed reconnaissance and surveillance assets—primarily small UAVs—as well as a qualitative leap in the volume and speed of intelligence information transmitted and received using these assets. As a result, near the line of contact, any covert transfer, deployment, concentration, and use of forces becomes 'impossible'.[69] This transparency in turn forces ground forces to be organized into small groups, subdivisions, and to operate from individual vehicles dispersed as much as possible across a given section of the front. This resulting inability to concentrate into larger formations and focus military forces and means on a mass-scale breakthrough 'overturns' many previous foundations of military affairs, forcing much

[64] 'Drones as 80% of the army: UK is radically changing its military doctrine based on Ukraine's experience', *DefenseExpress*, 28 May 2025. https://en.defence-ua.com/events/drones_as_80_of_the_army_uk_is_radically_changing_its_military_doctrine_based_on_ukraines_experience-14666.html.

[65] Dmytro Shumlianskyi, 'Ukrainian Drone Operators Defeated NATO Combat Group During Joint Exercises,' *Militarnyi*, 13 February 2026, https://militarnyi.com/en/news/ukrainian-drone-operators-defeated-nato-combat-group-during-joint-exercises/.

[66] CAST is one of the more authoritative Russian research organizations providing analysis of Russian and international military technologies, tactics, and concepts. http://cast.ru/.

[67] '"This is a different war." The drone revolution is radically changing the face of warfare and the military' ['"Это уже другая война." Дроновая революция кардинально меняет облик боевых действий и армии'], *RG*, 17 February 2025. https://rg.ru/2025/02/17/ruslan-puhov-v-hode-svo-sostoialsia-predskazyvavshijsia-triumf-bespilotnoj-aviacii.html.

[68] Ibid.

[69] Ibid.

smaller units and groups, as well as individual crews and fighters, to wage war on a large scale.[70]

He further explained that today, soldiers have no need to directly see the enemy or target—it can be seen theoretically from any distance and hit with high accuracy. Therefore, the survivability of remote dispersed platforms from closed positions and their crews is much higher than any weapon firing in line of sight.[71] He further reiterated that this situation therefore revolutionizes the entire system of strikes and destruction of enemy forces, and, in Ukraine today, these revolutionary means of destruction are enabled by small UAVs such as loitering munitions, munition-dropping quadcopters, and FPVs.

The CAST director also noted the evolving nature of tank warfare in such conditions, given the overarching importance of a tank as a key weapon in Soviet and Russian military operations stretching back to World War Two. He indicated that in Ukraine, the so-called 'tank crisis' is caused not so much by the vulnerability of tanks to UAVs strikes, but by the loss of the tank's significance as a means of fire in direct line of sight.[72] He explained that strike missions performed by a tank can now be performed by 'much cheaper, more effective, and more covert' means of high-precision destruction from closed positions—from FPV drones to guided artillery shells and tactical over-the-horizon missile systems.[73] Therefore, the experience of military operations in Ukraine shows that tanks now, as a rule, can be destroyed on the battlefield even before they are able to use their weapons, 'which, in fact, makes the very place of the tank in the armament system of the troops meaningless—if a tank is still capable of direct fires, it is only because the number of FPV drones and similar weapons is still insufficient everywhere at the front'.[74]

Thus, there is a crisis in Ukraine today of

> those forms and methods of warfare that many generations of soldiers were trained in—the very idea that the concentration of forces and resources in the traditional regulatory way or the concentration of tanks and armored vehicles in the current conditions is more likely a path to defeat rather than to success, and this still does not fit into the heads of many commanders on both sides in this war.[75]

He further commented that during the ongoing special military operation (SVO; a Russian term for its invasion of Ukraine), 'the long-predicted triumph of unmanned aircraft finally took place'.[76] This triumph was expressed not in the replacement of manned aircraft with heavier, more expensive medium-altitude, long-endurance (MALE) class drones, such as the Turkish-made Bayraktar TB2 in Ukrainian service

[70] Ibid.
[71] Ibid.
[72] Ibid.
[73] Ibid.
[74] Ibid.
[75] Ibid.
[76] Ibid.

and Russia's own Orion drones, but in the 'hordes' of small drones such as DJI Mavic quadcopters and FPV drones that roam the battlefield and can hunt any potential target, down to a single infantry soldier.[77]

Therefore, the war in Ukraine has become a war of drones, and the ongoing drone revolution is radically changing the face of combat operations—it has provided not only the previously mentioned 'unprecedented transparency' of the battlefield but has also begun to push out artillery and group infantry weapons: it is 'obvious that tanks, and in the future artillery, are losing their significance, with drones as a universal means of destruction that are already capable of operating at the full range of artillery; they also allow covert action and strikes from closed positions.'[78]

It is worth pointing out that the Russian military and defence industry are still not giving up on the idea of the tank as a forward-operating combat element, and attempt to combine legacy systems with new UAV technologies. In January 2026, Russia's Rostec defence-industrial enterprise discussed the role of tanks in current and future wars, arguing that tanks will retain their role as the primary breakthrough strike force on the battlefield due to new operational tactics and their integration with other forces and assets. One such current tactic involves one tank engaging the enemy from a standoff position while another executes a rapid manoeuvre to reach the front line, while the movements of these vehicles are coordinated with the aid of unmanned aerial vehicles for battlefield awareness and target spotting.[79]

At the same time, CAST director doubled down on his argument by noting that drones, unlike any other combat means, make it possible to identify all enemy targets and objects to tactical depth and destroy everything to tactical depth, 'and in the future, with further technology advancements, this will also become possible to do so at operational depth.'[80] For the CAST director, this is a 'revolution' that has drastically affected key elements of military affairs, including the concentration of forces and combat means; drones in Ukraine also make manoeuvring to ensure enemy defeat unnecessary, marking a fundamental change not only in tactics but also in operational art; and 'sooner or later, all this will lead to radical changes not only in the forms of combat operations, but also in the appearance and structure of armies.'[81]

According to the CAST director, FPV drones, given their 'unprecedented' cost-effectiveness, are becoming the main means of tactical destruction on the battlefield, and combatants in Ukraine continue to discuss their drone production in the

[77] Ibid.
[78] Ibid.
[79] 'Rostec has revealed what tanks will look like in the future following the transitional phase (Ростех рассказал, как после переходного этапа будут выглядеть танки в будущем)', *Tass*, 6 January 2026, https://tass.ru/armiya-i-opk/26096765.
[80] Ibid.
[81] Ibid.

millions.[82] Therefore, with such a scale of application, FPV-type drones could essentially displace 'all' group infantry weapons and anti-tank weapons, and will take on most of the tasks that are performed today by artillery.[83] This could lead to changes in the future appearance of military units, which can become a combination of squads of 'grenadiers' with FPV operators and assault units supported by reconnaissance UAVs and EW systems.[84] Moreover, he noted that the most important aspect of military affairs will be the fight against enemy FPV drones and one of the key components of future military unit structure should be forward target designators providing key guidance for high-precision weapons delivered by long-range systems and aviation.

Further, mass-scale UAV use and high-precision weapons could theoretically overcome positionality, including in military operations in Ukraine; therefore, the CAST director thinks that, in the foreseeable future, tactical battlefield success in Ukraine will depend on the parties' increasing production of FPV drones and loitering munitions (exemplified today by Russia's Lancet and Ukraine's Bulava drones), as well as increasing manufacturing of high-precision weapons, such as guided aerial bombs, missiles, long-range anti-tank guided systems, and others.[85] He predicts that if a military victory is won in a war of this type, it will be won by those who have more precision weapons at both tactical and operational levels. In such a scenario, more precision weapons would allow for the effective 'grinding down' and destruction of enemy troops and strongholds, as well as of their warehouses, communications, and command centres.[86]

Such tactics can also include using aviation as carriers of drones—a concept that is already making an appearance on the Ukrainian battlefield—which will lead to the isolation of the contested combat area. For the CAST director, the most decisive systems on the modern battlefield in Ukraine, such as FPV drones, DJI Mavic quadcopters, multi-rotor and fixed-wing UAVs, loitering munitions, guided bombs, and high-precision missiles, are the least dependent in production on Western sanctions.[87] Therefore, he thinks that scaling their production is a matter of vision and understanding of the main combat trends and war prospects by Russian leaders and commanders at all levels.

In a related Russian analysis, a military expert who is a corresponding member of the Russian Academy of Military Sciences writing about overcoming positional warfare in Ukraine has pointed out numerous Russian military analyses suggesting overcoming a positional deadlock by saturating breakthrough groups with reconnaissance and strike means, such as employing ZALA ISR drones, long-range Geran-2

[82] 'Ukraine has capacity to produce 5 million FPV drones per year, advisor says', *Kyiv Independent*, 29 March 2025. https://kyivindependent.com/ukraine-can-produce-5-million-drones-per-year-advisor-says/.

[83] '"This is a different war."'

[84] Ibid.

[85] Ibid.

[86] Ibid.

[87] Ibid.

UAVs, and loitering munitions such as Lancet, along with widespread use of tactical ground-based precision weapons for pinpoint destruction of enemy artillery and fortifications.[88]

Other suggestions listed in that particular analysis cites a Russian military expert pointing out that the success of the counter-battery fires in modern conditions can only be ensured by conducting reconnaissance and strike operations using high-precision guided missiles, artillery ammunition, reconnaissance UAVs, and loitering munitions.[89] Additional advice on overcoming the positional stalemate includes introducing a new support platoon to the organizational and staff structure of assault groups consisting of ammunition delivery and evacuation units, UAV operators, and counter-UAV units.[90]

Other proposals in the analysis for overcoming the positional stalemate in Ukraine include using FPVs and loitering munitions; and as this technology develops, to use UAVs and loitering munitions equipped with AI functions. Suggested tactics influenced by Ukrainian combat include a massive strike by FPVs and loitering munitions on the enemy's defensive lines with simultaneous mass destruction of the entire depth of the defensive lines of enemy manpower (down to individual servicemen), fortifications, control centres, communication nodes, as well as armoured vehicles and missile and artillery guns located on the defensive lines, with simultaneous isolation of the enemy in the breakthrough by preventing manoeuvre, regrouping, and pulling up enemy reserves to the breakthrough area.[91]

Additional proposed methods for overcoming a positional stalemate are described as follows: covert accumulation of small FPVs and loitering munitions in the planned breakthrough areas at a rate of two to five FPVs per unit of enemy manpower, or three to seven FPVs or one to two loitering munitions per unit of weapons and military equipment (guns, armoured vehicles, etc.).[92] The specified number of FPVs and loitering munitions is assumed based on the need to attack armoured vehicles and fortifications simultaneously with multiple UAVs to increase the probability of the target's successful destruction, as well as the advisability of forming a reserve of these weapons and military equipment to destroy undetected enemy forces and assets, as well as to target enemy reserves transferred to the breakthrough area.[93] Further, it is essential to conduct demonstration flights of UAVs for radio and electronic reconnaissance in order to identify the location of systems operating on electronic emission and to locate enemy radar stations that are part of the air defence systems,

[88] S. I. Makarenko, Overcoming the positional deadlock in modern combat operations through the massive use of unmanned aerial vehicles' ('Преодоление позиционного тупика в современных боевых действиях за счет массированного применения беспилотных летательных аппаратов'), [«Воздушно-космические силы. Теория и практика»]. *Aerospace Forces: Theory and Practice* 32 (December 2024). https://cyberleninka.ru/article/n/preodolenie-pozitsionnogo-tupika-v-sovremennyh-boevyh-deystviyah-za-schet-massirovannogo-primeneniya-bespilotnyh-letatelnyh.

[89] Ibid.

[90] Ibid.

[91] Ibid.

[92] Ibid.

[93] Ibid.

as well as radars that are part of the combat control system and radio communications systems used for tactical command and control.[94] This is where Russia's Rubicon Center is most effective today as it hunts Ukrainian drone command and control structures, systems and units, along with targeting Ukrainian logistics and supply efforts.

The advice further includes conducting individual training flights to understand the terrain and features of the defensive lines at the site of future breakthroughs. The following instructions are proposed—with one to two hours before the strike, deploy a network of signal repeaters on UAVs and balloons to control subsequent actions of UAVs and combat systems along the line of contact. Within 0.5 to 1.5 hours before the strike, conduct additional reconnaissance of the enemy defensive lines at the site of the future breakthrough using reconnaissance UAVs.[95] Finally, when delivering a massive strike, land- and air-based precision weapons are used against enemy EW, air defence radars, combat radars, and control systems; while combat aircraft and FPVs are used against enemy manpower, its armoured vehicles and artillery, and fortified defensive line facilities.[96]

Additional suggestions include the introduction of reconnaissance UAVs, strike FPVs, and loitering munitions reserves to the breakthrough location; providing support for assault operations by aviation and ground tactical precision weapons; and providing target designation and target illumination for tactical precision weapons by the combat formations of advancing troops; and the dispersal of reconnaissance UAVs at the rear of adversary lines for the purpose of detecting the signs of enemy regrouping and reallocation of reserves.[97]

For its part, Russia's Rostec is defending the future use of tanks in battlefield conditions outlined above, indicating that tanks have already seen significant improvements in their defensive capabilities, and that their operational deployment has become more sophisticated. In contrast to some of the CAST arguments outlined above, Rostec argues that the tank remains one of the most reliable combat platforms on the battlefield, capable of delivering precise direct fire even while under attack from a diverse array of enemy assets, including FPV drones.[98] Therefore, future tank designs will continue to incorporate the traditional hallmarks: a tracked chassis, robust armour, and a medium- or large-calibre cannon housed within a rotating turret. At the same time, these designs will be tailored to meet modern requirements when it comes to protection, firepower, and situational awareness on the battlefield.[99]

[94] Ibid.
[95] Ibid.
[96] Ibid.
[97] Ibid.
[98] 'Rostec has revealed what tanks will look like in the future following the transitional phase (Ростех рассказал, как после переходного этапа будут выглядеть танки в будущем)', *Tass*, 6 January 2026, https://tass.ru/armiya-i-opk/26096765.
[99] 'Rostec has revealed what tanks will look like in the future following the transitional phase (Ростех рассказал, как после переходного этапа будут выглядеть танки в будущем)', *Tass*, 6 January 2026, https://tass.ru/armiya-i-opk/26096765.

Thus for Rostec, which has a massive stake in current and future tank production for the MOD, 'for future armoured "fists"—concentrated armoured breakthrough formations—greater importance will be placed not so much on the individual specifications of a single tank, but rather on a new level of integration with other forces and assets: automated command-and-control systems, various classes of drones, ground-based robotic systems, artillery, electronic warfare and air defence assets, combat engineering units, and anti-tank assets.'[100]

Current Battlefield Tactical Realities

Many of the proposals described in the previous section are already implemented in whole or in part by both belligerents; in other instances, the ongoing combat is reflective of the proposal above for more effective operations. By 2026, Russian military commentators point out the growing importance of supply deliveries via aerial drones, especially when front-line conditions and open terrain without any natural or man-made shelter makes any movement very dangerous. In such circumstances, drones can carry food, bottled water, essential medicines, and other items. At the same time, these bloggers pointed out that a typical cargo drone's carrying capacity is limited, requiring multiple flights to bring supplies to the soldiers.[101] Some Russian military bloggers have argued that using these logistics and supply drones is an effective measure when every other solution is not available, that the UAV launch and landing points need to be changed periodically, and that drone pilots need to use camouflage to avoid Ukrainian counter-battery fires. They further suggested that cooperation and communication with neighbouring units during such UAV supply runs is crucial to avoid friendly fires.[102]

In 2025, Russian military commentators discussing best tactics for FPV units in assault operations pointed out that such units' main goals include taking out adversary equipment and fortifications in order to 'blind, deafen, and suppress' the adversary; striking the adversary's retreat and approach routes; suppressing firing points that prevent Russian units from advancing; and destroying running and retreating targets.[103] The FPV unit commander must be close to the assault group or observation post, using 'clear, short, tight' communications and code designations and pre-agreed markers on the map.[104] Each UAV crew must have at least four to six drones that are loaded, tested, and tuned to the required operating frequency.[105]

[100] 'Rostec has revealed what tanks will look like in the future following the transitional phase (Ростех рассказал, как после переходного этапа будут выглядеть танки в будущем)', *Tass*, 6 January 2026, https://tass.ru/armiya-i-opk/26096765.
[101] Urga_74, Telegram post, 23 April 2025. https://t.me/urga_74/4474.
[102] Ibid.
[103] Battle_unit, Telegram post, 22 April 2025. https://t.me/battle_unit/26311.
[104] Ibid.
[105] Ibid.

Russian military bloggers further described a 'carousel' drone tactic, which involves multiple drones: while the first UAV hovers over the target, its battery is gradually used up, and the second drone flies towards the first drone's position, preferably reaching the first UAV in less than ten minutes.[106] If the same operator simultaneously adjusts fires and controls with the first drone when the second UAV approaches, the operators then simply exchange drone remote controls for the second drone to resume the same functions. This allows the spotter pilot not to lose context and to continue working without interruption.[107] The carousel commentary also points out that, sometimes, drone piloting roles are divided: one controls the drone itself while the second one adjusts fire as an observer.[108] This particular approach is used when there are enough resources, and the task is to transmit data as accurately and quickly as possible.

Russian military bloggers have also lamented the advantage that Ukrainian drone forces currently enjoy, pointing out that when the Russian forces are attacking, the Ukrainian UAVs 'practically nullify' all Russian offensive potential, as mentioned above. In such cases, one well-trained FPV unit can hold back the attack of at least a platoon-sized formation, and in some cases, an entire advancing company.[109] As a result, as one Russian commentator notes, 'in many areas of the front, neither we nor the enemy have the opportunity to advance—the Ukrainian military cannot advance because of our [Russian] advantage in artillery, [and] we [Russian forces] cannot advance because of the enemy's advantage in FPV drones.'[110] They also pointed out available solutions to this stalemate that require prioritizing the destruction of identified enemy UAV crews by all available means and weapons, and creating special fighter UAV crews that, in cooperation with signal intelligence and EW units, can systematically destroy adversary aerial signal repeaters. Today, such tasks are increasingly taken on by the Rubicon Center units described earlier.

Russian military commentators writing about Ukrainian small unmanned aerial systems (sUAS) and counter-UAS(CUAS) tactics, techniques, and procedures also noted that up to 80% of the damage and destruction Russian forces experience today is inflicted by Ukrainian UAVs.[111] Those UAVs include not only the ubiquitous FPV, but also heavy Baba Yaga bombers that conduct remote mining and munitions drop operations. Commentators noted that in 2024, the priority target for an FPV strike was transport or infantry dugouts; today, drone operators often use multiple tactical UAVs to strike a single target.[112]

Furthermore, Russian commentary often notes the Ukrainian efforts to detect Russian personnel, identify equipment on rotation, and monitor delivery and evacuation routes via ISR UAVs followed by drone and artillery strikes.[113] A related Russian

[106] Urga_74, Telegram post, 17 April 2025. https://t.me/urga_74/4391.
[107] Ibid.
[108] Ibid.
[109] VictoryDrones, Telegram post, 21 April 2025. https://t.me/VictoryDrones/39624.
[110] Ibid.
[111] Boris Rozhin (Boris_rozhin), Telegram post, 30 April 2025. https://t.me/boris_rozhin/163037.
[112] Ibid.
[113] Ibid.

commentary noted that Ukrainian forces are actively using FPV drones, heavy Baba Yaga multi-rotor bombers, and quadcopters with grenade drops. They acknowledge that FPVs' real efficiency fluctuates within the range of 20–40%—the reason being the instability of UAV electronics, vulnerability to EW, as well as problems with accuracy and quality of drone-carrying warheads.[114] Often, a series of hits is required to bring down a single target, which sometimes reduces drones' economic and tactical efficiency.[115]

This commentary also reiterates that Ukrainian heavy drone bombers are used primarily at night to strike at concentrations of Russian equipment and against fortifications, while their high cost and vulnerability to small arms fires limit their large-scale use.[116] DJI Mavic quadcopters in Ukrainian service that conduct light munition drops are still effective against Russian infantry, but these light tactical drones require proximity to the line of contact, thus creating a higher risk for their drone operators.[117] The commentary further acknowledges that no matter how effective such Ukrainian drone operations are, they cannot replace the effects of artillery on the battlefield, since artillery retains a key advantage in speed and mass of fire.[118]

At the same time, the commentary noted that FPV drones—even with mass-scale use—are incapable of providing an equivalent density of destruction in a similar time frame; in addition, artillery is less susceptible to weather and EW limitations. The commentator also added that artillery can strike at tens of kilometres, making it indispensable for supporting the depth of defence and attack.[119] Thus, 'artillery is not disappearing from combat, but it is losing some of its versatility, and instead is turning into a specialized, rarer, but extremely important battlefield resource, and underestimating its role means waging war according to the wrong patterns.'[120]

In another analysis, Russian military commentators noted that some Ukrainian UAVs can 'freely' change their operating frequencies in a matter of seconds in order to evade persistent Russian EW counter-measures, along with using Baba Yaga heavy drones controlled by fibre optics. Specifically, this commentary acknowledged that as of early April 2025, the Ukrainian forces have increased their use of FPV applications 'at least threefold'.[121] When it comes to Ukraine's Baba Yaga heavy drones specifically, Russian commentators reported that it was used in Russia's Kursk region so often that such drones started to replace smaller DJI Mavic quadcopters, as these heavy multirotors performed half of the combat UAV tasks at the front as signal repeaters, drone carriers, mining and bombing UAVs.[122] They further argued that it is possible and

[114] Ibid.
[115] Ibid.
[116] Ibid.
[117] Ibid.
[118] Ibid.
[119] Ibid.
[120] Ibid.
[121] 'From Telegram about drones at the front' ['Из Telegram про дроны на фронте'], *LJ*, 12 April 2025. https://dandorfman.livejournal.com/4150678.html.
[122] Bayraktar1070, Telegram post, 21 April 2025. https://t.me/bayraktar1070/3918; Rusengineer, Telegram post, 21 April 2025. https://t.me/rusengineer/7021.

necessary to fight such heavy drones, but that the Russian forces lack a coherent system in place for such tasks, including (a lack of) 'interaction, coordination, exchange of information, and distribution of goals within a single information space.'[123]

Conclusion

The impact of tactical drones on Ukraine combat is one of the defining characteristics in the ongoing war, with both sides seeking to draw lessons from this confrontation to gain an advantage. This impact is felt far beyond this particular battlefront, as forces and militaries around the world are trying to, first, understand how such drones are impacting ground formations; and, second, to align their own capabilities to address such a potential threat. For example, the US military recently cancelled its reconnaissance helicopter programme, since much cheaper and numerous aerial drones can perform similar missions and tasks.[124] The aerial drones and UAVs are used in Ukraine today in such numbers against all kinds of targets that such tactics are pushing belligerents to change how they have to fight and adapt.

This in turn is impacting the debate and discussion of tactical warfare writ large, as legacy and new systems have to coexist in an increasingly versatile, rapidly changing, and competitive battlespace. This evolution of tactical combat is likewise influenced by new actors such as those who design and develop systems outside of the normal defence-industrial ecosystems, in turn pushing established defence industries to come to terms with rapidly innovating actors that have direct access to the military end users. The overall tactical situation in Ukraine remains similar to previous years, with incremental changes and slow, costly advances in an environment that favours the defending side equipped with reconnaissance and combat UAV and aerial drone assets.

At the same time, the arguments, statements, and discussions in this chapter allude to the need for a new organizational paradigm that will incorporate such tactical assets into existing formations and forces. Both belligerents are moving in that direction, with the Ukrainian and Russian militaries each establishing a USF branch to oversee the development and use of unmanned and robotic systems, with UAVs representing the bulk of such technologies. As Russian discussions in this chapter demonstrate, Ukraine's innovative and mass-scale UAV uses are effectively impeding the current Russian advances—with the Russian forces trying to develop their own tactics to overcome the current battlefield conditions.

The ubiquity of reconnaissance and combat tactical drones in Ukraine is a cause to rethink how close-quarters, urban, and short-distance fighting will potentially take shape in subsequent conflicts. Despite using long-range and advanced weapons, both sides in this war are still depending on infantry to take and hold ground—the very

[123] Ibid.

[124] Ashley Roque, 'Army cancels FARA helicopter program, makes other cuts in major aviation shakeup', *Breaking Defense*, 8 February 2024. https://breakingdefense.com/2024/02/army-cancels-fara-helicopter-program-makes-other-cuts-in-major-aviation-shakeup/.

force that bears the overwhelming brunt of drone and UAV attacks. It's evident that the Russian military is learning this lesson from drone combat in Ukraine, as are many forces around the world, including the US and NATO. It's also clear that tactical drone use will remain a feature in subsequent combats in one form or another—and it is up to future belligerents to draw the proper conclusions from this war to ready their forces for the battlefield full of unavoidable reconnaissance and tactical unmanned aerial assets.

12
'The Old Wars Are Back'
The IDF's Learning Process from the War in Ukraine

Eado Hecht and Eyal Berelovich

> We saw how the Russians fought in Ukraine and the mistakes they made. They fought there in a single-corps fashion, instead of using combined arms tactics, ... Israel's armoured corps has a different shape than it used to. We have been training in a combined-arms fashion at all levels for some years now. We no longer see the tank as being capable of doing everything. The battlefield has changed; it's much more crowded and built-up. There are many challenges to these big platforms and I expect the infantry and engineers to make up for my disadvantages. Our soldiers in all the courses and exercises are now accustomed to fighting in a combined-arms environment. (Brigadier General Hisham Ibrahim, commander of the IDF's Armoured Corps, October 2023)[1]

Introduction

The war between Russia and Ukraine found the Israel Defense Forces (IDF) at a crossroads: testing new concepts and preparing for a new five-year force generation plan, while simultaneously discarding old capabilities.

From the mid-1990s till the eve of the outbreak of the war in Ukraine, the IDF conducted an internal debate, assisted by academic researchers and analysts outside the IDF, on the changing political and strategic environment within which it was operating and the required changes needed in its operational doctrine. Though there were various contending concepts, the generally accepted understanding was that the entire world was changing and the 'old' type of wars were obsolete. Therefore, the war in Ukraine and the manner in which it was being conducted came as a shock to the IDF leadership. Initially, one response was to claim that this was because the two

[1] 'Israel's tank commanders are studying Russia's mistakes', *The Economist*, 16 October 2023. https://www.economist.com/briefing/2023/10/16/israels-tank-commanders-are-studying-russias-mistakes.

Eado Hecht and Eyal Berelovich, *'The Old Wars Are Back'*. In: *Russia's War in Ukraine and Modern Warfare*. Edited by: Mikael Weissmann and Niklas Nilsson, Oxford University Press.
DOI: 10.1093/9780198970903.003.0012

armies were backwards and therefore this war was not representative of current and future best practice as required by the existence of the latest technologies and political ideas developed in the West. The war was considered professionally interesting but not instructive. But gradually, as the war continued and information on the application of new technologies (especially drones and cyber warfare) that were supposed to revolutionize warfare came to light, their success, limited success, or failure, attitudes changed and the IDF began to take a deeper interest in learning from the war.

The Expectations

Over the past decades there has been much discussion in Western armies and academia as well as in Israel on the changing characteristics of warfare. Some proposed that the changes represented not merely an evolution but that actually a major revolution was occurring and that not only the characteristics of warfare were changing but perhaps even the very nature of war. Some of the discussion was purely theoretical and some based on an analysis of the wars conducted in the 1990s and early 2000s.

In the late 1990s and early 2000s the IDF leadership gradually adopted an extreme view of the transformation occurring in warfare and the necessary adjustments it needed to make to its own composition and operational art. Future wars, its senior commanders believed, would no longer include major manoeuvres of massed formations conducting high-intensity warfare (inaccurately dubbed: 'old warfare', 'classic warfare', or 'traditional warfare'), but would be purely low-intensity warfare—with the enemy invariably employing the methods of guerrilla warfare and terrorism, and that therefore the IDF should be reorganized and retrained to focus on counter-guerrilla and counter-terrorist operations (just as inaccurately dubbed: 'new warfare').[2] For example the former IDF chief of staff, Lieutenant General Aviv Kochavi referred to Hezbollah and HAMAS as a 'terrorist army'. A term that doesn't have a clear definition or a common understanding of its characteristics.[3] Conquering territory was irrelevant and even politically and militarily counterproductive—wars would be decided by an exchange of standoff fire with a minimum of manoeuvres or by small forces conducting counter-guerrilla raids, ambushes, and patrols.[4]

Furthermore, argued the proponents of this view, if by chance some enemy attempted to conduct war in the 'old warfare' style against Israel, they would be easily defeated by the new technologies of surveillance systems and precise long-range

[2] Any serious perusal of the history of warfare shows that the vast majority of wars have always been low intensity—i.e. focused on guerrilla and terror strategies and tactics. The so-called 'new wars' are in fact the oldest and most common type of warfare since humans began to fight each other.

[3] Tal Lev Ram, 'The chief of staff: We must stand fast against the terrorist armies that surround us' (in Hebrew). *Ma'ariv*, 12 June 2019. https://www.maariv.co.il/news/Article-703371.

[4] Eado Hecht and Eitan Shamir, 'The role of Israel's ground forces in Israel's wars', in Mikael Weissman and Niklas Nilsson (eds), *Advanced Land Warfare: Tactics and Operations* (Oxford: Oxford University Press, 2023), 301–19.

weapons. Mechanized formations attacking against these new capabilities would be rapidly decimated and therefore were no longer viable on the future battlefield. Statistical artillery weapons were virtually obsolete and should be almost entirely discarded by the IDF, as should the majority of armoured vehicles.[5] New non-lethal options such as offensive cyber operations would disarm and dislocate enemy forces and even the enemy as a political entity with a minimum of bloodshed, compelling them to desist from attacking Israel.

The IDF spent a number of years trying to adapt to its new vision of what the next war would look like.

The IDF Learning Process

As noted in the Introduction, the war between Russia and Ukraine found the IDF at a crossroads: testing new concepts and preparing for a new five-year force generation plan, while simultaneously discarding old capabilities.

Initially the learning process was focused on the 'new' technologies and factors that were expected to decide the war. It included both formal and informal studies, conducted both within the IDF and with the assistance of external researchers analysing and adapting the realities of the Ukraine war scenario to those of expected scenarios facing the IDF. The difference between the political, strategic, and tactical settings of the rivals in Ukraine and those of Israel and its enemies, required focusing on what was compatible and employing the lessons that could be gleaned to improve the force generation process then underway, improve training, and create a better understanding of what Israel's next war could look like. For two reasons the majority of these studies focused on the Russian perspective of the war—the first being that the Soviet and Russian armies had been studied by Israeli researchers over the decades because of their imprint (some major Arab armies modelled themselves on them) and actual physical presence in the Middle East, whereas the Ukrainian military had not been relevant and therefore had not been studied and was an unknown quality; the second reason was that though the IDF had shed most of its armoured forces, it still expected to employ tanks and armoured personnel carriers in a future battle and the Russian invasion was based on a similar force composition—so studying its successes and failures, especially against Ukrainian anti-tank weapons and other newer weapons, such as the armed drones, might provide insights on lessons for the IDF if, against expectations, it was compelled to attack across the border.

There were four common threads between the formal and informal learning processes:

[5] According to publicized figures over the past two decades the IDF has discarded the majority of its tanks and artillery and shut down a large number of formations. Eado Hecht and Eitan Shamir, 'Medium intensity threats: The case for beefed-up IDF ground forces', BESA Center for Strategic Studies, 16 October 2016, 41–2, note 11.

First, the knowledge community that was established spontaneously in the initial phase of the war included both IDF personnel and academics who all knew each other and had intercommunicated in the past; they debated among themselves and both formal and informal members of this community lectured and wrote papers for the IDF's various branches and corps. Furthermore, the IDF being an obligatory all-inclusive army, the external members had all served in the IDF and therefore had hands-on experience in addition to their academic training.

Second, lacking direct access to the battlefields, the members of this community collected information from both official and unofficial sources in Russia and in Ukraine—official statements; news reports; numerous snippets of videos, posts, and photographs published by participants in the war on social media; analysis published by various research institutes in North Atlantic Treaty Organization (NATO) states. However, the focus was less on accepting opinions and views, and more on attempting to collate the data presented and interpret it themselves. This in itself was a difficult task—sifting through tens of thousands of reports and videos and separating fact from fiction was difficult.

Third, the common conclusion reached by the separate researchers was that the future war imagined by the IDF as the basis for its future development was wrong. The common phrase in the initial phases of the war was: 'the old wars are back'. This led to the understanding that the force generation process must be changed.

Fourth, the perpetual worry was how to glean the correct lessons and not fall for mistaken ones—not everything from the Ukraine war is applicable in the Israeli context, not necessarily all the ideas of the new concept that seemed to have failed in the Ukrainian context were completely wrong.

The Formal Learning Process

The formal learning process, called in the IDF 'learning from the experience of others', did not begin with the war between Russia and Ukraine, it had started in the Syrian Civil War (2011–20)[6] and the Second Nagorno Karabah war (2020).[7] Major-General Amir Baram, then the commander of the Northern Command, wrote in his introduction to the *Ma'arachot*[8] special issue on the Syrian civil war: 'Military involvement of Hezbollah, Iran and Russia in the fighting alongside Regime forces during the

[6] Various writers, 'The Syrian Civil War', *Ma'arachot special issue*, 21 October 2020 (in Hebrew); two of the articles were translated to English: Eado Hecht, 'The Idlib Campaign 2019–2020', *Military Strategy Magazine* 7/2 (summer 2020): 10–18; Eyal Berelovich, 'The Syrian Civil War—Evolution of the Syrian army's way of war', *Military Strategy Magazine—MSM Exclusive*, February 2021.

[7] 'The New Wars—a seminar on the Nagorno Karabakh war' (in Hebrew), IDF GF Research, Doctrine and Concept and Ma'arachot, 17 January 2021, https://www.youtube.com/watch?v=w7zdLiV_h4Q

[8] *Ma'arachot* is the official professional journal of the IDF.

Syrian Civil War, created a kind of "battlefield laboratory" in which weapons, combat methods and operational concepts have been tested over the past decade.'[9] Baram noted the two pillars of the IDF method of learning from the war: understanding the wider context while delving deep into the details of the battles and applying the lessons learned to the IDF's unique operational problems.

In spring 2022 the IDF General Staff established a special learning team to monitor the fighting in Ukraine. The purpose of this team was to study the war in depth and acquire and adapt lessons relevant to the IDF.[10] In addition to the General Staff team, each professional branch established its own study team to focus on questions specifically interesting to it. The materials produced by the study teams were disseminated in writing to each other, to the commanders, and across the IDF. The teams intercommunicated with each other and with the external researchers continuously as the war developed.

Gradually the IDF began conducting one-day seminars about the war or single-lecture briefings by experts for senior and junior officers in courses and in units. These seminars didn't focus only on lessons from the ground campaign but also on other domains like the Russian air campaign.[11] The experts, nicknamed the 'travelling circus', included serving IDF officers, like Lieutenant-Colonel Shai Ventura from the IDF logistics corps, or retired officers like Lieutenant-Colonel (reserve) Oren Haas. The large numbers of these seminars were intended to disseminate the lessons learned from the war to all branches and units. Some of the seminars were conducted in unit gatherings (the staff of a particular division or brigade), some at professional branch headquarters, and others were integrated into the curriculum of various courses. While disseminating the knowledge acquired by the studies, these seminars simultaneously enabled the researchers to understand the interests of the various branches and units—what to look for in future studies.[12]

The IDF did not only conduct its studies in house and with Israeli researchers, it also invited leading researchers and analysts from other countries to Israel or sent IDF officers abroad. In these meetings the IDF senior officers were able to learn about the war from different perspectives including that of analysts who had visited Ukraine and learned directly from Ukrainian senior and junior officers.

Most of the studies and publications were not made available to the general public, however the main lessons of the IDF were not kept hidden. The IDF is a 'people's army'—most of the population serves initially as conscripts and then in the reserves, officers are enlisted from the ranks of conscripts, some serve for

[9] Amir Baram, 'The Syrian Civil War—Operational and tactical lessons' (in Hebrew), *Ma'arachot special issue*, 21 October 2020, 5.

[10] Amir Bohbot, 'Because of the war in Ukraine: IDF accelerates integration of Merkava Mark 5 tank' (in Hebrew), *Walla*, 10 March 2022.

[11] Lt Yaron Buskila on Twitter (in Hebrew). https://twitter.com/yaron1bu/status/1608163233830998017; https://x.com/DaganLauren/status/1611023885511839746?s=20; Lauren Dagan Amoss (in Hebrew). https://x.com/DaganLauren/status/1611023885511839746?s=20.

[12] Eyal Berelovich on Instagram (in Hebrew). https://www.instagram.com/p/Cv-h707MZA-/?utm_source=ig_web_copy_link&igsh=MzRlODBiNWFlZA==.

many years in the regular army but most serve only a few years in the regular army then transfer to the reserves where they have to maintain and update their military knowledge. Therefore, it is quite common for military affairs to be discussed in public forums and in the media. Senior officers are interviewed by the media or invited to lecture at public seminars on security affairs. Furthermore, some of the IDF researchers published their findings in open sources such as the IDF professional journal, *Ma'arachot*, open press, and social media—though of course these were vetted to prevent publication of information deemed confidential. These briefings to the public showed the scope of the subjects the IDF tried to learn from the war. For example, the subject of force survivability, especially of the ground forces, was broached in a news interview by the Ground Forces' Commander Major-General Tamir Yadai and Planning Division Commander Major-General Yaakov Benjo. They had concluded that more effort must be expended to improve the protective gear of the ground forces—both for individual soldiers (personal armour) and armoured vehicles (active protection systems), in addition to increasing purchases of passive or active protection they invested in radars to warn of incoming artillery fire.[13] Lecturing at the Reichman University Institute for Policy and Strategy, Yadai argued that the effectiveness of new forms of warfare that had been emphasized in the IDF's new war concept, such as cyber warfare and influence operations, had been overrated and in fact had been proven to have little effect on the conduct of war. Instead, he argued, the state of Israel and the IDF should rebuild their capability of ground manoeuvre as it had been shown that it was still a necessary capability.[14]

Informal Learning

The informal learning process was intertwined with the formal learning process. As mentioned, some articles, interviews, and lectures echoed the dilemmas and the internal debate inside the IDF. Others, mainly by retired officers, tried to explain the future course of action the state of Israel and the IDF should adopt due to the lessons each individual author had learned from the war. The difference between these two groups was that the first tried to tell a story about the war while the latter wanted to influence policy. All the authors and researchers used conventional media and especially social media to publish their work to conduct an open debate between them.

The core analysis of the informal learning community can be divided to a number of topics: first, telling the story of the war; second, examination of particular aspects of the war; third, how the Russian pre-war force generation process affected the initial

[13] Amir Bohot, 'The war in Ukraine has highlighted the failures the IDF is expected to encounter in Lebanon' (in Hebrew), *Walla*, 12 March 2022.

[14] Tamir Yadai, 'What can we learn from the war in Ukraine' (in Hebrew), Reichman University Institute for Policy and Strategy 2022 Herzliya Conference, 22 May 2022.

outcome of the war and, fourth, strategic aspects of the war; fifth, and of particular interest were all aspects of the logistics lessons that could be gleaned from the war.[15] The Russian logistics system is different from that of the IDF and it operated at distances that were extremely far beyond the average Israeli ground manoeuvres in all Israel's wars—hundreds of kilometres in Ukraine versus a few dozen kilometres in the regions around Israel.[16] Furthermore, the IDF has not manoeuvreed further than 15 to 20 kilometres since its 1982 Operation 'Peace for the Galilee' in Lebanon (approximately 100 kilometres to Beirut in six days). The methods, successes, and failures were therefore deemed important **not** because the IDF expects to conduct advances of hundreds of kilometres into its neighbours' territories, but to identify issues that might have been forgotten over the decades without IDF experience in long-range operations.

Telling the story of the war and how it evolved was the first priority because it established the context for the other issues. Scholars, analysts, and military bloggers followed the evolution of the war from day one. For example, Eado Hecht of the BESA Center for Strategic Studies, one of the authors of this chapter, published a daily description and preliminary assessment of the war that focused on the military aspects. This briefing was disseminated across the IDF from the most senior ranks in the General Staff down to the units, to both the regular and reserves officers of the various branches.[17] This preliminary description and analysis set the tone for the learning effort both in the formal and informal groups studying the war.

Reading the overall description enabled the researchers to pick the topics deemed relevant for the IDF. Thus, though the Russian invasion included five different fronts, in the first days of the invasion what interested the IDF analysts most were the battles *en route* to Kiev—the combination of an attempted special forces *coup de main*, an airborne deep operation and fast-moving mechanized forces were reminiscent of the 'old type' of wars and needed to be analysed to discern what had changed and what had not.[18] Yair Navot, a journalist and Russia analyst who had covered the war from its beginning in conventional media and social media, argued that the Russian attempt to seize Kiev was the pivotal effort to a fast win in the first days of the war.[19] The causes for the failure were analysed and described by Eado Hecht in his daily reports.[20] Already three days after the Russian air assault operation Yair Ansbacher

[15] Eyal Ziv, 'The Russian Ukraine war—The logistical failures' (in Hebrew), *Ma'arachot*, 30 May 2022; Alex Kafengauz on Twitter (in Hebrew). https://x.com/kafengauz/status/1746663985955344711?s=20; and numerous internal papers written by logistics officers in the IDF.

[16] The longest range major operation of IDF ground forces was less than 200 kilometres in four days in Sinai in the 1967 war.

[17] The reports were disseminated directly and virally via social media.

[18] It should be remembered that this was typical of Russian offensives in the past too: Hungary 1956, Czechoslovakia 1968, Afghanistan 1979, Chechnya 1994–5.

[19] His analysis can be found on Twitter (in Hebrew). https://x.com/newsisrael13/status/1497641217215148033?s=20.

[20] In his reports, debates in the forum of researchers and analysts and in lectures to IDF senior, formation, and unit commanders and staffs, especially the IDF airborne formations.

wrote a short paper on the battle in Hostomel airport. His main argument was that the inability to seize the airport was the main reason for the failure in Kiev.[21] Others argued that the failure of the Russian offensive against Kiev could not be put down to a single action or the failure of a single capability (in this case airborne forces)—one had to study a much wider scope of activities, successes, and failures to understand how this had occurred and what can be learned from that failure.[22] Two independent researchers, Oren Haas and David Lisovtsev, argued that since the failure of the initial war plan the Russian army was not trying anymore to imitate the initial course of action but planned and executed 'bite and hold' operations with the requisite tactical derivatives.[23] These and other researchers of the war, exchanged information and conducted debates on the factual analysis of the sources (given the imperfect, often contradictory information, and often deliberate misinformation in the sources) and on the relevant lessons for Israel and the IDF between themselves and with IDF personnel operating in the formal study process.

Though initially, for the reasons mentioned previously, the focus of the debates was on the Russian army's experiences, gradually more emphasis was placed on understanding the Ukrainian side—attempting to understand how it had acted to thwart the Russian invasion, what had worked and what had not, and what was relevant for Israel's situation. This emphasis increased when the Ukrainian army launched its first major and successful counter-offensive in late August 2022 and its second, failed, counter-offensive in June 2023. The successes and failures of these counter-offensives and the causes thereof were debated among the researchers and in papers written to provide lessons for the IDF.

Following its defeat near Kiev and the culmination of its successes in eastern Ukraine, the Russian army reverted from rapid deep manoeuvres to slow, methodical attritional attacks. This situation too was extremely interesting for the Israelis—what caused the shift in the operational method, what tactics were being employed to implement the new method, what were the causes of successes and failures.[24] The debate then turned to the question of whether the causes, methods, and tactics were relevant to Israel's situation and, if so, what, if at all, the IDF needed to change in its organization, equipment, and tactics to enable it to conduct such operations.

Gradually, the number of topics studied widened to include combat engineering,[25] new and old capabilities,[26] the naval aspect,[27] aerial operations, and long-range fires

[21] Yair Ansbacher, 'The battle for Antonov—An operational missed opportunity that became a strategic failure' (in Hebrew), *Ma'arachot—Between the campaigns*, 28 February 2022; David Lisovtsev, X (in Hebrew). https://x.com/david_lisovtsev/status/1749496398012166440?s=20.

[22] Eyal Berelovich, 'What stopped the Russian land manoeuvre in the first two months of the war' (in Hebrew), *Ma'arachot—Between the campaigns*, 16 June 2022.

[23] Oren Haas, 'The regrouping of the Russian forces indicates that the Russians analyze the situation of the war differently' (in Hebrew), *Haaretz*, 14 March 2022.

[24] Eyal Berelovich, 'The urban campaigns—Two case studies from the war between Russia and Ukraine' (in Hebrew), *Ma'arachot*, 22 February 2024.

[25] Gilad Avrelingy on Twitter (in Hebrew). https://x.com/GAvrilingi/status/1589708156527001601?s=20.

[26] https://x.com/AmitLidor/status/1628603431068786690?s=20 (in Hebrew).

[27] Shay Levi, 'The Ukrainan operation that terror organization can imitate—and Israel should study' (in Hebrew), *Mako*, 4 November 2022; Shaul Horev, 'We have a reason to worry about the naval warfare

capabilities.[28] To create a comprehensive picture of the war various scholars studied also the effect of the war on the civil society of both belligerents.[29]

The third theme was trying to understand how the force generation process created the conditions for the debacle of the initial Russian incursion. Shmuel Shmuel, director of the Institute of Military Studies in the IDF's Dado Center, published his research on the Soviet and Russian command and control system and its implications for the war. He argued that

> Allegedly, there is an inherent paradox in this approach that tries to produce flexibility at a higher echelon, while relying on rigidity at the lower levels. However, it is a result of the main limitation in which the Russian force operates, which is very large and relies on low-quality manpower in the tactical echelons. The Russians seek to square this circle by making tactical action as predictable as possible for the operational commander, as simple as possible for the tactical force, and at the same time reactive, rapid and surprising in a way that suits the modern pace of operations.[30]

Thus, in the different phases of the war the higher echelons had direct control over battalion tactical groups which caused some of the problems for the Russian army.[31]

Eyal Berelovich, senior defence researcher in the IDF Ground Forces Land Warfare Research Institute and one of the writers of this chapter, published two papers on how the Russians imagined future wars as a result of their participation in the Syrian Civil War. He argued that they had learned the wrong lessons from this war: they had learned that the battalion should be the focal point of the combined-arms organization in all high-, medium-, and low-intensity wars. They created combined-arms battalions too big and unwieldy for commanders to employ effectively when fighting a peer enemy such as the Ukrainian army and these failed in action during the first months, leading the Russians to abandon this concept and move the combined-arms focal point back to the brigades and divisions.[32]

The scope and the diversity of these and other formal research papers was intended to create a fact-based analysis of the war. The informal knowledge community tried to supplement the formal IDF learning endeavour by illuminating the topics that weren't in the scope of the IDF formal process. Furthermore, the external researchers who took part in studying the war were regular lecturers in the IDF, thus allowing the

in Ukraine' (in Hebrew), *YNET*, 24 April 2022; analysis by Dubi Raz on Twitter (in Hebrew). https://x.com/dubiraz?s=20.

[28] Omer Dank, 'Aerial warfare in the war between Russia and Ukraine—History, conclusions and the way ahead', *Strategic Question*, 18 November 2022; Shahar Shohat and Yair Ramati, 'Recommendations and first insights from the air campaign and stand-off capabilities in Ukraine', *Beyn HaKtavim*, vol. 37 Dado Center, 22 September 2022.

[29] Daniel Rakov and Sarah Fainberg, 'The growing impact of the civilian population on the modern battlefield: A glimpse into the Russia-Ukraine War' (in Hebrew), Elrom Center for Air and Space Studies, Tel Aviv University Research Report 0123E, May 2023; Shahar Heller, 'Ukrainian civilians in total war '22' (in Hebrew), *Ma'arachot*, 22 March 2022.

[30] Shmuel Shmuel, 'The Soviet and the Russian command and control system' (in Hebrew), *Ma'arachot*, March 2023, 32.

[31] Ibid., 33.

[32] Eyal Berelovich 'When a presentation meets reality' (in Hebrew), *Ma'arachot*, March 2023, 35–41.

officers to learn from others and to think on the adaptation they need to do in their units and organizations. However, the aim of this vast research endeavour was to start a public conversation about the war and its implication on the world order and on Israel and the IDF.[33]

Impact

The impact of the studies varied. IDF commanders at all levels agreed that many of the lessons were indeed relevant, even if needing adaptation to suit the needs of the IDF. However, actual implementation varied. Some of the lessons could be integrated rapidly into the IDF—these were especially those that were close to IDF thinking and merely required tweaks in the current doctrine, equipment, and organization. For example, the issues that caused the failure of the Russian airborne forces were not previously unknown to IDF doctrine which warned against them—so what was needed was merely to emphasize these weak points and risks in the relevant IDF units' training programmes, using the Russian example to prove the point to the trainees. Some lessons required only quick fixes—even copying techniques and technologies employed in Ukraine—thus, before entering Gaza in late October 2023, all the tanks of the IDF participating in this operation had anti-drone attack roofs added to them. Other anti-drone techniques were also adopted based on lessons from the Ukraine war or developed to suit local capabilities and limitations. Other lessons required retraining of officers to adjust from the focus on irregular warfare to regular warfare and therefore needed more time to implement, and some lessons required rethinking the composition and armaments of the IDF ground forces and will require many years to implement. These lessons are of course now being tested and affected by the Israel's current war with Hamas in Gaza and Hizbullah in Lebanon.

The first and foremost lesson of the Ukraine war was that the belief that high-intensity regular warfare, the force composition, and the combat doctrines relevant to it were no longer relevant for the IDF was wrong. The IDF had taken a wrong turn in its capabilities development, optimizing its ground forces for only low-intensity irregular warfare backed by a large precision fire air and ground force, and must turn back. Of course, not everything was completely wrong—but the basic concept, that the 'old' or 'traditional' force types, capabilities, and doctrine were no longer needed because of the political and cultural evolution in the world at large and the Middle East in particular and because modern technology had made them obsolete, needed to change. The 'new' had not replaced the 'old', it had only added another layer of capabilities. Some of the new technologies were providing exactly the same services as the old technologies in merely different ways, some were enhancing the capabilities of the old, and some were adding a new capability but not one able to fully replace the old.

[33] See: https://twitter.com/clarinetwoman2/status/1603312206078246912.

The prime example of such new technologies previously touted as replacing the old technologies but in fact merely adding this extra layer were the various remotely piloted aerial vehicles—able to fly in areas manned aircraft were too vulnerable to do so, they provided some of the same services provided by manned aircraft—surveillance, targeting, striking. However, their bomb loads are much smaller than those carried by manned aircraft, so they cannot yet fully replace them in the strike role. They add to the ability of ground forces to conduct all these roles without depending on the less available, more complicated to organize and to coordinate manned aircraft. They enable all levels of the force to conduct their own aerial missions, but cannot fully replace the old ground reconnaissance or ground weapons fire capabilities. This lesson was fully implemented in the war in Gaza and in Lebanon—the IDF employed numerous remotely piloted aerial vehicles, but these were attached to and used by the 'old' type ground forces, infantry, tank, engineer, and artillery, to enhance their actions rather than to replace them.[34]

Another lesson was the importance of statistical artillery. The IDF is one of the armies leading the incorporation of precision guided munitions and the reduction in statistical tube artillery. The war in Ukraine began to change the IDF outlook—precision guided munitions are of course a very useful weapon, but just as the sniper rifle cannot completely replace the machine gun so the guided munition cannot completely replace the 'dumb' artillery shell in all situations.[35]

Tanks have become much more vulnerable—this was proven again in Ukraine—but they are still an indispensable weapon system both in open terrain and in constricted terrain.

The combined-arms team must be the focus of all ground combat operations. However, the extreme Russian example of the 'battalion tactical group', which the IDF was studying before the war, was proven to be too unwieldy so that though battalions must fight as combined-arms teams, they must be smaller and some of the capabilities of the Russian 'battalion tactical groups' must be employed by the brigade or

[34] Eado Solomon, 'The quadcopter chaser: The 22 year-old who created a support unit for the fighting in Gaza' (in Hebrew), *Ulpan Shishi*, 27 January 2024. https://www.mako.co.il/pzm-soldiers/Article-4c4f57c49f74d81026.htm; Lilach Shoval, 'The quadcopter identifies—the force destroys: Watch video of of a 4th Brigade combat team fighting in Gaza' (in Hebrew), *Yisrael HaYom*, 30 January 2024. https://www.israelhayom.co.il/news/defense/article/15189071; Liran Antebi and Or Adar, 'FPV drones: From the Ukrainian battlefield to the Middle East', *INSS Insight No. 1825*, 8 February 2024. https://www.inss.org.il/publication/fpv/; 'The trainers who teach the entire IDF to hover another level above the enemy' (in Hebrew), IDF Ground Forces, 23 March 2024. https://www.idf.il/%D7%90%D7%AA%D7%A8%D7%99-%D7%99%D7%97%D7%99%D7%93%D7%95%D7%AA/%D7%96%D7%A8%D7%95%D7%A2-%D7%94%D7%99%D7%91%D7%A9%D7%94/%D7%9B%D7%9C-%D7%94%D7%9B%D7%AA%D7%91%D7%95%D7%AA/2022/%D7%9E%D7%93%D7%A8%D7%99%D7%9B%D7%95%D7%AA-%D7%A8%D7%97%D7%A4%D7%A0%D7%99%D7%9D-%D7%96%D7%A8%D7%95%D7%A2-%D7%94%D7%99%D7%91%D7%A9%D7%94-%D7%A7%D7%95%D7%A8%D7%A1-%D7%A8%D7%97%D7%A4%D7%9F/.

[35] Shai Levy, 'Tens of thousands of 155 mm shells have been procured for the IDF at a cost of 0.25 billion Shekel' (in Hebrew), *Mako*, 31 July 2023. https://www.mako.co.il/pzm-soldiers/Article-ae59ec7e51ba981026.htm; 'The background for the Elbit deal: The American withdrawal of shells from Israel for Ukraine', *ICE*, 31 July 2023. https://www.ice.co.il/finance/news/article/974340; Ami Rochkas Domba, 'The IDF is purchasing artillery shells from Elbit—Will the production rate suffice for a war?' (in Hebrew), *Israel Defense*, 1 August 2023. https://israeldefense.co.il/node/59104.

division and not become an organic part of the battalion. The brigade and the division, regarded over the past two to three decades by many (not only in Israel) as passé, are still essential rungs in the organizational hierarchy.

Last but not least—the concept of modern advanced fires capabilities making ground manoeuvre irrelevant was also debunked.

Summary

The IDF activated intensive research, both formal and informal, to study the war in Ukraine, to analyse it, and elicit lessons for its own development. The war reignited the debate conducted in the IDF that began in the mid-1990s and gradually adopted a new theory of how war should be conducted. This was because many of the lessons of that war contradicted that theory.

The new debate was still not resolved, though it was already beginning to affect the IDF's thinking, planning, and training, when on 7th October 2023 the Hamas terrorist organization ruling Gaza launched an all-out attack into Israel, aided by other armed Palestinian organizations located in Gaza. For the first time since 1982 the IDF was forced to conduct full-scale high-intensity regular warfare, as opposed to the routine low-intensity irregular warfare (counter-terrorist and counter-guerrilla) it was conducting constantly. The IDF counter-offensive into Gaza faced an enemy force equivalent to five typical infantry divisions (though officially termed brigades), organized and trained to conduct a regular prepared defence based on extremely well-fortified urban terrain.

Comparing the operational and tactical actions of the IDF in this war to the theories expounded by the IDF leadership over the past two and a half decades and the directions in which the IDF's development was heading, based on these theories, shows the extent to which the lessons elicited from the war in Ukraine affected some of the IDF's operational and tactical thinking. Thus, the quote opening this chapter expresses one of these lessons; others are the adoption of counter-drone methods observed in Ukraine (the most obvious being the erecting of counter-drone roofs on armoured vehicles) as well as updating methods of using drones (a field in which the IDF has been active since the late 1960s) and the usefulness of what were recently considered out-of-date weapons such as artillery and tanks.

As the war in Gaza progressed and a second major front opened on Israel's border with Lebanon, the people involved in studying the war in Ukraine were all mobilized to participate in Israel's own war. However, most of them managed to maintain some analytical contact with the war in Ukraine and conduct comparisons for better vetting of local lessons and those adopted from there.

13
Visibility of Women and Gender Equality Implementation in the Armed Forces of Ukraine

Tamara Martsenyuk

Introduction

The Russo-Ukrainian war has profoundly reshaped gender relations in the Ukraine armed forces and defence sector. Since the 2014 war in Donbas and, in general, more than ten years of the Russo-Ukrainian war, Ukrainian women have been actively participating in military resistance on the front line.[1] In response to Russia's full-scale military invasion of Ukraine, women remain active: results of a public opinion survey demonstrate that Ukrainian women are willing to resist with arms.[2] In fact, as of September 2024, as many as 68,000 women have enlisted in the armed forces of Ukraine (AFU), with over 48,000 of them being service members.[3] Close to 5,000 female military personnel are deployed in the combat zone.[4] There are more than 7,000 female officers, almost 12,000 non-commissioned officers, 23,000 female soldiers, and 1,300 female cadets.[5]

Since the Euromaidan protest of 2013–14 (called the 'Revolution of Dignity'), Ukrainian women challenged their traditional gender roles as caregivers and victims

[1] Invisible Battalion 5.0, *Monitoring of recommendations and research results from the series 'Invisible Battalion'*, ed. Tamara Martsenyuk (Berlin: NAKO, 2023); Olesya Khromeychuk, 'Experiences of women at war: Servicewomen during WWII and in the Ukrainian armed forces in the conflict in Donbas', *Baltic Worlds* 4 (2018): 58–70; Tamara Martsenyuk, 'Ukrainian women at war: From women's Sotnya to Invisible Battalion', Danyliw Research Seminar on Contemporary Ukraine, 10–12 November 2016, University of Ottawa, Canada. http://media.wix.com/ugd/ff1dca_b0fb19fcce7a44ac858816f7cd4965ad.pdf (accessed 10 January 2025).

[2] Invisible Battalion 5.0, *Monitoring of Recommendations.*

[3] According to the information from the state secretary of the Ministry of Defence of Ukraine, Liudmyla Darahan.

[4] Ministry of Defence of Ukraine, 'There are 68 000 women enlisted in the armed forces of Ukraine: Liudmyla Darahan highlights the Ministry of Defence's accomplishment in instituting systemic changes in gender policy', 27 September 2024. https://mod.gov.ua/en/news/there-are-68-000-women-enlisted-in-the-armed-forces-of-ukraine-liudmyla-darahan.

[5] Daryna Hrysiuk, 'How many women are defending Ukraine against Russia's invasion?', *War Ukraine*, 26 March 2024. https://war.ukraine.ua/articles/how-many-women-are-defending-ukraine-against-russias-invasion.

Tamara Martsenyuk, *Visibility of Women and Gender Equality Implementation in the Armed Forces of Ukraine*. In: *Russia's War in Ukraine and Modern Warfare*. Edited by: Mikael Weissmann and Niklas Nilsson, Oxford University Press.
 DOI: 10.1093/9780198970903.003.0013

of conflict, reclaiming visibility and recognition as revolutionaries and volunteers.[6] Victimhood of Ukrainian women in war is questioned: 'In reality, women also play a key role in resistance movements, both at the frontline and on the home front'.[7] Writing about Ukrainian women and war, scholars pay attention to the issues of 'freedom of expression' and women's agency in struggles for human rights.[8] Moreover, they also pay attention to the historical legacy of Ukrainian women's resistance.

This chapter will analyse the ongoing changes, asking how gender equality has been advanced—and where it still stalls—in the AFU. After a brief overview of the historical background, in this chapter, I will start with the institutional ('top-down') level of analysis of gender equality implementation in the AFU, especially during the last ten years of the Russo-Ukrainian war. As a member of the sociological team conducting research on women's visibility and integration of Ukrainian women into the military sphere, I summarize the successes and challenges of this process. Finally, I pay attention to the societal attitude (and its dynamic) towards women's integration in the AFU. I use primary (all-Ukrainian representative for the controlled territories polls) and secondary (international reports and previous studies) data, as well as results of five extensive sociological studies in our project and advocacy campaign 'Invisible Battalion'.[9]

A Brief History of Women in War

As touched upon above, the presence of women in the armed forces and defence sector has a more than century-long history. During World War I, women from western Ukraine found themselves in the war zone and survived the occupation. In contrast, women from the Ukrainian provinces of the Russian Empire suffered from famine, unemployment, and the mobilization of men for war.[10] Historian Dr Martha Bohachevsky-Chomiak noted:

> Women in Galicia were not as pacifists as a significant number of Ukrainian intelligentsia in the territory of Ukraine, which was under Russian rule. On the contrary, when World War I began, Ukrainians in Galicia believed that one could only fight for

[6] Martsenyuk, 'Ukrainian women at war'; Ioulia Shukan, 'Gender roles in the rear of the war in Donbas: Women's engagement in the care of wounded combatants', in Katharina Bluhm, Gertrud Pickhan, Justyna Stypińska, and Agnieszka Wierzcholska (eds), *Gender and power in eastern Europe: Changing concepts of femininity and masculinity in power relations* (Princeton NJ: Springer, 2021), 83–105.

[7] Feminist Initiative Group, 'The right to resist. A feminist manifesto', *Commons*, 7 July 2022. https://commons.com.ua/en/right-resist-feminist-manifesto.

[8] Jessica Zychowicz, 'Introduction', in Jessica Zychowicz (ed.), *Freedom taking place: War, women and culture at the intersection of Ukraine, Poland, and Belarus* (Wilmington, DE: Vernon Press, 2023).

[9] For more information, see our project webpage: https://invisiblebattalion.org/en/home-2.

[10] Maria Guk, 'Women of Ukraine in the First World War: Historiography' (in Ukrainian). Dissertation for obtaining the scientific degree of candidate of historical sciences in the specialty 07.00.06 'Historiography, source studies, and special historical disciplines', Taras Shevchenko Kyiv National University, 2000.

> their rights with weapons. That is why proportionally more women from Western lands joined the army than women from Eastern Ukraine.[11]

The researcher also mentioned that the first woman to enter military service, that is, to join the army of the Ukrainian Sich Riflemen,[12] was Olena Stepaniv.[13]

In Soviet historiography, the role of women during World War Two was shown in a relatively limited way due to the influence of ideology and censorship.[14] The gender strategies of the Soviet regime had a dual focus: on the one hand, they promoted the emancipation of women and, on the other, they used them as an economic resource in a totalitarian political regime.[15] The active participation of women in World War Two was partly a result of previous mobilization strategies that involved women in paid work in heavy industry, sports, and aviation.[16] Such measures were part of the overall project of women's emancipation. At the same time, women were required to learn defensive skills and first aid. The latter two areas, although important, were related to reproductive labour, which largely persisted after the war.[17]

Since 1991, Ukraine has achieved considerable progress in establishing gender equality in various spheres of Ukrainian society. For the last ten years of war, the military sphere has become a strategic one for women to fight for their rights, gender equality, and visibility. Understanding Ukraine and the fierce resistance of the Ukrainian society in the Russo-Ukrainian war can be based on analytically deploying the gender perspective.[18]

At the same time, scholars have labelled the Russo-Ukrainian war 'a war on gender order', particularly the war of Russia against gender equality in Ukrainian society.[19] For example, scholar of international politics Dr Jenny Mathers points out:

> While Russia's emphasis on the masculine character of its armed forces and the marked absence of women soldiers are important demonstrations of Moscow's commitment to a traditional social order, the presence of growing numbers of

[11] Martha Bohachevska, *The Duma of Ukraine is female gender* (in Ukrainian) (Kyiv: Resurrection, 1993), 60.

[12] It was a Ukrainian unit within the Austro-Hungarian army during World War One.

[13] Olena Stepaniv (1892–1963) was an Austro-Hungarian and Ukrainian soldier, public figure, and economist. She is popularly known as the first female officer in the Ukrainian army.

[14] Olga Bakovetska, 'Ukrainian women in the wars of the 20th–21st centuries', *Young Scientist* 91/3 (2021): 152–6 (in Ukrainian).

[15] Kateryna Kobchenko, 'Women as a military resource of totalitarian power: Soviet wartime gender strategies', in Gelinada Grinchenko, Kateryna Kobchenko, and Oksana Kis (eds), *Women of central and eastern Europe in the Second World War: Gender specificity of experience in times of extreme violence* (in Ukrainian) (Kyiv: Art Book, 2015), 61–78.

[16] Ibid.

[17] Ganna Hrytsenko, 'How women changed the Ukrainian army' (in Ukrainian), *Gender in Details*, 10 October 2019. https://genderindetail.org.ua/season-topic/gender-after-euromaidan/yak-zhinki-zminili-ukrainsku-armiyu-1341190.html.

[18] Tamara Martsenyuk, '(De-)gendering Ukraine: Understanding social change in a post-Soviet society', *Symposium Sociological Perspectives on Eastern Europe, part 2. Soziologie* 52/4 (2023): 425–36.

[19] Petr Kratochvíl and Míla O'Sullivan, 'A war like no other: Russia's invasion of Ukraine as a war on gender order', *European Security* 323/3 (2023): 347–66.

women in Ukraine's military demonstrates Kyiv's commitment to a very different type of state and society.[20]

As for Ukraine, results of the public opinion surveys of Ukrainian society demonstrate that after Russia's full-scale invasion, it is moving to more egalitarian values (i.e. gender equality, respect for diversity, and support of the registered partnerships for same-sex couples).[21]

Ensuring Gender Equality in the Armed Forces of Ukraine

Since Independence in 1991, the idea of equal rights and opportunities for women and men has been increasingly viewed in Ukraine as a key priority.[22] First of all, Ukraine has ratified major international documents on ensuring equal rights and opportunities for men and women: the United Nations (UN) Convention on the Elimination of All Forms of Discrimination Against Women (CEDAW), UN Global Sustainable Development Goals 2016–30, and UN Security Council Resolution 1325 on women, peace and security, and others. Ukraine also adopted the second National Action Plan (NAP) to implement UN Security Council Resolution 1325 for 2021–5.

Ukraine's commitments are not limited to UN frameworks. In 2020, Ukraine formally became a full member of the Biarritz Partnership for Gender Equality, a global coalition championed by the French government to empower girls and women world-wide.[23] In the framework of the partnership, Ukraine, as a full participant, undertakes commitments in five areas: (1) the development of a barrier-free public spaces, friendly to families with children and low-mobility groups; (2) teaching children the principles of equality between women and men; (3) prevention of violence; (4) reducing the pay gap between women and men; (5) creating greater opportunities for men to care for children.

Two years later, in the summer of 2022, the Council of Europe Convention on Preventing and Combating Violence against Women and Domestic Violence (the Istanbul Convention) was ratified.[24] As deputy prime minister for European and Euro-Atlantic Integration, Olha Stefanishyna, said: 'It was the first time in history that a state undertook obligations under such an important Convention in wartime.

[20] Jennifer G. Mathers, 'Ukraine recruits women soldiers: Why doesn't Russia?', *New Lines Magazine*, 13 June 2023. https://newlinesmag.com/argument/ukraine-recruits-women-soldiers-why-doesnt-russi.

[21] Tamara Martsenyuk, 'To be or not to be: Attitudes of Ukrainian society about gender equality and diversity after Russia's invasion of Ukraine', in Oleksandr Pankieiev (ed.), *Narrative of Russo-Ukrainian war: A look within and without* (Stuttgart: ibidem Press, 2025), 105–12.

[22] Martsenyuk, 'Ukrainian women at war'.

[23] President of Ukraine, 'Ukraine has become a full member of the Biarritz Partnership', 12 September 2020. https://www.president.gov.ua/en/news/ukrayina-stala-povnopravnoyu-uchasniceyu-ptheartnerstva-biarric-63521.

[24] Council of Europe, 'Ukraine ratifies the Istanbul Convention', 18 July 2022. https://www.coe.int/en/web/istanbul-convention/-/ukraine-ratifies-the-istanbul-convention.

This step was part of our fight for the same values we are fighting on the frontline, claiming our independence'.[25]

The same year, on 23 June, the European Council granted Ukraine the status of candidate for accession to the European Union, which means that Ukraine should develop its gender politics further. As the state secretary of the Ministry of Defence of Ukraine, Liudmyla Darahan, emphasized: 'Ukraine is demonstrating systemic success in incorporating inclusive strategies. Progress in addressing gender inequality, promoting women's active participation in diverse public domains, and incorporating inclusive principles, including within the Ministry of Defence, are essential for European integration.'[26]

As for gender equality implementation in the security and defence sphere, this process started after the 2004 Orange Revolution as a part of general state gender policy.[27] Natalya Dubchak, the captain of the AFU and former gender advisor to the minister of defence of Ukraine, wrote in 2008 that, traditionally, the AFU used to be the most conservative social institution on the issue of women staying in the military. In contrast, almost 10% of the AFU are women.[28] At this time, there were problems of both vertical and horizontal gender segregation in the armed forces as a labour market, with sexism and a paternalist perception of women in the security and defence sphere.

The situation with women's integration on the legislative level started to change faster after the annexation of Crimea and the beginning of the war in Donbas 2014. For example, in May 2015, the government of Ukraine adopted the Strategy of National Security of Ukraine, including the goals of affirming human and citizen rights and freedoms; ensuring economic, social, and humanitarian development; ensuring Ukraine's integration into the European Union; and meeting the conditions for entering North Atlantic Treaty Organization (NATO).[29]

In February 2016, Ukraine made further progress by adopting a NAP to implement UN Security Council Resolution 1325, 'Women, peace, security' by 2020.[30] As part of the implementation the aim of strengthening women's role in resolving conflicts and peace building in countries afflicted by wars and security issues was emphasized.

The Ukraine Ministry of Defence established a working group to implement Resolution 1325, and the corresponding action plans were developed. The same year, an action plan for implementing Resolution 1325 was developed and approved by

[25] Ukrainian Women's Fund, *Ukraine is not silent: Chronicles of fighting against war-related sexual violence (2022–2024)* (Kharkiv: Folio, 2025), 9.

[26] Ministry of Defence of Ukraine, 'There are 68 000 women enlisted'.

[27] Tamara Martsenyuk et al., *The 'Invisible Battalion': Women in ATO military operations in Ukraine. Sociological research* (Kyiv: Ukrainian Women's Fund, 2015), 18–19.

[28] Natalia Dubchak, 'Women in the armed forces of Ukraine: Problems of gender policy' (in Ukrainian), *Strategic Priorities* 9/4 (2008): 187–92.

[29] Verkhovna Rada of Ukraine, '"On resolution of the Council for National Security and Defence of Ukraine" as of May 6, 2015 "On Strategy of National Security of Ukraine": approved by the Decree of the President of Ukraine'(in Ukrainian), No. 287/2015. http://zakon3.rada.gov.ua/laws/show/287/2015.

[30] Cabinet of Ministers of Ukraine, 'On approval of the National Action Plan for implementing UN Security Council Resolution 1325 "Women, peace, security" for the period till 2020'(in Ukrainian), Decree as of 24 February 2016, No. 113-p. https://www.kmu.gov.ua/npas/248861725.

the national police. Since then, the Ministry of Internal Affairs, the national police, the state emergency service, and the National Guard have all been implementing the action plan for Resolution 1325.

In March 2017, the Cabinet of Ministers of Ukraine adopted a decree, assigning the deputy prime minister for EU and Euro-Atlantic integration to coordinate the state's gender equality policy. Furthermore, the position of the government commissioner for gender policy was introduced the same year. This position improved coordination between executive bodies to practically implement the principle of gender equality in all spheres of life and in the military.

The following year, Law #2523, 'On Amendments to Certain Laws of Ukraine Concerning the Provision of Equal Rights and Opportunities for Women and Men Throughout their Service in the Ukrainian Armed Forces and Other Military Formations', was adopted. This law enabled women to serve on equal grounds with men. In particular, it addresses enlistment by contract (voluntarily) and military draft, active service, reserve service, and observance of the military registration rules.[31]

For a long time, there was a problem with women's access to various officer positions. By 2022, the List of Military Occupational Specialties of Officers determined officer positions that women could assume. The position list was constantly updated but, from 10 February 2022, all officer positions in the UAF became available to servicewomen.[32] Moreover, women can now enter higher military educational institutions for any specialty. This huge step increases the army's professional level and brings it closer to world levels as Ukrainian women are given the same opportunities for a military career as men.

Increasing the ability of the defence sector to react to discrimination was a key priority of the government.[33] The institutional mechanism for gender equality implementation in Ukraine includes, among many state bodies, over ninety gender advisers, including about fifty in the Ministry of Defence system.[34] In general, the topic of women's access to the armed forces became part of the political agenda. The Equal Opportunities Inter-factional Caucus was one of the agents advancing it as part of the so-called 'top-down' level of solving the issue (via legislative mechanisms and by promoting political will).

On 24 February 2022, general mobilization of all men was announced in Ukraine. President of Ukraine Volodymyr Zelenskyy also declared martial law according to which, all men between the ages of 18 and 60 were considered liable for military service. Ukrainian men are forbidden to leave the country so that they can be called up to fight at any time. While many Ukrainian men have answered the call to serve,

[31] Tamara Martsenyuk et al., *The 'Invisible Battalion 2.0': Women veterans returning to peaceful life. Sociological research* (Kyiv: UN Women Ukraine, 2019), 6.

[32] Invisible Battalion 5.0, *Monitoring of* Recommendations, 22.

[33] CEDAW, 'Ninth periodic report submitted by Ukraine under article 18 of the Convention', 19 October 2022. https://www.ohchr.org/en/press-releases/2022/10/experts-committee-elimination-discrimination-against-women-commend-ukraine.

[34] NIRAS, 'Ukraine: Country gender profile', EU 4Gender Equality Reform Help Desk, European Union, 2023, 27. https://euneighbourseast.eu/wp-content/uploads/2023/09/eu4genderhelpdesk_ukrainegenderprofile_2023-cgp_v3.pdf.

some others have tried to evade conscription.[35] They dodge draft orders, hide, or escape the country. Some scholars have a rather critical perspective on Ukraine's mobilization policy after Russia's full-scale invasion. For example, Jessica Trisko Darden concludes that 'Ukraine's unequal approach to military mobilization harms women's position in society and does damage to men',[36] focusing on the gendered implications of Ukraine's mobilization policy, particularly regarding women's status as (non-)combatants. Women combatants still face gender stereotypes and discrimination also due to their mainly volunteer status and wish to serve, and lack of state politicy towards women's full integration into AFU. Jessica Trisko Darden underlines: 'The declaration of martial law reinforces the ambiguous position of women who want to serve. By holding men captive, martial law has undervalued women's potential contributions to the war effort and reinforced traditional gender norms that position women primarily as victims of, rather than participants in, armed conflict'.[37]

I here would like to pay attention to the fact that according to the law of Ukraine 'On Ensuring Equal Rights and Opportunities for Women and Men',[38] article 6, 'compulsory military service for men, set forth by law' is not considered as discrimination based on sex. So, according to this legislation, military service is a duty for men and a right for women.

Sociological Research and the Advocacy Campaign 'Invisible Battalion'

> The duty of every citizen is to defend the territorial integrity of the state. (Female respondent 23)[39]
>
> 'There is no place for women at war, but you're a brilliant exception.' It was said by some officers, it was said by other officers, it was said by all those with whom I had to interact. (Female respondent 38)[40]

When the war in Donbas began in 2014, women actively joined the volunteer battalions, including occupying several combat positions. During the first year of the war (2014–15), some military women, including aerial reconnaissance drone operator and National University of Kyiv-Mohyla Academy (NaUKMA) graduate Maria Berlinska, decided to draw attention to the issue of gender discrimination in

[35] Constant Méheut, 'As Ukraine expands military draft, some men go into hiding', *New York Times*, 21 June 2024. https://www.nytimes.com/2024/06/21/world/europe/ukraine-war-draft-dodgers-conscription.html.

[36] Jessica Trisko Darden, 'Ukrainian wartime policy and the construction of women's combatant status', *Women's Studies International Forum* 96 (January–February 2023). https://doi.org/10.1016/j.wsif.2022.102665.

[37] Ibid.

[38] Verkhovna Rada of Ukraine, 'Law of Ukraine No. 2866-IV "On ensuring equal rights and opportunities for women and men"', 8 September 2005. http://gender.cawater-info.net/publications/pdf/ukr1.pdf.

[39] Martsenyuk et al., *The 'Invisible Battalion'*.

[40] Ibid.

the AFU. In the spring of 2015, Maria Berlinska approached me with a proposal to conduct a sociological study on this topic.[41] The sociological research team also included two other graduates of the NaUKMA Department of Sociology—Anna Kvit and Hanna Hrytsenko.

As a result, since 2015, this research team (since 2018, together with lawyers from the non-government organization (NGO), Legal Hundred)[42] has conducted five sociological studies on the situation of women in the AFU:[43]

- 'Invisible Battalion: Women's Participation in Military Operations in the Anti-Terrorist Operation' (ATO) (2015),
- 'Invisible Battalion 2.0: Return of Female Veterans to Peaceful Life' (2019),
- 'Invisible Battalion 3.0: Sexual Harassment in the Military Sphere in Ukraine' (2021),
- 'Invisible Battalion 4.0: Women's Access to Military Education in Ukraine' (2022),
- 'Invisible Battalion 5.0: Monitoring Recommendations and Research Results from the "Invisible Battalion" Cycle' (2023).

The study in 2015 conducted, an empirical survey of forty-two women fighting in the ATO confirmed the main problems: horizontal and vertical gender segregation, the paternalist approach towards women's integration into the military sphere of Ukraine, and lack of gender sensitivity.[44] Women were almost not allowed to make decisions in the AFU. Many were not officially employed and could not receive remuneration for their work. Consequently, they will be deprived of the benefits the state grants, including the status of combat participant and all the benefits that go with it. Another problem was that some officially employed women were registered as holding positions different from those they held. They faced limitations, and the explanation was that the staffing table did not include these positions for women. Moreover, some issues not discussed by the state were everyday issues: lack of gender-sensitive medical care for women, appropriate sizes of uniforms and shoes, and satisfactory living conditions. The infrastructure of the AFU was adjusted to the needs of men, and women with specific needs were excluded from the army. That proved the lack of gender sensitivity in women's integration and the lack of understanding of the needs of military women.

Even the first study's results were supported by military women who decided to fight for their rights and promote gender equality in the AFU. These studies exemplify public sociology and the so-called 'bottom-up' in regulating societal gender relations. Together with another mechanism—'top-down'—legislative changes managed to change the situation in Ukraine in terms of ensuring equal rights and opportunities in the AFU.

[41] More information about Maria Berlinska is at: https://en.wikipedia.org/wiki/Maria_Berlinska.
[42] See: https://legal100.org.ua/en.
[43] All texts can be found at the project's webpage: https://invisiblebattalion.org/en/invisbat-2/.
[44] Martsenyuk et al., *The 'Invisible Battalion'*.

In 2016, the 'Invisible Battalion' advocacy strategy for 2016–17 was developed to achieve gender equality in the AFU and the Ministry of Internal Affairs of Ukraine. Military women actively advocated for their rights in the AFU. For example, in January 2016, in Kyiv, activists for equal rights and opportunities for women and men in the AFU marched to the Ministry of Defence of Ukraine as part of the 'March of the Invisible Battalion' campaign.

Among the first results of this campaign was the Decree of the Ministry of Defence of Ukraine No. 292 as of 3 June 2016, amending the temporary list of staff positions of soldiers and petty officers.[45] It considerably broadened the list of combat positions for servicewomen and increased them by more than a hundred combat military vocations. This allowed women to openly serve in the Ukrainian army, in roles such as scouts, snipers, and commanders of combat vehicles.

In November 2016, Maria Berlinska also created an NGO called the Institute of Gender Programs, which focuses on the issues of gender equality in the security and defence sector.

In 2017, a Ukrainian documentary film almanac about women in combat conditions in eastern Ukraine, 'The Invisible Battalion', was filmed.[46] It is about six stories of female soldiers, told by three Ukrainian directors: Iryna Tsilyk, Svitlana Lishchynska, and Alina Gorlova. Another documentary film about women in war and rehabilitation problems after the ATO is 'No Obvious Signs', directed by Alina Gorlova.[47]

These and other successes in ensuring equal rights and opportunities in the AFU, as well as particular challenges, are summarized in Table 13.1 .

Awareness of gender equality and the ability to recognize sexual harassment, counteract it, and deal with the consequences are essential aspects of countering gender-based violence in the security and defence sector. Dignity and professionalism should be the values of the armed forces, and they are promoted through education and training.

Organizations that represent the 'bottom-up' level play an important role in working with gender-based violence in the AFU. For example, statistical information about cases of sexual harassment in the military service is recorded on their hotlines by public organizations, such as the All-Ukrainian Human Rights Organization—Legal Hundred NGO, the Ukrainian Women Lawyers Association JurFem, and La Strada-Ukraine NGO.[48] The Ukrainian Women Lawyers Association 'JurFem' has developed a memo for servicewomen in case they encounter sexual violence during military service. The memo was distributed in cooperation with the Ukrainian Women Veteran Movement NGO.

[45] Martsenyuk et al., *The 'Invisible* Battalion', 32.

[46] 'Invisible Battalion', film, 2017, I. Tsilyk, S. Lishchynska, and A. Horlova (dirs.). https://invisiblebattalion.org/mediahub/#films.

[47] 'No obvious signs', film, 2018, A. Horlova (dir.). https://takflix.com/films/no-obvious-signs.

[48] 'Invisible Battalion 3.0', in Tamara Martsenyuk (ed.), *Sexual harassment in the military sphere in Ukraine: Sociological research* (Kyiv: Institute of Gender Programs, 2021), 51.

Table 13.1 Ensuring gender equality in the AFU: Successes and challenges

Aspect	Successes	Problems That Remain Unresolved
Institutional mechanism	The presence of a gender policy and a system of gender advisors in the AFU. Legislative attention ensures gender equality in the military sphere.	Sustainability of the mechanism and realism of implementation. In some situations, there is a lack of political will for successful implementation.
The AFU as a labour market	Overcoming vertical gender segregation in the AFU will result in more women in status positions at the decision-making level. Overcoming horizontal gender segregation in the AFU, opening combat military specialties for women. Moving away from paternalistic policies towards women, the list of 450 prohibited professions for civilian women was eliminated.	There is a lack of women at the military decision-making level and among the leadership of the AFU. The maximum lifting standards for women remain relevant. Implementing a more gender-sensitive approach to integrating women into the labour market is worth it.
Gender-based violence in the AFU	Ratification of the Istanbul Convention. Training programmes and online courses have been introduced (e.g. the online course 'Gender Equality and Combating Sexual Harassment in the Military'). Civil society organizations provide information to victims through a hotline mechanism.	Available statistics on sexual harassment in the AFU do not reflect the real number of cases, as victims do not always report incidents of harassment and violence against them, and there is a lack of political will to investigate these cases. The issue receives little media coverage.
Representation of female military personnel and veterans in the media	Increased visibility of women in the AFU and increased representation of their current problems. The topic of female servicewomen began to be raised in cinema. A shift away from sexist rhetoric by Ukrainian public figures regarding female military personnel. Public figures (including the president of Ukraine) increasingly recognize women's contribution to the war.	There is a lack of women among the military experts consulted by the media. There is little coverage of women's experiences returning from war.

Note: Based on the results of five sociological studies from the 'Invisible Battalion' project, 2015–23.

In 2021, the Prometheus platform introduced a course called 'Gender Equality and Counteracting Sexual Harassment in the Military Sector'.[49] The NGO 'Institute of Gender Programs' developed the online course within the information campaign 'Invisible Battalion 3.0: Sexual Harassment in the Military Sphere in Ukraine'. It is designed for the personnel of the security and defence sector of Ukraine, female and male veterans, female and male military journalists, female and male employees of the military education area, and all those interested in gender equality promotion in the military sector. It is essential to add that the Department of Military Education and Science of the Ministry of Defence recommended that higher military educational institutions and military structural units of higher education institutions include it in their curricula for cadets.[50]

War-related gender-based and sexual violence is a special kind of violence that is not only directed at one person but that affects entire communities and society as a whole. To provide comprehensive assistance to survivors, a network of relief centres was established with the support of international partners. These centres provide various psychological, social, and legal services. A Platform for Assistance to Survivors was established to guarantee everyone who needs support access to services. It is the first and only online platform in Ukraine with full information on all the main services assisting people affected by the war.[51]

Public Opinion on Ensuring Gender Equality in the Armed Forces of Ukraine

To understand the attitudes to the involvement of women in the security and defence sectors and issues of equal rights and opportunities, it is important to monitor public opinion constantly. Even though political will is crucial to implementing gender equality at an institutional level, societal attitudes towards diversity and dignity are also essential factors.[52]

In this chapter, I analyse quantitative data to assess public opinion on equal rights and opportunities in the military during the last six years of the Russo-Ukrainian war. The data were obtained as a result of three representative surveys by Kyiv International Institute of Sociology (KIIS) in 2018 and the research agency 'Info Sapiens' in 2023, 2024 (see Table 13.2).

The limitations of this study are the lack of information from territories not controlled by Ukraine, as well as the departure of a significant number of the population abroad, and large-scale movements within Ukraine after 24 February 2022. In addition, this is not a panel survey where attitudes are asked from the same people.

[49] 'Gender equality and counteracting sexual harassment in the military sector, Online course on the Prometheus platform, basic level. https://courses.prometheus.org.ua/courses/course-v1:Prometheus+GE101+2021_T2/about.

[50] 'Invisible Battalion 3.0', 54.

[51] Ukrainian Women's Fund, *Ukraine is not silent*, 8–9.

[52] Martsenyuk, 'To be or not to be', 105–12.

Table 13.2 Empirical data on the attitude of Ukrainian society to equal rights and opportunities in the military sphere

Year	2018	2023	2024
Research agency	KIIS (Kyiv International Institute of Sociology)	Info Sapiens	Info Sapiens
Dates	8–23 September	11–23 January	11–19 February
General population	Population of Ukraine aged 18 and over, except for the Autonomous Republic of Crimea and the occupied territories in Donetsk and Luhansk regions	The population of Ukraine aged 16 and over, except the Autonomous Republic of Crimea, Luhansk Oblast, and certain temporarily occupied territories of Donetsk, Kherson, Zaporizhia, Mykolaiv, and Kharkiv Oblasts, where there is no connection.	
Sample	2026	1,000	
Representativeness error	< 3.3% with a probability of 0.95%.	< 3,1% with a probability of 0.95%.	
Method	Personal formalized interview (face-to-face)	CATI (computer-assisted telephone interviews)	

Thematic questions were selected from the 2018, 2023, and 2024 surveys, and questions from sociodemographic blocks were also used: gender, age, and level of education, which are most often used to test their influence on attitudes towards a certain phenomenon. IBM SPSS Statistics software was used for further data processing.

In 2018, more than half of those surveyed in Ukraine expressed support for providing women with equal opportunities with men to serve in the AFU and other military units (see Table 13.3). However, support has increased significantly over the past four and a half years. In 2023, after the full-scale invasion, almost half of those surveyed fully supported this idea. According to data from January 2023, half of the surveyed respondents fully agreed with the statement that women should be given equal opportunities with men to work in the AFU and other military formations.

This suggests that support for egalitarian principles becomes more common in military conflict, as all human resources are considered essential to military activity. Support for such a view strengthens the military readiness of society while simultaneously promoting military and democratic ideals.[53]

[53] Julie Collins-Dogrul and Jaimis Rebecca Ulrich, 'Fighting stereotypes: Public discourse about women in combat', *Armed Forces & Society* 44/3 (2018): 436–59.

Table 13.3 Dynamics of the attitude of Ukrainian society to the statement that women should be given equal opportunities with men to work in the AFU and other military formations

Variants of Answers	KIIS, N=2026, September 2018 (%)	Info Sapiens, N=1,000, January 2023 (%)	Info Sapiens, N=1,000, February 2024 (%)
I agree completely	24.0	50.0	26.0
I rather agree than disagree	29.3	33.8	31.6
I rather disagree than agree	18.9	8.4	22.6
I completely disagree	12.2	5.8	16.5
Difficult to say	14.3	1.6	2.7
Refusal to answer	1.4	0.4	0.6
Total	100.0	100.0	100.0

In 2024, most Ukrainians continue to support equal rights and opportunities in the military sphere, albeit to a lesser extent than in 2023. Instead, we see that after a full-scale invasion, the share of those who find it difficult to answer this question significantly decreases. The realities of a full-scale war and active public discussions about mobilization, in particular, encourage women to decide on this issue.

Thus, Ukrainian society changed its level of support for equal rights and opportunities in the military from 53% in 2018 to 84% in 2023 and 58% in 2024. However, from 2023 to 2024, there is a decrease in the level of support for egalitarian ideas regarding the involvement of women in the military. This may be because the 2023 survey was conducted approximately a year after the full-scale invasion, when anger and fear had converted into general resistance, and most people were ready to fight for democracy. In 2024, there is an increase in fear for life, pessimism about war, and the so-called phase of 'war fatigue'. The constant threat to security, human losses, air raids, shelling, power outages, victories, and defeats in both the political and military spheres act as factors in fatigue.

The results of the latest survey in 2024 are convincing that gender differences in attitudes towards the fact that women should be given equal opportunities with men to work in the AFU and other military formations are not significant (see Table 13.4). The vast majority—about 59% of surveyed men and 56% of women—agree with this statement. On the other hand, 37% of men and 41% of women disagree. Generally, the differences in agreement or disagreement (if the answers are combined) are no more than 5%. The most significant differences are visible in that men are 10% more likely to completely agree with the egalitarian attitude towards involving women in the AFU.

Education level is the only sociodemographic factor significant in the studies (2023 and 2024). The higher the level of education of the respondents, the greater the level of support for the statement that women should be given equal opportunities with men

Table 13.4 Attitudes by gender to the statement that women should be given equal opportunities with men to work in the AFU and other military formations

Variants of Answers	Men (%)	Women (%)	Total (%)
I agree completely	30.9	21.7	26.3
I rather agree than disagree	28.2	34.7	31.5
I rather disagree than agree	19.6	25.4	22.5
I completely disagree	18.0	15.3	16.7
Difficult to say	2.4	2.6	2.5
Refusal to answer	0.9	0.4	0.7
Total	100	100	100

Note: Info Sapiens, N=1,000, February 2024.

to work in the AFU and other military formations. For respondents with a higher level of education, it is more critical to support egalitarian values of equal rights and opportunities. In 2024, more than 62% of Ukrainians with higher education and about 50% with complete secondary education supported the proposed statement.

An important question is also about what kind of the army should be introduced in Ukraine (see Table 13.5). Regardless of the survey year, the most popular answer is the conscription of both men and women only on a voluntary, contract basis. Over 60% of Ukrainians chose this option in 2018, 2023, and 2024. By 2024, the share of those polled who were inclined to support the conscription of only men decreased to 9% from 17%.

The current situation in Ukraine—the draft of all men and only some women—is supported by only around 11–14% of Ukrainian society. The least popular option in all three periods is the total conscription of all men and all women. In 2018, only 1.6% of Ukrainians supported it. In 2024, it rose only to 3.8%. So, support remains robust for enlistment over conscription and equal opportunity for women and men to enlist.

Table 13.6 presents gender-specific attitudes towards the appropriate army option that are not large and insignificant. In 2024, women were more likely than men (68% vs 63%) to choose to draft both men and women voluntarily (under contract). Ukrainian men, on the other hand, are more likely to choose to draft all men and all women (5% vs 3%), i.e. the Israeli army option.

Finally, it is advisable to learn about the dynamics of the (un)willingness of Ukrainians and Ukrainian women to engage in armed resistance in recent years after the full-scale invasion. As seen from Table 13.7, in 2024, there is an overall decrease in readiness among men and women.

If in 2023, 77% of men and 64% showed readiness to engage in armed resistance, then a year later, in 2024, the share of those ready dropped to 60% of men and 44% of women. However, it is important to note that in 2024, the vast majority—51% of

Table 13.5 In your opinion, what kind of army should Ukraine introduce?

Variants of Answers	KIIS, N=2026, September 2018 (%)	Info Sapiens, N=1,000, January 2023 (%)	Info Sapiens, N=1,000, February 2024 (%)
Enlistment of both men and women only voluntarily (enlistment by contract)	61.9	65.9	65.7
Conscription of all men and only of some women	10.6	12.8	14.2
Conscription only of men	17.0	12.0	8.9
Conscription of all men and all women	1.6	3.3	3.8
Other	0.4	4.0	1.4
Difficult to say	7.4	1.9	3.6
Refusal to answer	1.1	0.1	0.9
Conscription of all men and women upon request			1.5
Total	100	100	100

respondents—are ready for armed resistance. The share of both men and women who are already resisting has also increased; in 2024, there will be more than 4% of respondents. It should be noted that the choice of these answers by male and female respondents does not mean that they will necessarily act accordingly. However, we see a decrease in readiness for armed resistance to the enemy, which can also be considered a manifestation of the 'war fatigue' phenomenon mentioned above.

Conclusions

Ukraine ensures equal rights and opportunities in society and the military sphere. Faced with the challenges of war, the country has taken on international obligations to improve gender policy, including in the AFU. The integration of women into the Ukrainian armed forces has been significantly influenced by the broader Women, Peace, and Security (WPS) agenda, which emphasizes the importance of women's roles in security forces. The government of Ukraine demonstrates the political will to ensure the rights of women in the military sector by approving the second NAP for the Implementation of UN Security Council Resolution 1325, 'Women, Peace, and Security', and integrating gender equality issues into national programmes to adapt the security and defence sector of Ukraine to the norms and principles of NATO, etc. After Russia's full-scale invasion, more attention has been paid to the issue of

Table 13.6 In your opinion, what kind of army should Ukraine introduce? Distribution of responses by gender

Variants of Answers	Men (%)	Women (%)	Total (%)
Enlistment of both men and women only voluntarily (enlistment by contract)	62.7	68.1	65.7
Conscription of all men and only of some women	14.2	14.3	14.2
Conscription only of men	9.9	8.1	8.9
Conscription of all men and all women	5.0	2.9	3.8
Conscription of all men and women upon request	1.9	1.2	1.5
Other	2.0	0.9	1.4
Difficult to say	3.1	4.0	3.6
Refusal to answer	1.2	0.6	0.9
Total	100	100	100

Note: Info Sapiens, N=1,000, February 2024.

Table 13.7 Are you personally ready or not ready to put up armed resistance to stop the Russian occupation of Ukraine?

Variants of answers	Info Sapiens, N=1,000, January 2023 (%)			Info Sapiens, N=1,000, February 2024 (%)		
	Men	Women	Total	Men	Women	Total
Ready[a]	77.41	64.29	70.23	59.68	44.18	51.21
Not ready[b]	15.78	30.03	23.59	27.67	48.26	38.92
I am already resisting	5.00	0.12	2.33	7.89	1.48	4.39
Don't know / Difficult to say	1.82	5.54	3.86	4.76	6.08	5.48
Total	100	100	100	100	100	100

Notes: [a]Readiness refers to the combined response options 'Completely ready' and 'Rather ready'.
[b]Not readiness refers to the combined response options 'Completely not ready' and 'Rather not ready'.

combating war-related sexual violence, to which all categories of the population are vulnerable, among them servicewomen.

A series of sociological studies and the advocacy campaign 'Invisible Battalion' are examples of grassroots activist and research initiatives to improve the position of women in the AFU and their greater visibility in society. The results of five studies conducted during the last eight years of the war draw attention to the successes and challenges of ensuring gender equality in the AFU.

Among the successes, it should be noted the legislative attention to ensuring gender equality in the military sphere, the gradual overcoming of the phenomenon of vertical (more women hold higher military ranks) and horizontal (more women in combat specialties) gender segregation in the military sphere, the departure from paternalistic gender policy, and the provision of formal access for women to all combat specialties and military education.

Instead, it is worth solving such urgent problems as combating and counteracting gender-based violence and sexual harassment in the AFU, investigating these cases, and developing a sustainable professional mechanism to deal with these phenomena, as well as infrastructural integration and greater visibility of women's needs in the AFU (a gender-sensitive approach). The lack of sustainability of gender policy mechanisms and, in some cases, political will to implement them, the relevance of the 'glass ceiling' issue in the AFU as a labour market, and the lack of visibility of women among the military elite and expert military environment are other important issues that require attention.

If we summarize the results of the public opinion poll, we see that after the invasion, egalitarian tendencies towards equal rights and opportunities for women and men in the military sphere, as well as public expectations regarding the involvement of women in the AFU, have intensified. As of the beginning of 2023, the most popular option remains a professional contract army with the conscription of men and women voluntarily. In addition, we have the readiness of Ukrainian society (not only the majority of men but also women) for armed resistance to the enemy. At the same time, the results of 2024 demonstrate the phenomenon of 'war fatigue' that is quite expected after two years of full-scale war.

In the mobilization challenges that modern Ukrainian society and the authorities face, it is essential to consider the desire of some women to defend Ukraine. These women should have decent conditions in service, taking into account a gender-sensitive approach to the integration of women in the AFU, which still has unresolved issues.

PART III
CASE STUDIES

14

Russia's War on Ukraine

Lessons in Hybrid Defence and Conventional Warfare (2014–2025)

Omar Ashour and Hanna Shelest

Outgunned but Unbreakable: Ukraine's Roadmap to Hybrid Defence and Strategic Resilience

In an era defined by asymmetric warfare, understanding the mechanisms of survival and success of the disadvantaged, smaller forces is critical for military strategists and policy makers alike.[1] This chapter delves into operational art, sampling tactics deployed in the Russian war on Ukraine from 2014 to 2024. It primarily aims to extract lessons and offer critical insights for combat effectiveness, military effectiveness, and defence policies, particularly for the small(er) states in Europe and beyond.

The chapter argues that the Ukrainian Security and Defence Forces[2] employed a 'hybrid defence'[3] strategy (HDS), featuring tactics, techniques, and procedures

[1] Sections of this chapter are a part of a broader research agenda of Professor Omar Ashour focused on investigating how small(er) forces adapt, innovate, and persevere against superior adversaries across the globe. See for example: Omar Ashour, 'Hybrid defence for small(er) states: Lessons in combat effectiveness from Ukraine to Gaza', Keynote address presented at the Strategic Studies Unit Fifth Annual Conference, 'Defence strategies of small states and smaller forces: Tactical adaptations and operational innovations', Centre for Research and Policy Studies, Doha, Qatar, 23 February 2025. https://www.youtube.com/watch?v=__djCAJidL0&t=3755s; Omar Ashour, 'Hybrid defence for small states: Strategic lessons and operational implications from Ukraine and Gaza (2014–2024)', *Siyasat Arabiya* 12/71 (November 2024): 31–51. On non-state force employment of hybrid defence, see Omar Ashour, 'Security, oil, and internal politics: The causes of the Russo-Chechen conflicts', *Studies in Conflict & Terrorism* 27/2 (2004): 127–43; Omar Ashour, *How ISIS fights: Military tactics in Iraq, Syria, Libya and Egypt* (Edinburgh: Edinburgh University Press, 2021).

[2] The term 'Security Forces' is used here to encompass institutions under the Ministry of Interior—such as the national guards of Ukraine (NGU)—as well as forces affiliated with the Security Service of Ukraine (SBU) and other institutions, including the State Border Guard Service of Ukraine (DPSU).

[3] 'Hybrid Defence' is defined in the next section of this chapter. For more on the definition of the concept and its grounding in the literature, see Ashour, 'Hybrid defence for small(er) states'; Omar Ashour, 'How Hamas fights: Hybrid defence and combat effectiveness in Gaza (2023–2024)', *Studies in Conflict & Terrorism* (forthcoming in 2025). In doctrinal terms, hybrid defence (HDS) and asymmetric warfare intersect with—yet are not identical to—the North Atlantic Treaty Organization (NATO)-adjacent 'Resistance Operating Concept' (ROC) manual attributed to Otto Fiala (2020). HDS (Ashour) is a denial-centric, society-in-arms posture: multi-domain intelligence, surveillance, and reconnaissance

Omar Ashour and Hanna Shelest, *Russia's War on Ukraine*. In: *Russia's War in Ukraine and Modern Warfare*. Edited by: Mikael Weissmann and Niklas Nilsson, Oxford University Press.
DOI: 10.1093/9780198970903.003.0014

(TTPs) in four or more domains of operations. This has enhanced both their military and combat effectiveness and compensated for asymmetries in force sizes, force equipment, and initially in force quality (during specific operations). The overall impact of the HDS on attrition remains inconclusive in this particular case. Whereas the strategy permitted the smaller force to militarily (not just politically) attrite the larger one, the overall results of attrition remained acceptable to the larger force up to the time of writing of this chapter.

The chapter is divided into five sections. The following section provides a brief history of the war as a background for the analysis. The second and third sections reintroduce the concept of *hybrid defence* and relates it to the two concepts of *military effectiveness* and *combat effectiveness*. It analyses the trajectory of Ukraine's hybrid defence adaptations based on the phases of the war between 2014 and 2025. The fourth section focuses on selective tactics employed by Ukraine and Russia in multiple domains of operations within an asymmetric war. The fifth and final section concludes with key lessons for the smaller forces, thus outlining strategic implications for the defence policies of Ukraine and, more generally, of small(er) states.

A History of the War

While most media date the Russian war against Ukraine to February 2022—after largely eschewing the term 'war' for nearly a decade—this is incorrect from both political-legal and military-operational perspectives. The absence of a classical declaration of war is immaterial; the conflict's onset is February 2014 in Crimea and April 2014 in eastern Ukraine, as recognized in proceedings before the International Court of Justice (ICJ)[4] and the European Court of Human Rights.[5]

Ukrainian and Russian periodization of the 2014–25 war diverge. From the Ukrainian military perspective, the conflict comprises three phases: 2014–18 (Anti-Terrorist Operation, or ATO); 2018–22 (Joint Forces Operation, or JFO); and 2022–present (full-fledged invasion/war). From the Russian perspective, two periods are asserted: 2014–22—officially framed as 'limited involvement', i.e. a *de facto* proxy

(ISR)-to-fires integration, Counter-Unmanned Aerial Systems (C-UAS) / electronic warfare (EW) dominance in the air-littoral, dispersion/deception, hardened redundant command and control (C2), and mobilization of reserves/territorials and partisan networks to break an invader's kill web and hold terrain. Asymmetric warfare is the broader method of war available to strong or weak leveraging dissimilarity to exploit adversary seams and invert cost exchange, including offensive action. The ROC (Fiala) operationalizes the societal component of this ecology—before and under occupation—by specifying structures (underground-auxiliary-guerrilla), pre-crisis preparation (legal authorities, continuity of government, caches, comms redundancies), special operations forces (SOF) enablement, decentralized / mission command, clandestine, logistics, and information/psychological operations.

4 'Application of the international convention for the suppression of the financing of terrorism and of the international convention on the elimination of all forms of racial discrimination (Ukraine v. Russian Federation)', Judgment, International Court of Justice, 31 January 2024. https://www.icj-cij.org/sites/default/files/case-related/166/166-20240131-jud-01-00-en.pdf.

5 'Ukraine v. Russia (re Crimea) [GC]—20958/14 and 38334/18. Judgment 25.6.2024 [GC]', Legal summary, European Court of Human Rights, June 2024. https://hudoc.echr.coe.int/eng?i=002-14347.

war without formal acknowledgement of Russian regulars in the occupied territories; and 2022–present, the so-called 'special military operation', with explicit acknowledgement of direct Russian military involvement.

In April 2014, in response to Russia's annexation of Crimea, Ukraine's acting president, Oleksandr Turchynov, announced an 'Anti-Terrorist Operation with involvement of the Armed Forces' to prevent Russia from 'repeating the Crimean scenario in the Eastern region of Ukraine'.[6] It was launched under the Law on Fighting Terrorism, with the command of the security services' Anti-Terrorist Centre. As such a situation did not allow them to utilize all capabilities and comprehensive use of the defence forces, in 2018, the anti-terrorist operation was replaced with a Joint Forces Operation (JFO) under military command, through a new 'Law on De-Occupation'.[7] In part, this was a recognition that the adversary were primarily regular Russian military personnel, as opposed to so-called 'separatists', with an advanced command and control system.[8] The goal of the JFO was defined as to liberate the occupied territories.[9] JFO officially ended on 24 February 2022. The reason—it had a limited geographical cover, and the Russian full-fledged invasion demanded a lifting of this frame.

The Russian periodization since 2014 will see a certain dichotomy. On the one hand, all its actions fall under the 'special military operation' term, which significantly influenced the command and control structure, mobilization of personnel procedures, and deployment. On the other hand, till 2022, Russian officials had not recognized Russian forces' direct involvement; thus, covert operations and a proxy war dominated regular warfare. The stationing of the Russian forces in Crimea in the winter of 2014 made it legally more difficult to identify the exact start date of aggression. However, a number of its actions clearly fall under the conditions of the United Nations (UN) General Assembly resolution 3314 (XXIX)[10] dated 14 December 1974, in particular: (a) The invasion or attack by the armed forces of a state on the territory of another state, or any military occupation, however temporary, resulting from such invasion or attack, or any annexation by the use of force of the territory of another state or part thereof; (c) The blockade of the ports or coasts of a state by the armed forces of another state. The act of annexation was officially recognized by President Putin and confirmed by the Treaty on Acceptance of Crimea to the Russian

[6] 'РНБО розпочинає масштабну АТО із залученням Збройних Сил—Турчинов [NSDC start full-scale anti-terrorist operation involving the armed forces—Turchynov]', *Ukrainian Pravda*, 13 April 2014. https://www.pravda.com.ua/news/2014/04/13/7022274/.

[7] 'On particularities of state policy and guarantee of the state sovereignty of Ukraine at the temporarily occupied territories of Donetsk and Luhansk regions', Embassy of Ukraine in the United Kingdom of the Great Britain and Northern Ireland, 27 February 2018. https://uk.mfa.gov.ua/en/news/63253-please-read-a-basic-ideas-of-a-law-of-ukraine-on-peculiarities-of-state-policy-on-ensuring-state-sovereignty-of-ukraine-in-the-temporarily-occupied-territories-in-donetsk-and-luhansk-re.

[8] 'Сергій Наєв: «Об'єднані сили мають поставити крапку в російській агресії проти України» [Sergiy Nayev: Joint forces should mark a decisive end to the Russian aggression against Ukraine]', *Tyzhden*, 2 August 2018. https://tyzhden.ua/Society/217692.

[9] 'АТО й операція Об'єднаних сил триватимуть паралельно,—юрист ГШ, [ATO and Operation of the Joint Forces will continue simultaneously]', *Novynarnya*, 27 April 2018. https://novynarnia.com/2018/04/27/ato-y-operatsiya-ob-yednanih-sil-trivatimut-paralelno-yurist-gsh/.

[10] UN General Assembly resolution 3314 (XXIX), 14 December 1974. https://docs.un.org/en/A/RES/3314(XXIX).

Federation[11] signed in Moscow on 18 March 2014, and the best example of the ports blockade is the events in Donuzlav in March 2014.[12] Eastern Ukraine events of spring 2014 added yet another clause—(g) The sending by or on behalf of a state of armed bands, groups, irregulars, or mercenaries, which carry out acts of armed force against another.

Ukraine's Hybrid Defence and Combat Effectiveness: A Multi-Domain Analysis

Building on previous works,[13,14] a 'hybrid defence' strategy is defined as 'a defensive strategy that embraces the combined employment of techniques, tactics and procedures of both "hard" and "sharp" powers in four or more domains of operations, usually with the aim of enhancing the combat or the military effectiveness of the disadvantaged force during a defensive, info-kinetic manoeuvre.'[15] Hybrid defence can be employed in almost any domain or terrain. However, as a strategy, it is much more effective when simultaneously[16] employed in at least four domains of operation: land, information/intelligence, cyber, and electronic.[17] The air, sea, and space domains of operations remain important, but the initial four are essential for a minimum level of military effectiveness.[18] The maximum impact of an HDS on an adversarial force usually takes place in the urban terrain. Hybrid defence strategies require connectivity, a sizable segment of the population with technological and technical education, urban infrastructure, and a complex geography.[19] These structural conditions are likely to be found in cities and city suburbs more than in towns, and in towns more than in villages and remote settlements. Hybrid defence operational plans and TTPs are likely to

[11] 'Договор между Российской Федерацией и Республикой Крым о принятии в Российскую Федерацию Республики Крым и образовании в составе Российской Федерации новых субъектов [Treaty between the Russian Federation and the Republic of Crimea on the admission of the Republic of Crimea to the Russian Federation and the formation of new subjects within the Russian Federation]', the Kremlin, 18 March 2014. http://www.kremlin.ru/acts/20605.

[12] 'Konstantin Olshansky: History of Ukrainian ship stolen by Russia in Crimea in 2014', *RBC*, 26 March 2024. https://newsukraine.rbc.ua/news/kostiantyn-olshansky-history-of-ukrainian-1711448715.html.

[13] The section is based on: Omar Ashour, 'Hybrid defence for small states: Strategic lessons and operational implications from Ukraine and Gaza (2014–2024)', *Siyasat Arabiya* 12/71(2024): 31–51.

[14] Ibid.; Ashour, 'How Hamas fights'.

[15] Ashour, 'Hybrid defence for small states', 4.

On sharp power see: Christopher Walker and Jessica Ludwig, 'The Meaning of Sharp Power: How Authoritarian States Project Influence,' *Foreign Affairs*, 16 November 2017, https://www.foreignaffairs.com/articles/china/2017-11-16/sharp-power-china.

[16] An HDS can also be employed sequentially. However, it is likely to be less effective if employed sequentially.

[17] The seven domains are the land, maritime, air, space, informational, electronic, and cyberspace realms. However, the United States (US) Department of Defence consolidates them into five principal operational domains: land, sea, air, space, and cyberspace. The latter—cyberspace—subsumes both the informational and electronic/electromagnetic spectrums, wherein the functional dimensions and operational effects of these 'subdomains' may converge and overlap.

[18] See for example David G. Perkins, 'Multi-domain battle: Driving change to win in the future', *Military Review* 97/6 (November–December 2017): 8–15.

[19] For the impact of these structural factors on urban warfare, see David Kilcullen, *Out of the mountains: The coming age of the urban guerilla* (Oxford: Oxford University Press, 2015).

be well over 1,000. They are incessantly dynamic and constantly developing. Several empirical examples from the war on Ukraine are given below. Hybrid defence strategies usually weaponize the urban civilian terrain via manipulating the infrastructure and designing (or employing existing) multi-layered, subterranean defence-in-depth operational plans. This in addition to the employment of multi-domain, combined arms manoeuvres (to include the information domain / intelligence functions).

Having briefly introduced the concept, the following two subsections overview the employment of hybrid defence in Ukraine.

How Ukraine Fights: An Overview of Hybrid Defence (2014–2025)

From the very beginning of the war in February 2014, Ukraine suffered from a series of interrelated challenges that have led to territorial losses and repetitive risks to national security. These challenges stemmed from seven variables: force size (shortages in manpower), force quality (limited training, low morale, and low capability to conduct and scale up combined arms manoeuvres), force generation (limited full-, partial- and tiered-mobilization capacities), limited defensive lines,[20] weaknesses in critical domains of operations to include the sea, the air, the cyber, and the intelligence/information domains (weaknesses in the latter domain led to multiple infiltrations in the security and the military higher echelons), allies' commitment problems (low level of commitment, resulting in slow and hesitant decision making and late and limited support), and loyal non-state actors with limited capabilities (low-to-no armament and low level of organization). None of Ukraine's non-state loyalists operated beyond the 1991 borders of Ukraine. Within Ukraine's developing HDS, all of these variables have been addressed between 2014 and 2025—with major successes, several challenges, and intermittent pauses. Below is an overview of some military implications of these challenges.

On 20 February 2014, the Ukrainian military and security forces were outnumbered and outgunned in every domain of operations. In February and March 2014, the forces of the Russian Federation executed a hybrid offensive operation that ultimately led to the occupation and the annexation of Ukraine's Crimean Peninsula. Russia's military intelligence and security forces used multiple techniques and tactics of sharp power to include the employment of false-front organizations, intensive propaganda and disinformation campaigns, psychological warfare, and infiltrations of senior ranks within Ukraine's military and security institutions with the aim of destabilization.[21] To secure the strategic gains, hard power was necessary. Russia executed

[20] In urban areas, Soviet civilian subterranean infrastructures—to include industrial areas like Azovstal Steel Plant in Mariupol and the underground metro stations in Kyiv—were utilized as potential bunkers and urban defensive lines. However, that was not the case in many of the rural areas especially in the South (across from Crimea) and in the East.

[21] Kent De Benedictus, *Russian 'hybrid warfare' and the annexation of Crimea* (London: I.B. Tauris, 2021), 264–71; Omar Ashour, 'How Putin's armies fight like ISIS in Ukraine', *Middle East Eye*, 9 May

a large-scale air assault and amphibious operations, with soldiers wearing unmarked uniforms—earning the infamous moniker 'little green men'—for plausible deniability.[22] The Russian operation highlighted Ukraine's weaknesses in four of the seven above-mentioned factors, namely: force quality and armament; essential domains of operations, especially in the intelligence/information domain (weaknesses in counter-intelligence and counter-propaganda); commitment problems among the allies; and unarmed (non-state actors and civil society) loyalists for rapid reaction and popular resistance.

After Crimea, Moscow's proxy forces repeated similar hybrid offensive operations in several cities in eastern and southern Ukraine, including in the Donbas region in April 2014 and afterwards. These operations were led by both former and active Russian security and military personnel, among the most infamous of whom is Colonel Igor 'Strelkov' Girkin.[23] The outcomes of these 'hybrid offensives' were far from Russia's outright victory in Crimea. In Slovyansk, Kramatorsk, Donetsk, Luhansk, Mariupol, Kharkiv, Dnipro, and Odesa the outcomes ranged between successes and failures. The clearest failures were in Odesa and Kharkiv. 'Kharkiv People's Republic' did not last fifteen hours between 7 and 8 April 2014. Ukrainian Russian speakers in Ukraine's second largest Russian-speaking city rapidly brought down Russia's proxy 'republic' by employing a combination of civil resistance and armed action, both state and non-state.[24] 'Odesa People's Republic' was such an utter failure that it was never even declared.

In 2014 and afterwards, Ukraine was reactively adapting and gradually addressing some of the above-mentioned weaknesses—including arming local non-state formations and enhancing counter-propaganda and counter-intelligence efforts.[25] The surprise element—a major force multiplier in Crimea in February and March of 2014—was reduced in Kharkiv in April 2014. That impacted Ukraine's effectiveness within a 'hybrid defence' strategy (HDS). Additionally, Russia's hybrid offensive in Kharkiv failed because it was hardly 'hybrid' (as defined above). Unlike the Russian grouping of forces that exceeded 20,000 soldiers in Crimea, 'hard power'—whether in the form of a conventional or an irregular force—was almost negligible in Kharkiv.

2022. https://www.middleeasteye.net/opinion/how-putins-army-fights-islamic-state-group-ukraine. For the level of infiltration, its depth and breadth, see: 'New head of Ukraine's navy defects in Crimea', *BBC*, 2 March 2014. https://www.bbc.com/news/world-europe-26410431; Ellen Nakashima, 'Inside a Russian disinformation campaign in Ukraine in 2014', *The Washington Post*, 25 December 2017. https://www.washingtonpost.com/world/national-security/inside-a-russian-disinformation-campaign-in-ukraine-in-2014/2017/12/25/f55b0408-e71d-11e7-ab50-621fe0588340_story.html.

[22] See for example: The US army Special Operations Command, '"Little green men": A primer on modern Russian unconventional warfare Ukraine 2013–2014' (Fayettevile, NC: Fort Bragg, 2022). https://www.jhuapl.edu/sites/default/files/2022-12/ARIS_LittleGreenMen.pdf.

[23] Lawrence Freedman, *Command: The politics of military operations from Korea to Ukraine*, (London: Allen Lane, 2022), 371–86.

[24] Oleksandr V. Danylyuk, former chief advisor to the minister of defence of Ukraine, interview by Omar Ashour, Tallinn, 17 May 2024. For the history, see Natalia Shapovalova and Balázs Jarábik, 'How eastern Ukraine is adapting and surviving: The case of Kharkiv', *Carnegie Europe*, 12 September 2018. https://carnegieendowment.org/research/2018/09/how-eastern-ukraine-is-adapting-and-surviving-the-case-of-kharkiv.

[25] Lt Colonel S. I., NGU, interview by Omar Ashour, Kyiv, July 2023; Colonel V. H., air-assault forces of Ukraine (DShV), interview by Omar Ashour, Kyiv, October 2023.

In the Donbas cities and towns, the hybrid offensive that was launched by Russian-led irregular forces achieved mixed results. In Luhansk and Donetsk cities, it was successful, whereas initial successes were followed by ultimate failures in Sloviansk, Kramatorsk, and Mariupol. These cities and other towns were all liberated during Ukraine's first counter-offensive in the summer of 2014.

Between April 2014 and February 2022, Ukraine invested in enhancing its hybrid defences in all domains of operations: land, sea, air, space, electronic, cyberspace, and information/intelligence. That included the initial coordination and then the integration of volunteer armed activist formations within the framework of the national guards of Ukraine (NGU)[26] and other structures of the armed forces or security services of Ukraine. This is in addition to the augmentation of the territorial reservist forces (territorial defence forces, or TDF),[27] to significant increases in defence and security budgets,[28] equipment upgrades and boosting training. In terms of the latter, the Joint Multinational Training Group-Ukraine (JMTG-U) was established in April 2015 and, until 2021, it had trained over 20,000 soldiers, mainly in the Yavoriv Combat Training Centre near Lviv.[29] Since 2015, almost every infantry, artillery, and armour brigade[30] has gone through at least three special training programmes under the United States (US), United Kingdom (UK), or Canadian instructors. These training courses were interrupted by combat tours in the Donbas region and then followed by more advanced training.[31] Between 2014 and 2022, Ukraine also invested in its cyber and electronic defence capabilities,[32] and vetted the intelligence service to remove or prosecute some of the pro-Kremlin elements and corrupt security officials. The reforms and capability enhancement efforts were necessary to design and develop Ukraine's HDS, and to protect sovereignty and the remaining territorial integrity. However, while this was necessary it was insufficient to deter the Russian Federation.

In February 2022, Ukraine had to defend over 3,000 kilometres of land and maritime borders with Russia, Belarus, and occupied Crimea. Like in 2014, Ukraine remained outgunned in all domains of operations in 2022. Moreover, due to Ukrainian miscalculations and successful Russian force concentration and relative surprise, Russia's forces managed to outnumber Ukraine's, achieving manpower ratios that exceeded 5 to 1 in the south and parts of the north. These ratios reached

[26] Lt Colonel S. I., interview, July 2023.

[27] Andriy Zagorodnyuk, former defence minister of Ukraine, interview by Omar Ashour. Warsaw, 3 October 2023; for more details, see Taras Tarasiuk, Petro Burkovskyi, and Olexiy Haran, 'From underdogs to masters of modern warfare: The transformation of Ukrainian nationalist movements', *Strategic Papers* 16 (7 March 2024). https://www.dohainstitute.org/en/Lists/ACRPS-PDFDocumentLibrary/from-underdogs-masters-modern-warfare-transformation-ukrainian-nationalist-movements-2022-2023.pdf.

[28] Zagorodnyuk, interview, 3 October 2023; Andriy Zagorodnyuk, former defence minister of Ukraine, interview by Omar Ashour, Kyiv, 15 July 2023.

[29] Ben Hodges, former commanding general of the US army in Europe, interview by Omar Ashour, Warsaw, 1 October 2024.

[30] With the exception of air force and naval units, and air defence; Hodges, interview, 1 October 2024; Lt Colonel S. I., interview, July 2023.

[31] Hodges, interview, 1 October 2024.

[32] Danylyuk, interview, 17 May 2024.

10 to 1 in Chernihiv Oblast (northeast) during the famous battle, marked by a heroic combat performance from the Ukraine's 1st Tank Brigade.[33]

Between February 2022 and November 2024, the war on Ukraine has had eight phases. In February 2022, Russian armoured formations broke through Ukrainian defences in the north and the south. By April—the end of the first phase of this war—Ukraine had liberated four northern regions (Zhytomyr, Kyiv, Chernihiv, and Sumy), as well as parts of the occupied Kharkiv region (northeast). The investments in the regular forces, territorial reserves, and the NGU, the strategic communication strategy employed by the political leadership, the intelligence-vetting processes as of 2014, the combat-experienced activist-volunteers and their rapid deployment in the capital have all paid off.

By February 2022, a TDF platoon of twenty-seven soldiers in the Kyiv region had been trained and equipped with eleven anti-tank guided missiles (ATGMs), including American Javelins and British NLAWs; a major upgrade compared to the armament and the training of reservists in February 2014.[34] Soldiers of the NGU's 4th Rapid Reaction Brigade, the Georgian legion (Georgian-Ukrainian volunteers), and other units managed to attrite and delay Russia's elite air-assault (VDV) and special forces (*spetsnaz*) units during the battle for Antonov airport in Hostomel, before artillery units from the regular 72nd Mechanized Brigade entered the fray, cratered the airport runway, and further degraded the attacking force.[35] The Ukrainian victory in the bigger Battle of Hostomel significantly reduced the threat from the northwest to the capital, Kyiv. In the northeast, the investment in conventional armoured units also paid off. In general, tanks were critical to Ukraine's defence-in-depth strategy. Tanks provided mobility, protection, and firepower, while Ukrainian forces were trading space for time and mobility for attrition. Ukraine's 1st Tank Brigade (also known as the Severia Brigade) fiercely resisted units from Russia's 1st Guards Tank Army (GTA) and 41st Combined Arms Army (CAA), while being ridiculously outnumbered. The Severia Brigade not only repelled the invasion attempt on Chernihiv City but also undermined the Russian advance towards Kyiv from the northeastern vector. Ukrainian resistance in the northeast rendered Russia's elite 1st GTA temporarily combat ineffective.[36]

The first phase of the war also showed how critical the subterranean (civilian) infrastructure and its weaponization is for urban defence operations. That was primarily demonstrated in the defence of Mariupol, led by the 36th Separate Marine Brigade, the Azov Regiment, and other smaller combat units.[37] Despite losing the city, a force of about 4,000 Ukrainian soldiers tied down a Russian division-sized

[33] See for example David Axe, 'How Ukraine's 1st tank brigade fought a Russian force ten times its size—and won', *Forbes*, 25 December 2022. https://www.forbes.com/sites/davidaxe/2022/12/25/how-ukraines-1st-tank-brigade-fought-a-russian-force-ten-times-its-size-and-won/.

[34] Captain O. S., officer in territorial defence forces, interview by Omar Ashour, Kyiv, 2 November 2023.

[35] Ibid.

[36] This outcome has degraded the threats posed by Russia's Western military district to smaller NATO countries in eastern Europe, to include the Baltic States and Poland.

[37] Mariia Chekh, 12th Separate Special Forces Brigade (Azov) of the NGU, interview by Omar Ashour, Zagreb, 11 October 2024.

formation mainly from the 58th CAA, arguably the best performing unit in the entire Russian groupings of forces during the first phase of the full-scale invasion.[38] Due to the effective employment of civilian subterranean infrastructure and general force quality of the Ukrainians, over 15,000 Russian soldiers were tied down in Mariupol, fighting a much smaller force during the first phase—the most critical phase of the war—and well into the second phase, until May 2022.

The second phase of the war took place during the late spring and summer of 2022. It was dominated by attritive artillery duels in the Donbas and relatively limited manoeuvre warfare. The phase has shown the critical importance of reliable military alliances and sustained investments in the defence industries, including the local manufacturing of artillery shells. During this phase, Russian artillery outmatched and overwhelmed Ukraine's defenders, resulting in the fall of Lysychansk, followed by the fall of the entire Luhansk region in July 2022. Luhansk was the only region to (temporarily) fall entirely under Russian occupation since the fall of Crimea—a Ukrainian autonomous republic—in 2014.

The third phase of the war was fought in autumn 2022. It featured a series of stunning Ukrainian counter-offensives, liberating most of the Kharkiv region as well as Kherson City, the only provincial capital captured by Russia during the war.[39] The liberation of Kherson city highlighted the critical role of Western guided rocket-artillery systems, such as the High Mobility Artillery Rocket Systems (the M142 HIMARS). The Kharkiv 2022 counter-offensive operation showed the critical need for larger quantities of armoured personnel carriers (APCs), armoured fighting vehicles (AFVs), infantry fighting vehicles (IFVs) and main-battle tanks (MBTs), as well as close air support (CAS). Given that Ukraine's armoured formations were attritted and undersupplied by this phase, Ukraine's mechanized brigades[40] executed Kharkiv's counter-offensive with a hybrid force. These brigades combined conventional armour with wheeled civilian 4x4 sport utility vehicles (SUVs) and pickup trucks with makeshift armour. Despite the hybrid nature of the armoured force, the thunder runs were successful. Overall, the third phase highlighted the importance of extensive and long-term investment in training and equipment needed for combined arms' manoeuvres. It also demonstrated the superiority of some of the Western-made weapon systems—specifically in rocket artillery—in terms of range, mobility, and precision firepower.

The third phase, like all the upcoming phases, rehighlighted the cruciality of manpower / force size, early recruitment, and rapid mobilization capabilities. To a lesser extent, it rehighlighted the importance of force quality in terms of equipment and

[38] This is an assessment based on Omar Ashour's field observations and interviews in 2022 and 2023. It is based on breakthroughs and two-pronged thrusts in southern Ukraine, into the southwest (Kherson and Mykolaiv Oblast) and into the southeast (towards Melitopol and then Mariupol).

[39] Mariupol was the provisional capital of Donetsk Province, since the fall and the occupation of Donetsk City in the summer of 2014 by Russian-led irregular forces.

[40] Namely, the 92nd and the 93rd Mechanized Brigades and the Kraken regiment. The rest of the units were mainly motorized and light infantry brigades, including units from the NGU and the TDF as well as from the International Legion, the Dudayev Battalion, and others.

training. Lacking primarily manpower,[41] Ukraine's units could not fully exploit the route and liberate all of the Kharkiv region nor several relatively vulnerable positions in the northern Luhansk region. Hence, the operations stalled in occupied Luhansk, ending the third phase with a Ukrainian operational, but not strategic, victory.

The fourth phase of the war (winter 2022–3) featured a brutal Russian air campaign that caused significant damage to Ukraine's civilian infrastructure. It rehighlighted the general importance of mobile, dense, and multi-layered air defences, given the geographical size of Ukraine. The fifth phase was fought in spring 2023. It showed the importance of non-state forces in an overall hybrid offence strategy. The Russian Wagner Group has had a lasting impact despite almost entirely leaving Ukraine after its June 2023 insurrection.[42] The organization—and its supporters in the airborne, artillery, and air force units—attrited some of the most experienced Ukrainian brigades during the Battle of Bakhmut. Although manpower attrition ratios have favoured Ukraine in Bakhmut—with an estimated loss exchange ratio (LER) of 5 to 1 for Ukraine[43]—the battle gave the Russian forces the time needed to bolster their defensive positions in the south. At the end of that phase, the Wagner Group and its allies occupied Bakhmut, achieving an operational victory while Russia's regular forces were struggling in the rest of the fronts.

The sixth phase featured Ukraine's summer 2023 counter-offensive, which has not been successful for various reasons.[44] However, it gave Ukraine—and any small(er) state force paying attention—invaluable lessons in both hybrid and conventional warfare. While the land operation stalled in this phase, Ukraine was more successful in the sea domain. Russia's Black Sea fleet command and assets were forced out of Crimea by a combination of short-range precision strikes and innovative unmanned aerial vehicles (UAVs), unmanned surface vehicles (USVs), and unmanned underwater vehicles (UUVs) tactics. Without a conventional navy and with a hybrid 'mosquito fleet' and a combination of deep battle strikes and remote naval warfare, Ukraine has not only inflicted major losses but also forced a strategic realignment within Russian naval operations.[45] The operational defeat of the Black Sea fleet and the withdrawal of its assets from occupied Crimea to Novorossiysk (Russia) and Ochamchire (separatist Abkhazia) has allowed Ukraine to export millions of metric tons of grain and oils during the 2023–4 season via the Western part of the Black Sea, thus breaking the naval blockade.

[41] The armed forces of Ukraine (AFU) also lacked proper armour, sufficient ammunition, and airpower. However, they managed to successfully execute the 2022 Kharkiv counter-offensive. Manpower remained the main reason for the limited exploitation of the success.

[42] On the insurrection, see 'Wagner: World leaders react to Russian insurrection', *Middle East Eye*, 24 June 2023. https://www.middleeasteye.net/news/wagner-world-leaders-react-turmoil-russia.

[43] The number of Russian soldiers killed for every fallen Ukrainian soldier. See for example 'NATO estimates Russia lost 5 times more soldiers in Bakhmut than Ukraine', *CNN*, 6 March 2023. https://edition.cnn.com/europe/live-news/russia-ukraine-war-news-03-06-23. On LERs, see Omar Ashour, 'From counteroffensive to counter-invasion: How did the Ukrainian armed forces invade the Russia Federation?' (in Arabic), *Strategic Analysis*, no. 9, Arab Centre for Research and Policy Studies, 17 September 2024. https://tinyurl.com/3bs84stm (accessed 26 January 2025).

[44] Ibid., 4–7.

[45] Sidharth Kaushal, 'Ukraine's naval campaign and the sinking of the Moskva: Implications for the war at sea', *RUSI Commentary*, April 2022; Igor Delaneo, 'Russia Black Sea fleet in the "special military operation" in Ukraine', *Foreign Policy Research Institute Report*, February 2024. https://www.fpri.org/wp-content/uploads/2024/02/russias-black-sea-fleet-.pdf.

Finally, the seventh and eighth phases of the war were fought in the spring and summer of 2024. They featured Russian advances mainly in the southern Donbas, followed by a Ukrainian multi-brigade combined arms offensive that successfully occupied parts of the Kursk region in western Russia. Like the counter-offensives of 2022 and 2023, the armed forces of Ukraine (AFU)'s multi-domain counter-invasion of August 2024 offered major lessons for the small(er) side, some of which are illustrated in the final sections of this chapter.

The Strategic Logic of Asymmetric Warfare after 2022

While an HDS primarily concentrates on denial-oriented defensive efforts, the concept of asymmetric warfare extends beyond purely defensive functions. It rests on the same premise as forces asymmetry, yet encompasses offensive capabilities, and recognizes that it is not exclusive to the weaker actor. Originally centred on force disparity and smaller states tactics,[46] the later concept developed towards the recognition that

> asymmetrical warfare then refers to unconventional strategies, tactics and arms systems adopted by a force when its military capabilities towards its enemy are not simply unequal but are so significantly different that it cannot appeal to the same sorts of attacks... The essential characteristic of asymmetric strategies is the attack of vulnerabilities not appreciated by the adversary, using different operational concepts, doctrines, tactics, and weapons from those used in symmetric conventional confrontations.[47]

Considering the size of forces, ability to produce ammunition, and other parameters, Russia and Ukraine applied different strategies to their choice of warfare methods. The Russian Federation concentrated on classical tactics which relied on dominance in manpower. In this case, wider fronts, multiple assaults, and a combination of air and land assaults with a significant number of personnel involved were present. Still, there are three elements that are on the edge of classical and asymmetric warfare that can be analysed. The use of the private military company—the Wagner Group, maritime blockade of Ukrainian ports, attacks against critical infrastructure, and, first of all, energy.

The Wagner Group's[48] involvement surprised a lot in the beginning as the personnel was not an issue to the Russian armed forces. Nevertheless, the Wagner Group enabled the Kremlin to address several critical needs: it facilitated the recruitment of prisoners and mercenaries, thereby circumventing the politically sensitive

[46] Andrew Mack, 'Why big nations lose small wars: The politics of asymmetric conflict', *World Politics* 27/2 (1975): 175–200.

[47] Lautaro N. Rubbi, Victoria Álvarez Magañini, Lucas Nascimento, and Dana Sager, 'Asymmetric warfare: Exploratory study on the non-conventional military developments of the People's Republic of China in the period 2012–2018', *Revista SAAP* 14/1 (2020): 180–200.

[48] Régis Genté, Rodolphe Oberle, Hennadiy Maksak, Yan St-Pierre, and Anastasiya Shapochkina, 'Wagner as a tool of Russia's geopolitical strategy', Foreign Policy Council 'Ukrainian Prism', 2023. https://prismua.org/en/english-wagner-as-a-tool-of-russias-geopolitical-strategy/.

option of general mobilization while still fielding disciplined and tactically competent forces. Simultaneously, it allowed for the redistribution of financial resources—bypassing entrenched corruption within the Russian Ministry of Defence—while also serving as a financial incentive for loyal actors who might otherwise pose a challenge to the existing regime.

Attacks on Ukrainian energy infrastructure[49] could be considered regular warfare, despite the limitations that rules of war usually set regarding such objects if the logic behind it were purely military. *De facto*, the massive scope of the attacks and the geography of coverage (the whole of Ukraine) aimed first and foremost at bringing fear and dissatisfaction to the Ukrainian population. These attacks were accompanied by coordinated information attacks that intricate narratives undermining trust in the central government, calling for 'quick peace' with Russia, and bringing discord between different regions of Ukraine. Such a combination allows us to assume that the primary goal was to influence the population to go against its own government and to trigger domestic instability.

The maritime blockade of the Ukrainian ports *de facto* started earlier than the full-fledged invasion. The ports of the Azov Sea had been regularly blocked and navigation of them was disturbed since the construction of the Crimean bridge, and especially after the 2018 incident with the capture of the Ukrainian ships. Odesa's port witnessed an attempt of the blockade two weeks before the February 2022 attack and a full blockade afterwards until the deblockade later in 2023. While creating a maritime blockade is one of the traditional methods, it is usually used to prevent supply to a blockaded territory. In the case of the Russian-Ukrainian war, in addition to its military reasoning, Russia has also satisfied its economic ambitions, as it has prevented Ukrainian grain export to vulnerable states, thus excluding their main competitor to those markets in Asia and Africa, but also Russia used it as a manipulative instrument in relations with the so-called 'Global South',[50] presenting itself as a main guarantor of the food security.

Considering the limited resources Ukraine has had, asymmetric warfare became a chance for balance against a bigger power. Here, it involved tactical decisions, differences in command and control (North Atlantic Treaty Organization, or NATO) principles vs Soviet doctrine), motivation (value of defending the own land vs constantly changing and badly explained reasons for the 'special military operation'), involvement of the territorial defence forces (described in a previous section), as well as technological advances and techniques. While in some cases we can argue over

[49] 'Attacks on Ukraine's energy infrastructure: Harm to the civilian population', UN Human Rights Monitoring Mission in Ukraine, September 2024. https://ukraine.ohchr.org/sites/default/files/2024-09/Attacks%20on%20Ukraine%E2%80%99s%20Energy%20Infrastructure-%20%20Harm%20to%20the%20Civilian%20Population.pdf.

[50] Omar Ashour, 'Defence diplomacy revisited: Ukraine and the "Global South"', *New East Europe* 61/3 (April/May 2024). 67–75: https://neweasterneurope.eu/2024/04/11/defence-diplomacy-ukraine-and-the-global-south/.

whether all this creates a coherent strategy, the use of drones and, especially maritime drones, has changed the future of warfare significantly.

The Russian-Ukrainian war has already been named a mix of nineteenth-and twenty-first-century wars. While the role of the infantry and artillery has not diminished, technologies and, particularly, unmanned systems (UAVs or drones) have shaped the battlefield. Their introduction *en masse* has impacted reconnaissance (allowing full coverage of the battlefield without the necessity to send a single soldier) and striking capabilities (kamikaze drones and drones with warheads, from 'upgraded' Mavic to Palianytsia missile drones,[51] and hundreds of other variations that are constantly upgraded, developed, and improved), especially with the introduction of artificial intelligence (AI) in the decision-making process and command and control system (with the Delta system[52] and Defence Mapping Software Kropyva as examples[53]). War is not static. The side which reacts quickly to changes in the external environment, documents them, digitizes them, and transfers them to their command centres has a critical advantage.[54]

The most vivid asymmetric warfare has been happening in the maritime domain. After the annexation of Crimea in 2014, Ukraine lost 70% of its navy, and since 2022, additional losses have occurred, which have led to the situation of its naval forces being without any serious battleship. The Russian navy, on the contrary, had militarized Crimea and has increased both personnel and ship numbers there since 2014. Compared to other types of military equipment and the closure of the Bosporus to battleships by Turkey in 2022 (according to the Montreux Convention), Ukraine needed an extraordinary solution to prevent Russian landing operations, jeopardizing the safety of navigation, and even being stationed in Crimea and initiating strikes from there. Maritime drones became a solution.

The paradox of this war is that due to limited resources, the expected actor does not always come up with the solution. It was not the navy but the security services of Ukraine and military intelligence[55] that presented two of the most well-known maritime drones—Sea Baby[56] and Magura.[57] These drones, aimed for

[51] Daria Svitlyk, 'Everything we know about Ukraine's new Palianytsia missile-drone', *Kyiv Independent*, 1 September 2024. https://kyivindependent.com/everything-we-know-about-ukraines-new-palianytsia-missile-drone/.

[52] 'Battlefield innovation: Ukraine's DELTA system paves the way for allied interoperability at CWIX24', NATO Allied Command Transformation, 12 July 2024. https://www.act.nato.int/article/delta-system-cwix/.

[53] Defense Mapping Software, ArmySOS. https://armysos.com.ua/defense-mapping-software/.

[54] 'On the importance of drones on the battlefield and on the prospects of Ukraine in military drone production', *Ukraine World*, 5 June 2023. https://ukraineworld.org/en/articles/analysis/ukraine-military-drone-production.

[55] 'Naval drones of the defence intelligence of Ukraine dominate the Black Sea', Defence Intelligence of the Armed Forces of Ukraine. https://gur.gov.ua/en/content/morski-drony-hur-dominuiut-u-chornomu-mori.

[56] 'Updated Sea Baby 2024 naval drone: One ton of explosives with a range of over 1000 km', *Defence Express*, 6 March 2024. https://en.defence-ua.com/weapon_and_tech/updated_sea_baby_2024_naval_drone_one_ton_of_explosives_with_a_range_of_over_1000_km-9748.html.

[57] Sofiia Syngaivska, 'Ukrainian Magura V5 drone swarms break through, overloading Russian defenses with new upgrades', *Defense Express*, 19 March 2025. https://en.defence-ua.com/weapon_and_tech/

both reconnaissance and strikes, developed their range of distance up to 1,000 kilometres and capabilities that came from initial kamikaze (above-sea torpedo-style) drones—carriers of munitions capable of targeting helicopters and jets.[58] As a result of sophisticated operations that were conducted in cooperation with the navy, due to the use of maritime drones, the country without a naval fleet managed to destroy 30% of the Russian Black Sea fleet[59] and to push others to leave Crimea and anchor in Novorossiysk.

UAV strikes against Russian refineries[60] are another example of asymmetric warfare. Attacking refineries deep in the Russian territory—well beyond the forward edge of the battle area—allowed Ukraine to achieve three effects: (1) induce a fuel shortage, degrading military logistics; (2) generate negative domestic perceptions by demonstrating that the Russian integrated air and missile defence (IAMD) is unable to protect critical national infrastructure; (3) 'impose' economic costs as a functional substitute where allies were reluctant to apply severe sanctions on the Russian energy sector.

Hence, in 2023–4 Ukraine employed a calibrated combination of HDS and asymmetric warfare, contingent on its aims and operational priorities. During phases of Russian offensive action, the HDS was prioritized; during counter-offensive phases—and in theatres with minimal prospects for Russian advance (e.g. the Black Sea)—Ukraine prosecuted asymmetric operations that combined HDS and unconventional offensive action.

Conclusion: Lessons Learned for Small State Defence Policies

We conclude here with comparative observations to extract lessons for defence policies and strategies of small(er) states. These lessons mainly address force size/manpower, force quality, and force structure and reconfirm the cruciality of the hybrid nature, including multi-domain and combined arms operations, for military effectiveness. The first lesson is to invest in reserves. Generally, reservists play a pivotal role in any HDS of a smaller force, state or non-state. As shown in Ukraine, reserve forces have provided a cost-effective means to augment active military personnel, enhance rapid mobilization capabilities and scalability, strengthen national resilience, proffer flexible and responsive force structures, leverage civilian

ukrainian_magura_v5_drone_swarms_break_through_overloading_russian_defenses_with_new_upgrades-13890.html.

[58] 'Ukrainian drones take out two Russian Mi-8 helicopters near Crimea on New Year's Eve', *The New Voice*, 3 January 2025. https://english.nv.ua/nation/ukrainian-naval-drones-destroy-two-russian-mi-8-helicopters-near-crimea-on-dec-31-50478654.html.

[59] Dinara Khalilova, 'These are the most important Russian ships destroyed by Ukraine', *The Kyiv Independent*, 30 January 2024. https://kyivindependent.com/these-are-most-important-russian-ships-destroyed-by-ukraine/.

[60] 'White House confirms Ukraine destroyed 20% of Russian oil refineries in August', *Ukrinform*, 29 August 2025. https://www.ukrinform.net/rubric-economy/4030619-white-house-confirms-ukraine-destroyed-20-of-russian-oil-refineries-in-august.html.

professional/specialist skills in multiple domains of operations, and multiply the force size to offset asymmetrical disadvantages.

The second lesson is to enable the infrastructure. The subterranean, hybrid defence-in-depth tactics—employed by the defenders of Mariupol[61]—have had a major impact on combat effectiveness against superior forces, as outlined above. Hence, pre-combat preparations and the organization of the civilian infrastructure are critical; not only for the protection of the defending smaller force but also for the protection of the urban centre *per se*, whether a town or a city.

A third lesson is that there is no escape from multi-domain, combined arms for creating upsets against a larger force. That means that the smaller force must endorse multi-domain, combined arms operations as a pillar of a HDS. A fourth lesson is about remote warfare, characterized in Ukraine by the deployment of UAVs, USVs, UUVs, unmanned ground vehicles (UGVs), and other 'autonomous'[62] or semi-autonomous systems. Arguably, the scale, lethality, intensity, and overall effectiveness of these systems may represent a paradigm shift in force employment and may further lead to major updates in doctrines and concepts of operations. At the very least, these systems have enhanced combat effectiveness, if not overall military effectiveness. Hence, they are worth investing in, in terms of procurement, training, equipping, reorganizing, scaling up, and revising doctrines. However, their effectiveness may be short lived as electronic warfare develops, and they require constant development and upgrades, which brings us to the need for a separate legislative framework distinct from classical munitions testing and defence procurements. Second, changes to the normative basis would allow quicker interactions between developers, producers, and final users. As modern technologies are changing rapidly and constantly, to be effective on the battlefield, you should be ready to upgrade your capabilities respectively. This involves real-time testing and feedback from users, shorter periods of governmental certification, and state support for the innovation.

The final lesson is the importance of asymmetric response. Given contemporary technologies, numerical disadvantages in personnel and ammunition can be offset through technological solutions (UAV / Unmanned Combat Aerial Vehicle (UCAV) swarms, loiterers, AI-enabled intelligence, surveillance, and reconnaissance (ISR)-to-fires, a devolved and distributed command and control architecture (e.g. Delta; delegated decision rights to mid-level commanders without constant top-down approval), and prioritization of target sets. This asymmetric approach should

[61] The subterranean infrastructure encompassed by the Azovstal Steel Plant, the Ilyich Iron and Steel Works, Soviet era civil defence bunkers, as well as the frequently utilized smaller existing structures—such as municipal sewer lines, drainage tunnels, and building basements—for covert movement and shelter. In contrast to Gaza—as an example—very few elements of this overground and underground infrastructure were constructed by Ukrainian units themselves. Chekh, interview, 11 October 2024; Lt Colonel S. I., interview, July 2023.

[62] Most of these systems require significant human intervention (including, but not limited to, 'human-in-the-loop' and 'human-on-the-loop' tasks) to run them properly and effectively. Hence, the term 'autonomous' remains a controversial adjective, subject to intense debates in the defence and the security studies literature.

also incorporate elements not detailed above but operationally decisive: societal support for the defence effort—from volunteer networks capable of supplying kits up to and including military equipment, public donations to defence spending, civic monitoring and cueing (e.g. e-air defence / E-PPO applications). These instruments collectively enhance denial capacity, accelerate kill-chain closure, and sustain operational tempo, thereby converting material inferiority into combat-credible resilience consistent with a HDS.

15

The Russian Invasion of Ukraine

How Putin Lost in Ten Days

Ed Arnold

Introduction

President Putin's 'Special Military Operation' (SMO)—to subjugate Ukraine in ten days—failed. Poor intelligence, preparation, planning, and execution of the invasion set the conditions for a protracted attritional war which continues to draw in external powers on both sides. The Russian politico-military leadership was both complacent about Ukraine's ability and will to resist and overconfident in Russian military power. After an underwhelming performance against Georgia in 2008, the armed forces of the Russian Federation (AFRF) had undertaken a period of modernization and reform and had experienced successes in Crimea and the Donbass in 2014 and in Syria from 2015. However, the full invasion of Ukraine was an operation on a completely different scale and neither the political leadership nor the AFRF was prepared for the mission. This dangerous combination of complacency and overconfidence meant that no plan B was developed or tested. Therefore, when the SMO began to fall apart within the initial few days—as Kyiv held, Hostomel airport was denied to Russian forces, and Ukrainian air defence (AD) was still operational—Russian commanders were unable to gain the initiative. Put simply—it was a bad plan which was poorly executed.

However, the story of the invasion is not only one of Russian failures, but also of critical Ukrainian successes. Ukraine's remarkable initial resistance—while far from perfect—converted a series of tactical victories on the battlefield into political and strategic wins. Ukraine's early defiance in the face of the Russian onslaught secured its centre of gravity by winning the significant backing of the Western international community on which the country has become wholly reliant to continue to fight. Therefore, Ukraine's heroic resistance has put European security on a fundamentally different path to the one should the SMO have succeeded. Ukraine has simultaneously exposed Russia's significant military weaknesses and enhanced the defence of Europe by degrading and fixing the Russian military within Ukraine.

Ed Arnold, *The Russian Invasion of Ukraine*. In: *Russia's War in Ukraine and Modern Warfare*. Edited by: Mikael Weissmann and Niklas Nilsson, Oxford University Press. © Ed Arnold (2026). DOI: 10.1093/9780198970903.003.0015

It is important to analyse why Russia squandered such a dominant position. Russia successfully achieved surprise by convincing Ukraine that the SMO would be limited with its 'main effort' being in the Donbass. This deception delayed Ukrainian national mobilization and left Kyiv less well defended, giving Russia a 12:1 force ratio advantage on its main axis of advance towards Kyiv.[1] Despite this overwhelming physical advantage the AFRF was unprepared and consistently outfought by the armed forces of Ukraine (AFU).

This chapter will examine the origins of Russia's war against Ukraine and why Putin's expectation of a swift subjugation of Ukraine within ten days was so far from the reality of the attritional war which has developed. It seeks to understand and explain: why the SMO failed, how close it came to succeeding, Russian military weaknesses, the strength of Ukrainian resistance, and the geopolitical repercussions of Russia's failure to subjugate Ukraine.

How Putin Lost in Ten Days

Drawing on preliminary assessments of the war's opening days, expert commentaries, and international media reports, this chapter argues that Russia's early failures were comprehensive and systemic. Three core factors best explain the outcome of the so-called SMO.

First, differences at a political level. Putin's personal views on Ukraine infected the Russian system, which created a disconnect between the political and strategic levels of warfare on one hand and the operational and tactical on the other. Specifically, Putin—supremely confident in the swift subjugation of Ukraine—failed to prepare or mobilize the Russian Federation for war. In direct contrast, Ukraine's levels of warfare were harmonized—resist the Russian invasion at all costs—which enabled significant agility and flexibility and allowed Ukraine to seize the initiative and deny Russia a quick victory. In this sense, the war was the AFRF versus the whole of Ukrainian society.

Second, there was a deficient Russian—and superior Ukrainian—moral component of fighting power. Assessments of a rapid Russian victory and swift Ukrainian capitulation were made with reference to the physical and conceptual components of fighting power. The moral component—often overlooked and difficult to assess—was the main difference between the AFRF and AFU and provides the strongest explanation for early Ukrainian battlefield successes.

Third, Ukraine was much better than Russia at applying the principles of war (selection and maintenance of the aim, maintenance of morale, offensive action,

[1] Mykhaylo Zabrodskyi, Jack Watling, Oleksandr V. Danylyuk, and Nick Reynolds, 'Preliminary lessons in conventional warfighting from Russia's invasion of Ukraine: February–July 2022', *RUSI Special Report*, 30 November 2022, 28.

security, surprise, concentration of force, economy of effort, flexibility, cooperation, sustainability) to both the planning and execution of the war.[2]

Therefore, despite the significant advances in modern warfare that the war in Ukraine has established, the result can best be explained through the application and misapplication of traditional military tenets.

The rest of the chapter is divided into three parts which align with the military operational phases of 'Shaping', 'Decisive', and 'Sustaining'. The shaping covers the preparation and planning of the SMO between the annexation of Crimea in 2014 to the large-scale invasion on 24 February 2022. The decisive phase covers the fighting between the invasion on 24 February and 25 March—when Russian forces were given the redeployment order to the Donbass, and Kyiv was safe—and focuses on the pivotal Battle of Hostomel for the control of Antonov airport. Finally, the sustaining section concludes the chapter and examines how the initial ten days of the invasion created the conditions for the war of attrition which has developed and its wider implications for European security.

Shaping—2014 to 24 February 2024

On paper, and in the minds of most military planners in Europe, the 2022 Russian army was the second most powerful in the world.[3] Russia's military expenditure was ten times that of Ukraine—US$66 billion versus US$6.9 billion and boasted a total of 900,000 regular soldiers, to the AFU's 196,600.[4] The physical component (the means to fight) of fighting power unsurprisingly dominated the assessments of the strength of the AFRF.[5] The conceptual component (the ideas behind how to fight) was assessed as healthy under the false assumption that it had significantly improved through the post-Georgian war reforms from 2008. A detailed assessment of the moral component (the ability to get people to fight) consisting of leadership, motivation, moral cohesion, and an ethical foundation was overlooked, through a mix of negligence, a lack of curiosity, and a lack of available data.

It is interesting that it was not only Putin—with his false reading of history and hubris—who severely miscalculated the balance of power. Defence analysts were near unanimous in predicting a rapid Russian victory and an immediate Ukrainian

[2] For British army definitions of the principles of war, see British Army, 'Army doctrine primer', May 2011. https://assets.publishing.service.gov.uk/media/5a79b6c540f0b63d72fc7eac/20110519ADP_Army_Doctrine_Primerpdf.pdf, pp. 4–1 to 4–2 (accessed 6 June 2024).

[3] Jack Watling, 'Russia's callousness towards its own soldiers is undermining its combat power', *The Guardian*, 4 March 2022. https://www.theguardian.com/commentisfree/2022/mar/04/russias-callousness-towards-its-own-soldiers-is-undermining-its-combat-power (accessed 15 October 2024); and 'Global Firepower Ranking 2022', Global Firepower. https://www.globalfirepower.com/global-ranks-previous.php (accessed 15 October 2024).

[4] '2021 figures', SIPRI Military Expenditure Database. https://milex.sipri.org/sipri (accessed 3 January 2025); and 'The military balance 2022', International Institute for Strategic Studies, 14 February 2022.

[5] For a full explanation of fighting power, see British Army, 'Army doctrine primer', 1–4.

collapse.[6] Such was the strength of conviction of the analytical community in the outcome that it likely directly influenced the United States (US) government and intelligence community, with then chairman of the Joint Chiefs of Staff reportedly commenting that Kyiv could fall in just seventy-two hours.[7] While US intelligence correctly assessed that Russia would undertake a full country-wide invasion of Ukraine, it was incorrect that the AFRF was highly capable of achieving its aims against a weak AFU that would not be able to resist. Shamiev summarizes the situation as:

> From the beginning of 2000 to the end of 2021, the Russian defence budget almost tripled, from $22.84 billion to $65.91 billion (in 2020 US dollars). In those 22 years, Russia spent over $1.1 trillion on the military, had three ministers of defence and four chiefs of the general staff who oversaw three military reforms, a short war with Georgia, an intervention in Syria, and several smaller deployments in former Yugoslavia and the former Soviet republics. Yet, when that military was finally tested in a full-scale conflict, almost nothing appeared to have changed.[8]

Therefore, the moral component was the decisive factor between the AFRF and the AFU in the first month of the war.[9] The AFRF had a deficient moral component which was heavily influenced by the Russian leadership and military culture. In a dictatorial and hierarchical system, the confused political leadership impacted all levels of command. It was rigid, inflexible, and could not adapt to changes on the ground. In contrast, the Ukrainian moral component was superior, with stronger political leadership which exploited the flexibility and adaptability which had been developing in the AFU since 2014. Critically this political leadership brought together the Ukrainian military and civil society, creating a whole society resistance against the existential threat that Russia posed. These differences had a profound effect on the command structures and cultures. Lawrence Freedman states that 'From the start, the war has provided a remarkable contrast in approaches to command. And these contrasts may go a long way towards explaining why the Russian military has so underperformed expectations'.[10] As such, the causes of the Russian poor performance and Ukraine

[6] Samuel Charap and Scott Boston, 'The West's weapons won't make any difference to Ukraine', *Foreign Policy*, 21 January 2022. https://foreignpolicy.com/2022/01/21/weapons-ukraine-russia-invasion-military/ (accessed 5 November 2024).

[7] Eliot A. Cohen and Phillips O'Brien, 'The Russia-Ukraine War: A study in analytic failure', Center for Strategic and International Studies, September 2024, 17–18. https://csis-website-prod.s3.amazonaws.com/s3fs-public/2024-09/240924_Cohen_Russia_Ukraine.pdf?VersionId=1YNnRnwS.6DkrwNcAkdb5DbsfjclgOJR (accessed 5 November 2024).

[8] Kirill Shamiev, 'Brass tacks: Why Russia's military fails to reform', European Council on Foreign Relations Policy Brief, May 2024, 3. https://ecfr.eu/wp-content/uploads/2024/05/Brass-tacks-Why-Russias-military-fails-to-reform-1-1.pdf (accessed 5 November 2024).

[9] Ed Arnold, 'Slava Ukraini: Assessing the Ukrainian will to fight', *RUSI*, 23 March 2022. https://rusi.org/explore-our-research/publications/commentary/slava-ukraini-assessing-ukrainian-will-fight (accessed 10 July 2024); and C. Patrick Welsh, 'Fighting power and the war in Ukraine', Wavell Room, 16 November 2022. https://wavellroom.com/2022/11/16/fighting-power-and-the-war-in-ukraine/ (accessed 10 July 2024).

[10] Lawrence D. Freedman, 'Why War Fails', Foreign Affairs, 14 June 2022, https://www.foreignaffairs.com/articles/russian-federation/2022-06-14/ukraine-war-russia-why-fails (accessed 5 November 2024)

defying expectations can best be explained by their respective approaches to planning and preparation.

Russian Planning and Preparation

Putin—whose sole decision it was to invade Ukraine—bears most responsibility for failing to prepare the Russian military to achieve his political objectives. His July 2021 essay 'On the Historical Unity of Russians and Ukrainians' argued that Russians and Ukrainians were 'one whole', that Ukraine 'is Europe's poorest country' for which he firmly puts the blame on the 'Ukrainian authorities'.[11] As such, Putin's designs on Ukraine are maximalist—and have not changed even after the SMO failed—he wants the complete subjugation of Ukraine including recognition of a Russian sphere of influence, no prospect of North Atlantic Treaty Organization (NATO) membership, and the AFU disbanded.[12]

This argument established the complacency that infected the AFRF and special services. Putin failed to apply the master principle of war of 'selection and maintenance of the aim' and in doing so obfuscated his case for war, simultaneously arguing for 'de-nazifying the political leadership of Ukraine', that Ukraine, alongside the US, was developing weapons of mass destruction and biolaboratories, and that Ukraine was preparing to invade Russia and therefore his action was pre-emptive.[13] This confused rationale for war meant that his political goals could not be converted into achievable military objectives and tasks.

The Russian and dictatorial culture of top-down directives meant that political mistakes trickled down through the strategic, operational, and tactical levels. In addition, because of Putin's false assumptions, he also did not consider the need to prepare the Russian people for a long war. Edmonds argues that 'It would be difficult to find a better example of how the faulty views of one person could so straightforwardly bring about the initial failures of an operation than Putin and his invasion of Ukraine'.[14]

The complacency was matched by a false confidence in the abilities of the AFRF for several reasons. First, after a poor performance in the 2008 Georgian war the AFRF had undertaken a radical reformation and modernization programme under Defence Minister Anatoliy Serdyukov. However, civil-military relations did not see similarly

[11] Vladimir Putin, 'On the historical unity of Russians and Ukrainians', Kremlin, 12 July 2021. http://www.en.kremlin.ru/events/president/news/66181 (accessed 5 November 2024).

[12] Andrew A. Michta, 'Russia's imperial mindset: Understanding the roots of the Ukraine war', *1945*, 2 Jan 2025. https://www.19fortyfive.com/2025/01/russias-imperial-mindset-understanding-the-roots-of-the-ukraine-war/ (accessed 3 Jan 2025).

[13] Rachel Treisman, 'Putin's claim of fighting against Ukraine "neo-Nazis" distorts history, scholars say', *npr*, 1 March 2022. https://www.npr.org/2022/03/01/1083677765/putin-denazify-ukraine-russia-history (accessed 10 January 2025); and Olga Robinson, Shayan Sardarizadeh, and Jake Horton, 'Ukraine war: Fact-checking Russia's biological weapons claims', *BBC News*, 15 March 2022. https://www.bbc.com/news/60711705 (accessed 10 January 2025).

[14] Jeffrey Edmonds, 'Start with the political: Explaining Russia's bungled invasion of Ukraine', *War on the Rocks*, 28 April 2022. https://warontherocks.com/2022/04/start-with-the-political-explaining-russias-bungled-invasion-of-ukraine/ (accessed 17 December 2025).

effective reform and problems persisted. Shamiev argues that the 'Ministry of Defence policies simply could not solve the military's long-standing quality problems, such as unsatisfactory service conditions, intra-military brutality, corruption, waste, and poor tactical command and control skills'.[15] The programme stalled in 2012 when Sergei Shoigu succeeded Serdyukov as defence minister.[16] Second, the 2014 Russian annexation of Crimea was an incredibly successful operation with only two fatalities and the West reacted with limited sanctions which likely gave Putin the belief that he was a 'shrewd supreme commander'.[17] Third, Russian military operations in Syria from 2015 had been successful in achieving Russian strategic ambitions to become more involved in the Middle East and establish military facilities within the country. However, combat operations were principally in the air domain which were supported by limited ground forces, mainly irregulars, with limited crossover to the context of the SMO. Furthermore, the quality of the militia opposition that Russia faced was far lower than that of the AFU.[18] Fourth, the Western international community had not reacted to the initial build-up of Russian forces towards Ukraine's borders from March 2021 and therefore the strong belief was that the West collectively would not decisively act against Russia on Ukraine's behalf. Finally, the size of Ukraine and the maximalist objectives of the SMO was without modern precedent in Europe and was qualitatively and quantitively of a completely alien scale to post-Cold War limited operations.

Consequently, from July 2021, the 9th Section of the 5th Service of Russia's Federal Security Service (FSB) were selected to shape and plan for the full-scale invasion of Ukraine.[19] All military plans are based on intelligence. If the intelligence is flawed, then so is the plan.

The failure of Russian special services to build an accurate intelligence picture is thus a critical—and surprising—factor in the overall failure of the SMO. It is surprising because Russia had started the operation to destabilize Ukraine in 2014 with the annexation of Crimea and incursions into the Donbass, allowing Russian special services a significant opportunity for intelligence collection and preparation of the battlefield. The assessment that Ukrainian society and the AFU would not resist was completely wrong and likely because the FSB was feeding Putin exactly what he wanted to hear. The resulting confirmation bias fed the sense of complacency. Zabrodskyi et al. argue that there was a tension between 'diffusing Ukraine's political unity and deceiving Ukraine as to the intent' with the former requiring greater shaping activities and the latter speed. The optimism bias concluded that speed could defeat Ukraine's will to resist and therefore 'opted to undertake a shock and awe campaign with little preliminary shaping'.[20] Selecting speed as

15 Shamiev, 'Brass tacks', 4.
16 Ibid., 27.
17 Freedman, 'Why war fails'.
18 Seth G. Jones, 'Russia's ill-fated invasion of Ukraine', CSIS Briefs, June 2022, 7.
19 Zabrodskyi et al., 'Preliminary lessons', 7.
20 Ibid., 8.

the main planning assumption—coupled with complacency and overconfidence—provided several weaknesses in the plan. Furthermore, these weaknesses also meant a misapplication of the principles of war.

First, 'Flexibility' and 'Offensive Action'. No plan B was developed. Therefore, once the opening stages of the offensive started to go wrong, the AFRF had no pre-planned and wargamed courses of action to fall back on and were unable to adapt to a rapidly changing situation.

Second, 'Economy of Effort'. The force was too small to achieve the mission. For the SMO, Russian amassed approximately 190,000 soldiers—regular and irregular—task organized in approximately 120 battalion tactical groups (BTGs) to invade a country with a pre-war population of forty-four million within 600,000 square kilometres, or a ratio of only four Russian soldiers per 1,000 of the Ukrainian population.[21] This force ratio was totally inadequate both for the anticipated warfighting phase and also the long-term occupation that would have been necessary had the SMO succeeded.

Third, 'Surprise' and 'Maintenance of Morale'. The Russian deception plan was successful in making Ukraine think the main effort would be in the Donbass and thus achieved surprise. However, this came at a high cost. The operational security surrounding the SMO was so tight that it has been reported that Sergei Lavrov, the Russian foreign minister, did not even know that the invasion was happening on 24 February.[22] Consequently orders were not issued to units until approximately twenty-four hours before H-hour, with some soldiers even believing that the troop build-up was an exercise, or a limited operation, not a full-scale invasion, and felt they were 'duped' into participating in the war.[23] Therefore, commanders were unable to do detailed ground studies or mission rehearsals with little appreciation for their part in the overall plan.[24] Politicians and commanders cannot expect soldiers to do their jobs if they deliberately choose not to tell them what they are to do. This military dereliction of duty certainly contributed to the high Russian death toll.

Fourth, 'Sustainability'. The FSB failed to identify Western support for Ukraine as its vital ground and centre of gravity. As such, there was no action to interdict and block the main supply routes from Poland to Kyiv. Many countries, including the US, United Kingdom (UK), Poland, and the Baltic States had accelerated the supply of lethal aid to Ukraine in early 2022 which increased after the first few days of the war when the AFU were successfully repelling the AFRF in multiple engagements. With heavy fighting, it was too risky for Western countries to supply arms and ammunition by air and had to use road or rail to move supplies from Poland. However, despite this vital lifeline, there appears to have been no plan whatsoever to exert control over these routes by Russia. The E373/M07 highway that connects Poland to Kyiv is only 34 kilometres from the Belarussian border at its shortest, and to the south, the E40

[21] Jones, 'Russia's ill-fated invasion', 7.

[22] Robbie Gramer and Amy Mackinnon, 'The undignified fall of Russia's once-dignified diplomatic corps', *Foreign Policy*, 31 August 2022. https://foreignpolicy.com/2022/08/31/putin-ukraine-war-lavrov-diplomacy-russia-diplomatic-corps/ (accessed 16 January 2025).

[23] Luke Harding, 'Demoralised Russian soldiers tell of anger at being "duped" into war', *The Guardian*, 4 March 2022. https://www.theguardian.com/world/2022/mar/04/russian-soldiers-ukraine-anger-duped-into-war (accessed 16 January 2025).

[24] Zabrodskyi et al., 'Preliminary lessons', 26.

through Zhytomyr, is approximately 100 kilometres away. Had the AFRF exerted control over these routes, which would have been relatively straightforward to achieve in the opening hours of the war, then it would have denied a significant source of Ukrainian strength and further isolated Kyiv. That an operation of this kind was not even considered shows the overconfidence that the Russian's had in plan A.

Fifth, 'Concentration of Force' and 'Security'. The invasion plan had four axes of advance: to Kyiv from the north, a northeast axis towards Kharkiv, a push from occupied Donbass from the east, and finally a breakout from Crimea and an invasion from the Black Sea fleet in the south. This concept of operations divided logistics support across the whole country which further exacerbated sustainment problems. Russian military logistics relies heavily on internal rail networks and was slow to adapt and establish effective sustainment operations within Ukraine. There were numerous accounts of combat units overrunning their logistics lines and the 'Russian army failed to secure its critical lines of communication'.[25]

Finally, 'Cooperation'. Shaping activities and destabilization were not coordinated with the offensive actions. Significant cyber and destabilization attacks were perpetrated against Ukraine by the whole Russian security apparatus, including the main directorate of the General Staff of the armed forces of the Russian Federation (GRU), foreign intelligence service (SVR), and FSB. The lack of coordination failed to make the most of these actions and therefore had limited impact.[26] There were also disagreements on the intelligence picture and readiness of the AFRF between the Russian General Staff and special services which reportedly led to delays in the invasion date and Putin making the ultimate decision in isolation, away from his arguing security apparatus.[27]

Ukrainian Planning and Execution

In contrast to the confused political leadership in Russia, Ukraine's political objective was simple, clear, and amplified by the communication skills of President Zelensky—'all Russians out of Ukraine'. This simplicity allowed Ukrainian units to operate with full mission command and without direct and overbearing orders and communications. Moreover, the Ukrainian preparation had not just focused on the AFU but on the wider mobilization of Ukrainian society to contribute to its fighting power.

The AFU, like Russia, had also undertaken a period of reform through three cycles since 1991 and understanding it is vital to explaining the successes Ukraine experienced in 2022. The latter cycle, from 2016 was instrumental in enabling flexibility and adaptability from the political down to the tactical level—allowing the

[25] Jones, 'Russia's ill-fated invasion', 6.
[26] Ibid., 7.
[27] 'Ukrainian intelligence: Putin postponed Ukraine invasion date three times', *The Kyiv Independent*, 19 December 2022. https://kyivindependent.com/ukrainian-intelligence-putin-postponed-ukraine-invasion-date-three-times/ (accessed 17 December 2024).

Ukrainians to seize the initiative—which focused on a transition from a Soviet hierarchical and inflexible command philosophy, towards a NATO comparable mission command philosophy.[28]

The experience provided by the AFU's Anti-Terrorist Operation (ATO—latterly renamed the Joint Forces Operation or JFO) against Russian forces in the Donbass helped it to identify deficiencies and develop solutions and 'reform in contact'. In 2016 the AFU undertook the 'Strategic Defence Bulletin for Ukraine' which was an ambitious four-year transformation programme. Sanders argues that it was a highly tailored programme which adopted a 'hybrid approach' which kept the best of the Soviet organization and mindset, including mass and a large standing army, with more precise elements of Western modernization. This modernization focused on improved command and control (C2)—including fostering a mission command culture—the establishment of a non-commissioned officer (NCO) corps, and developing special forces, which raised combat power and helped absorb Western military equipment and training.[29] The implementation of the 'Strategic Defence Bulletin for Ukraine' was supported by the 2017–20 'State Program for the Development of the Armed Forces' and the 2018 'Law on National Security' simplified Ukraine's defence planning and strengthened civilian control over the military.[30] As the AFU was approaching military reform from a lower baseline to that of Russia, they had to focus on doing the basics of military operations very well, which obviously paid off in the confusion of the early days of the invasion.

A specific ten-year procurement plan prioritized special forces, the growth and modernization of artillery and missile systems, including advanced radars, improved C2, drones for artillery spotting, army aviation, drones—such as the procurement of the Bayraktar TB2 with Turkey (which was used to great effect in the opening stages of the war), coastal artillery, and AD systems. The Ukrainians also procured Javelin anti-tank systems from the US and complemented this with the domestic development of Stugna P and Korsar systems. Again, these would prove very effective against poor Russian anti-armour tactics.

Finally, the AFU reserve system was completely transformed in 2020 with the creation of three components: the operational, mobilization, and civilian reserves. When AFU regulars were let go in the Donbass due to pay issues, they went back into the civilian population and provided a source of trained and combat-experienced soldiers ready to fight.[31]

Although the reforms to the AFU were impressive given the scale of the task, the timeframes, and financial constraints they also had help. In 2020 Ukraine issued an updated national security strategy which identified Russia as the central long-term

[28] Deborah Sanders, 'Ukraine's third wave of military reform 2016–2022—Building a military able to defend Ukraine against the Russian invasion', *Defence and Security Analysis* 39/3 (2023): 317. https://doi.org/10.1080/14751798.2023.2201017.

[29] Ibid., 317.

[30] Arda Mevlutoglu, 'Ukraine's military transformation between 2014 and 2022', *Politics Today*, 7 April 2022. https://politicstoday.org/ukraine-military-transformation/ (accessed 10 January 2025).

[31] Zabrodskyi et al., 'Preliminary lessons', 14.

threat to Ukraine's security and sought closer ties with NATO, the European Union (EU), and the US.[32]

From 2014, a variety of training programmes delivered by the US, UK, Canada, Lithuania, and Germany supported the transformation of the AFU including a $2.7 billion combat training centre. Each NATO country took responsibility for training the AFU at a different echelon of command (the US, Canada, Lithuania, Germany, the UK) making the overall package comprehensive and it established strong links with the AFU which were immediately drawn upon during the invasion. The close coordination with Western partners allowed for very specific training for the most dangerous course of action—a full Russian invasion. Therefore, the AFU General Staff had extensively wargamed and exercised for this exact scenario. Moreover, key commanders such as Valerii Zaluzhnyi, commander in chief of the AFU during the invasion and Colonel General Olexandr Syrskyi—who led the defence of Kyiv and would succeed Zaluzhnyi as commander in chief—had significant experience of fighting the Russians on the ATO. They had been promoted from tactical command positions to senior commanders during those years where the ATO was reforming its command structure and philosophy. Therefore, they were both very attuned to the capabilities of their soldiers and the ability to apply mission command to defend against Russian advances without overbearing orders from Kyiv.[33]

Western involvement was not only in improving tactics, training, and procedures (TTPs). Valuable senior relationships were also formed. From 2018, then-Secretary of Defence Ash Carter appointed General Abizaid as a senior defence adviser to Ukraine for defence reform. He was supported by a team of allies, including UK General Nick Parker (former UK land forces commander), Jill Sinclair (senior Canadian defence official), and General Jonas Andriškevicius (former Lithuanian chief of defence).[34] This group regularly visited Ukraine and provided advice on reform and civilian control over the military as it was happening and forming very strong bonds between seniors. This supported the cultural changes necessary to move towards a more Western command philosophy and a national culture of volunteerism to become increasingly compatible with NATO and the EU.[35]

However, Ukraine's preparations were far from perfect. Russia's success at surprise and deceiving Ukraine as to the true objective of the SMO delayed the mobilization of Ukrainian society. Zelensky had believed that issuing the orders for mobilization ahead of the invasion could precipitate it. In addition, as Ukraine believed the main attack would be against the JFO forces in the Donbass, many of Ukraine's best troops

[32] Taras Kuzio, 'The long and arduous road: Ukraine updates its national security strategy', *RUSI*, 16 October 2020. https://rusi.org/explore-our-research/publications/commentary/long-and-arduous-road-ukraine-updates-its-national-security-strategy (accessed 19 December 2024).

[33] Simon Shuster and Vera Bergengruen, 'Inside the Ukrainian counterstrike that turned the tide of the war', *Time*, 26 September 2022. https://time.com/6216213/ukraine-military-valeriy-zaluzhny/ (accessed 19 December 2024).

[34] 'A conversation with Col. Liam Collins', *Fletcher Security Review*, March 2020. https://www.fletchersecurity.org/post/a-conversation-with-col-liam-collins (accessed 19 December 2024).

[35] Liam Collins, 'In 2014, the "decrepit" Ukrainian army hit the refresh button. Eight years later, it's paying off', *The Conversation*, 8 March 2022. https://theconversation.com/in-2014-the-decrepit-ukrainian-army-hit-the-refresh-button-eight-years-later-its-paying-off-177881 (accessed 20 January 2024).

were not defending Kyiv. As an example, despite Central Intelligence Agency (CIA) Director William J. Burns warning Ukraine in January 2022 that Hostomel was key terrain for Russia, the Ukrainians were still taken by surprise.[36]

Overall, the AFU transformation was more successful in preparing them for the war that was about to unfold.

Decisive—24 February to 25th March 2024

The decisive stage of the SMO began in the early morning of 24 February 2024 with a country-wide strike operation to degrade the AFU, especially its AD and infrastructure, as Russian troops crossed their line of departures and advanced on their four main axes.

Ukraine had inherited a large ground-based AD network from the Soviet Union. Therefore, the task of the Russian air and space forces (VKS) was to neutralize the network to enable air assault by their airborne forces (VDV) and then support ground combat manoeuvres from the air. However, following initial country-wide strikes on ground-based early warning radars and long-range surface-to-air missile (SAM) batteries, the VKS failed to exploit the situation and destroy the Ukrainian air force and approximately 300 modern VKS combat aircraft remained grounded in the first four days of combat.[37] Bronk offers several explanations, including limited stocks of precision-guided munitions (PGMs), a lack of confidence in Russian air-to-ground fire control measures and concerns of fratricide, and a lack of time on pilot training.[38] The surprising inability of the VKS to achieve air superiority and provide widespread support to ground operations denied the Russians a key strength and was a striking example of broader coordination problems.

Russia was also unable to integrate combined arms warfare at a suitable echelon to support large-scale ground manoeuvres. The BTG was an old Russian force structure to be used to mitigate poor manning problems in brigades. In combat they are better suited to low-intensity conflicts and counter-insurgencies, such as the fighting Russia experienced in the Donbass from 2014 and in Syria from 2015. It was completely the wrong task organization for the SMO and exhibited weaknesses in unity of command, C2, and sustainment.[39] Therefore a common theme was that despite overwhelming

[36] Erin Banco, Garrett M. Graff, Lara Seligman, Nahal Toosi, and Alexander Ward, '"Something was badly wrong": When Washington realized Russia was actually invading Ukraine', *Politico*, 24 February 2023. https://www.politico.com/news/magazine/2023/02/24/russia-ukraine-war-oral-history-00083757 (accessed 10 January 2024).

[37] Justin Bronk, 'The mysterious case of the missing Russian air force', *RUSI*, 28 February 2022. https://rusi.org/explore-our-research/publications/commentary/mysterious-case-missing-russian-air-force (accessed 5 November 2025).

[38] Ibid.

[39] Amos C. Fox, 'Reflections on Russia's 2022 invasion of Ukraine: Combined arms warfare, the battalion tactical group and wars in a fishbowl', Land Warfare Paper No. 149, Association of the United States Army, September 2022, 4–8. https://www.researchgate.net/publication/364039849_Reflections_on_Russia27s_2022_Invasion_of_Ukraine.

physical quantitative advantages the Russians were repeatedly unable to make the most of them.

As a result of a combination of planning defects, multiple tactical problems compounded to operational mistakes on each axis. First, the mapping that was issued to Russian commanders was either provided late or was out of date. Therefore, the available global positioning system (GPS) systems did not correlate with the maps, and in some cases the GPS had been sold by the commanders. Consequently, Russian armoured and mechanized units stuck to roads to advance and choked up routes and junctions, avoiding the February thaw and marshy ground, and, as they were not expecting stiff Ukrainian resistance, they advanced in administrative columns. In addition, there were reports that cheap Chinese tires failed on vehicles, there was not enough fuel available, or even that fuel was sold by commanders, all reducing vehicle availability and creating the need for recovery.[40] As Russian forces were task organized as BTGs, they had no organic recovery assets, meaning that damaged, broken down, or stuck vehicles could not be recovered, further clogging up routes and junctions and slowing the advance.

This situation gave the AFU considerable opportunities for anti-armour and infantry ambushes, for which they had been directly trained with Western partners and supplied with the necessary weapons for success, such as the next generation light anti-tank weapons (NLAW). As a result of poor Russian C2 and a lack of functioning secure communications, the AFU, supported by Ukrainian intelligence, showed particular skill in targeting Russian commanders, which further compounded AFRF command and coordination.[41] In the first weeks of the war, twelve Russian flag officers were reported killed on the battlefield. Moreover, several senior military officers were also suspended or fired for a variety of reasons, further disrupting effective C2.[42]

The Battle of Hostomel

The first pivotal battle of the war was fought at Hostomel for the control of Antonov airport. Twelve miles to the northwest of Kyiv, it was key terrain to quickly seize the Ukrainian capital and overthrow the political leadership. The capture of the airfield, which was capable of receiving Russian strategic lift, would open an air line of communication (ALOC) with Pskov airfield in Russia where 18 IL-76 transporters were ready to deliver substantial mechanized follow-on forces to the outskirts of Kyiv.

Antonov airfield was home to the Ukrainian national guard's 4th Rapid Reaction Brigade as part of the 72th Mechanized Brigade which was responsible for the defence of Kyiv. A delayed mobilization and successful Russian deception had left the defenders with approximately 200 troops, including some intelligence and special

[40] Lily Russell-Jones, 'Cheap Chinese tires blamed for Russian convoy unable to reach Kyiv', *CITYAM*, 6 March 2022. https://www.cityam.com/cheap-chinese-tires-blamed-as-russian-convoy-unable-to-reach-kyiv/ (accessed 3 January 2025).

[41] Jones, 'Russia's ill-fated invasion', 7.

[42] Ibid.

forces augmentees, but with the contracted professional soldiers and the bulk of the heavy equipment deployed to the JFO in the east. Almost a quarter of the force was clerical staff, they had not received any Western weapons and had never fired the main weapon that they did have, such as the Igla man-portable air defence systems (MANPADS).[43]

Between 200 and 300 air assault forces from the 31st Guards Air Assault Brigade and 45th Separate Guards Spetsnaz Brigade (VDV) loaded onto thirty-four helicopters (Mi-8s, Mi-24s, and KA-52 Alligators) launched from Bolshoy Bokov airport in Belarus, approximately 170 kilometres to the north.[44] Two helicopters were downed by Ukrainian defenders using the Igla MANPADs in the assault but the VDV were able to land and secure the airfield. Despite the seizure of an airfield being an implied task of VDV forces, Kofman et al. argue that there was no evidence that the soldiers had rehearsed the task or knew their part in the plan.[45] Reports suggest that the Ukrainian defenders were similarly surprised.[46]

Having secured the airfield, the VDV became isolated and so Ukrainian forces counter-attacked. Zelensky reportedly directly gave the order to 'show no mercy. Use all available weapons to wipe out every Russian thing that's there'—demonstrating effective mission command.[47] In recognition of Hostomel's significance to the Russian SMO and denying Russia the ALOC, crack troops assigned to the defence of Kyiv were sent north including elements of the 80th Air Assault Brigade, 95th Air Assault Brigade, 72nd Mechanized Brigade, and the 3rd Special Purpose Regiment of the special operations forces (SSO).[48] This motley Ukrainian force fought back and was able to dislodge the understrength VDV forces that could not hold the airfield without the support of heavier weapons. The 72nd Brigade's artillery was brought up from their base at Bela Tserkva, 100 kilometres south of Kyiv, and targeted the VDV. The artillery fire also crated the runway, denying an ALOC being secured with Pskov. Consequently, the Russian's called off the planned reinforcement, due to the Ukrainian artillery fire, the inability of the VDV to hold the airfield, that Ukrainian AD was still functioning, and because they had lost six to seven helicopters in the assault.[49] Despite this loss, Russian forces continued to attempt to land at Antonov airfield after 24 February and fighting continued at the airfield until early April.[50]

[43] Oksana Torop and Svyatoslav Khomenko, 'The fight for Hostomel airfield: How the gates to Kyiv stayed locked', *BBC News Russian*, 29 February 2024. https://bbcrussian.substack.com/p/ukraine-war-the-fight-for-hostomel-airfield (accessed 7 November 2024).

[44] Michael Kofman, Liam Collins, and John Spencer, 'The Battle of Hostomel airport: A key moment in Russia's defeat in Kyiv', *War on the Rocks*, 10 August 2023. https://warontherocks.com/2023/08/the-battle-of-hostomel-airport-a-key-moment-in-russias-defeat-in-kyiv/ (accessed 5 November 2024).

[45] Ibid.

[46] Torop and Khomenko, 'The fight for Hostomel airfield'.

[47] Simon Shuster, *The showman: The inside story of the invasion that shook the world and made a leader Volodymyr Zelensky* (New York: William Morrow, 2024): 28.

[48] Kofman et al., 'The Battle of Hostomel airport'.

[49] Ibid.

[50] Torop and Khomenko, 'The fight for Hostomel airfield'.

It is hard not to overstate the significance of the Battle of Hostomel when evaluating the SMO and the course of the war to date. Kofman et al. argue that 'The actions of individual commanders, soldiers, and citizens can and did have a profound impact in a pivotal battle that would help decide the course of the war'.[51]

Indeed, the Battle of Hostomel pitted some of the very best and most well-equipped Russian units against a depleted and uncoordinated Ukrainian national guard unit.[52] Viewing the battle through the lens of the physical and conceptual components (the VDV specialize in air assault) the Russians should not have lost the battle. The moral component is therefore the best explanation for the outcome. So is luck.[53] Shuster recalls a conversation between the US Joint Chief of Staff Milley and General Zaluzhny where the former 'did not ascribe the success to clever Ukrainian planning ... it looked more like a military miracle' and said 'Son, you just got lucky'. The latter said the remark 'stung'.[54]

Had the VDV been able to secure the ALOC between Pskov and Hostomel then thousands of Russian forces, with mechanized support, would have been delivered to the outskirts of Kyiv in the opening days of the war. Under this outcome then the SMO could have been successful. Instead, it was a seismic defeat for the Russians. There are no precise numbers of Russians killed at Hostomel. However, Oryx confirmed the loss of seven armoured fighting vehicles, twenty-three infantry fighting vehicles, three armoured personnel carriers, two towed artillery pieces, six helicopters, one jamming system, and sixty-eight support vehicles at Hostomel by the time the Russians withdrew on 29 March.[55]

This was not only a lost battle but a failure of the SMO. With no plan B developed, Russia was unable to seize the initiative, or Kyiv. With the Antonov airport ALOC denied, the AFRF reactively decided to undertake an 80 mile mechanized ground move from the Belarussian border to Kyiv. The 40 mile long convoy of approximately 15,000 troops, stalled eight days into the war due to Ukrainian action, a lack of leadership, fuel, and supplies, with the lead elements approximately 19 miles from Kyiv city centre.[56] Indeed, the convoy was a tangible metaphor for the logistics and command troubles that plagued the AFRF across the whole of Ukraine.[57] On 3 March, the deputy commander of the 41st Combined Arms Army of the Central Military

[51] Kofman et al., 'The Battle of Hostomel airport'.

[52] Torop and Khomenko, 'The fight for Hostomel airfield'.

[53] Tim Mak, 'The counteroffensive: Russia's failure to take Kyiv was luck and timing', *The Kyiv Independent*, 22 February 2024. https://kyivindependent.com/opinion-russias-failure-to-take-kyiv-was-pure-luck/ (accessed 20 January 2024).

[54] Shuster, 'The showman', 82.

[55] Jakub Janovsky et al., 'Destination disaster: Russia's failure at Hostomel airport', *Oryx*, 13 April 2022. https://www.oryxspioenkop.com/2022/04/destination-disaster-russias-failure-at.html (accessed 5 November 2024).

[56] Luke McGee, 'Here's what we know about the 40-mile-long Russian convoy outside Ukraine's capital', *CNN World*, 3 March 2022. https://edition.cnn.com/2022/03/03/europe/russian-convoy-stalled-outside-kyiv-intl/ (17 December 2024).

[57] 'Why a huge Russian convoy remains stalled north of Kyiv', *The Economist*, 4 March 2022. https://www.economist.com/europe/2022/03/04/why-a-huge-russian-convoy-remains-stalled-north-of-kyiv (17 December 2024).

District, Major General Andrei Sukhovetsky, was killed by a Ukrainian sniper at the front of the convoy and was (by then) the highest ranking Russian officer killed.[58]

The AFU targeted the vanguard of the convoy which blocked progress and fixed it to the road. The AFRF had to dismount and dig into the roadsides, to defend against the Ukrainian special forces, who used ambushes, artillery, and drones to attrite the convoy.

Indeed, while the Battle of Hostomel has become the focal point of assessing the SMO, Ukrainian counter-attacks in other areas of the Kyiv Oblast periphery also contributed to the success. Fighting in the Kyiv suburbs of Bucha, Irpin, and Hostomel (wider than the airfield) successfully held up the AFRF advance. AFU special forces were specifically trained for this role and worked closely with Ukraine's artillery units to blunt advancing Russian mechanized forces, causing heavy Russian losses and disrupting the whole operation.[59] Zabrodskyi et al. argue that 'Despite the prominence of anti-tank guided weapons in the public narrative, Ukraine blunted Russia's attempt to seize Kyiv using massed fires from two artillery brigades' to the north.[60] The pre-war focus on improving the quantitative and qualitative strength of its artillery systems was decisive in the battle for Kyiv, where the Ukrainians held the advantage, and the defence of Hostomel airport.[61]

Sustaining—25th March 2022 and Beyond

Vladimir Putin's SMO failed in the opening days of the 2022 invasion. Without a plan B to quickly fall back on, the AFRF were unable to press home their significant physical advantages and gain the initiative. They were consistently outfought at the operational and tactical levels by the AFU. Consequently, on 25 March 2022, the Russian military announced a withdrawal from Kyiv—to redeploy to the Donbass—and, by 6 April, Russian forces had completely withdrawn from Kyiv Oblast.[62] The first month of the fighting was critical to the survival of Ukraine as a sovereign state. With the AFU winning key battles at Hostomel, Bucha, and Irpin, Kyiv was saved and forced Russia into pursuing modest military objectives in the Donbass, which, as Western support has increased, has become a challenge for the AFRF.

[58] Tom Batchelor and Jane Dalton, 'Russian Major General Andrei Sukhovetsky killed by Ukrainians in "major demotivator" for invading army', *The Independent*, 7 March 2024. https://www.independent.co.uk/news/world/europe/andrei-sukhovetsky-russian-general-killed-b2029363.html (accessed 20 January 2024).

[59] Federico Borsari, 'Hunting the invader: Ukraine's special operations troops', *CEPA*, 15 March 2022. https://cepa.org/article/hunting-the-invader-ukraines-special-operations-troops/ (accessed 20 January 2024).

[60] Zabrodskyi et al., 'Preliminary lessons', 2.

[61] David Axe, 'Ukraine's artillery did the most killing around Kyiv, ultimately saving the city from Russian occupation', *Forbes*, 26 December 2022. https://www.forbes.com/sites/davidaxe/2022/12/26/ukraines-artillery-did-the-most-killing-around-kyiv-ultimately-saving-the-city-from-russian-occupation/ (accessed 3 January 2024).

[62] Peter Dickinson, 'Russia in retreat: Putin appears to admit defeat in the battle for Kyiv', *Atlantic Council*, 29 March 2022. https://www.atlanticcouncil.org/blogs/ukrainealert/russia-in-retreat-putin-appears-to-admit-defeat-in-the-battle-for-kyiv/ (accessed24 January 2025).

In the opening days of the invasion, the US offered to evacuate President Zelensky. However, he has been widely quoted as refusing, declaring that 'The fight is here; I need ammunition, not a ride'.[63] It was a pivotal moment and an expression of leadership, clarity, and a superior moral component which was absent within the Russian system. Of equal significance, one week before the invasion, President Zelensky travelled to the Munich Security Conference at great personal risk to himself and to Ukraine, as his absence could have precipitated the invasion. This journey was instrumental in personifying the Ukrainian dilemma to Western backers—including US Vice President Kamala Harris—and convince them that he was a person that they could do business with.[64] The heroic resistance of the AFU and Ukrainian society, not just around Kyiv but across the county, such as the Azovstal steelworks defenders in Mariupol and the defenders of Snake Island, was amplified by social media and made a direct connection with Western societies. Moreover, early reporting of war crimes committed by the Russians in Bucha highlighted the lack of ethical foundation in the AFRF. Therefore, the superior Ukrainian moral component was central in replenishing its physical component through significant Western military support and its conceptual component through greater NATO input into developing Ukrainian strategy and operations.

Conclusion

The Russian political, strategic, operational, and tactical failures with the SMO was comprehensive and systematic. Not only has this failure weakened Russia, through the AFU exposing the AFRF as a paper tiger and unworthy of the 'second most powerful military in the world' label, it has strengthened both Ukraine and NATO. Ukraine is now central, not peripheral, to European security, has significant and growing European support, and is on an 'irreversible' path to NATO membership. Meanwhile, the alliance has two new members in Finland and Sweden, a transformation of its Russia facing defence plans, and an agreement to spend 5% of gross domestic product (GDP) on defence by 2035. In this context, it is hard not to overstate the significance of the initial ten days of fighting and how consequential they were to the current situation in Europe.

Had the SMO succeeded in quickly seizing Kyiv, supplanting the Ukrainian leadership, and defeating the AFU in the field, then European security would look very different today. In this scenario, Russian military power would have successfully subjugated Ukraine and would now pose a significant threat to Europe and NATO. Instead, an attritional war has developed which both sides view as existential, with neither having the offensive potential to regain the initiative and make decisive gains on the battlefield. Despite the significant technological innovation on both sides in

[63] Sharon Braithwaite, 'Zelensky refuses US offer to evacuate, saying "I need ammunition, not a ride"', *CNN*, 26 February 2002. https://edition.cnn.com/2022/02/26/europe/ukraine-zelensky-evacuation-intl/index.html (accessed 20 January 2024).

[64] Arnold, 'Slava Ukraini'.

the war in Ukraine, the best explanation for the outcome of the first ten days of fighting remains traditional military tenets, specifically the principles of war and the moral component of fighting power. Indeed, one of the greatest lessons for the intelligence and analytical community, and defence planners, is to develop methodologies to accurately assess the morale component and its interaction with the physical and conceptual components.

This is a critical requirement as there are strong indications that Russia is institutionalizing it learning from the initial failure and subsequent attritional war. In 2022, Russia launched a systematic review of its performance and has been optimizing its approach to war, including better relationships between the military and defence industry.[65] The development of new tactics, more effective weapons and changes in command culture all demonstrate signs of the AFRF being capable of 'reforming in contact' and therefore readying itself for a potential war with NATO. Therefore, it is a strong assumption for NATO defence planners that Russia is capable of learning from its myriad mistakes and the Russian military that emerges from the war in Ukraine will be stronger and more experienced than the one that lost the SMO.

[65] Dara Massicot, 'What the Kremlin is learning from the war in Ukraine', *Foreign Affairs*, 8 October 2025. https://www.foreignaffairs.com/russia/how-russia-recovered (accessed 8 October 2025).

16
Ukrainian Counter-Offensives in Kharkiv and Kherson June–November 2022

Dara Massicot

Introduction

This chapter examines two Ukrainian counter-offensive operations in the Kharkiv and Kherson regions during the summer and autumn of 2022. It surveys the strategic context for the operations and force dispositions of Ukrainian and Russian forces beginning in June 2022 through the Russian withdrawal across the Dnipro River in November 2022. This analysis considers the factors that contributed to the operational successes and failures of the armed forces of Ukraine (AFU) and Russian forces, respectively. In the summer of 2022, Russian forces in Ukraine were at the nadir of their power, having taken high casualties and equipment losses with no ready strategic reserves available.

Ukrainian forces were able to exploit Russian forces' weakness by first luring Moscow's forces to the south, and then attacking the most sparsely occupied parts of the front line elsewhere in Kharkiv. Next, Ukrainian forces launched a second counter-attack to take advantage of Russia's unfavourable positions on the western side of the Dnipro River in Kherson, forcing them to withdraw under fire by November.

However, this chapter cautions that the circumstances that contributed to successful Ukrainian counter-offensives in 2022, namely understrength Russian units and Russia's decision to hold unfavourable terrain, would not be present in the years that followed, particularly after Russia mobilized 300,000 personnel and its economy to support the war effort. As subsequent chapters will discuss, the AFU counter-offensive in the summer of 2023 and surprise attack on Kursk would not find the same success.

Strategic Context and Status of Forces as of June 2022

At the start of June 2022, the AFU withstood the initial Russian invasion of their country and pushed back multiple Russian axes of attack on Kyiv and northeastern

Dara Massicot, *Ukrainian Counter-Offensives in Kharkiv and Kherson June–November 2022*. In: *Russia's War in Ukraine and Modern Warfare*. Edited by: Mikael Weissmann and Niklas Nilsson, Oxford University Press. © Dara Massicot (2026). DOI: 10.1093/9780198970903.003.0016

Ukraine. By June, Russian forces occupied around 20% of Ukrainian territory.[1] The AFU swelled in size thanks to mobilization and large numbers of volunteers. Their main operational tasks during this time were continuing active defence operations against Russian attacks in the Donbas and southern Ukraine, second, stabilizing areas in the north recently liberated from Russian occupation, and third, securing state borders with Belarus and Russia.

By June, Ukraine had depleted most of its own artillery stockpiles, and was increasingly dependent on artillery support that had recently begun flowing in larger quantities from the West.[2] Western partners initially focused their weapons resupply efforts in this phase of the war on Soviet-origin equipment from abroad, as these equipment types were already in use by Ukrainian forces and were easier to maintain. When this effort reached its limits, Western partners of Ukraine began supplying 155 millimetre ammunition and Western artillery systems and other armoured vehicles by late spring. Ukraine would receive systems like the M777 Howitzer, which used global positioning system (GPS) guided Excalibur rounds. Ukraine also received the M270 self-propelled multiple launch rocket system (MLRS). On 23 June, the first US high-mobility artillery rocket system (HIMARS) arrived in Ukraine and a second set of launchers was delivered in early July.[3] Both the M270 MLRS and HIMARS systems can fire M31A1 Guided Multiple Launch Rocket System (GMLRS); Ukraine was originally provided with the 70 kilometre range variant.

In July, Ukraine received National Advanced Surface-to-Air Missile Systems (NASAMS) and IRIS-T air defence systems from Western partners. By August, Ukraine received additional self-propelled artillery systems like the Caesar and PanzerHaubitze 2000 systems, helicopters, M113 armoured personnel carriers and Maxxpro armoured vehicles, Switchblade loitering munitions, and other critical systems. While Bayraktar drones (manufactured in Turkey) were useful during the opening months of the war, Russian forces adapted and increasingly targeted the Bayraktars with air defence and jamming; this platform would be increasingly held back from front-line missions from summer 2022 onwards.[4]

American and European support to Ukraine expanded and became formalized through various governmental agreements (like the Lend-Lease Act) and establishment of new working groups (like the Defense Contact Group, which meets in Ramstein, Germany). During this time, Ukraine was granted candidate status for European Union (EU) accession. Crucially, in the spring of 2022, the Western

[1] Matt Murphy, 'Ukraine war: Zelensky says Russia controls a fifth of Ukrainian territory', *BBC News*, 2 June 2022. https://www.bbc.com/news/world-europe-61675915.

[2] Isobel Koshiw, 'We're almost out of ammunition and relying on Western arms, says Ukraine', *The Guardian*, 10 June 2022. https://www.theguardian.com/world/2022/jun/10/were-almost-out-of-ammunition-and-relying-on-western-arms-says-ukraine.

[3] 'U.S. long-range rocket systems arrive in Ukraine—Minister', *Reuters*, 23 June 2022. https://www.reuters.com/world/europe/us-long-range-rocket-systems-arrive-ukraine-minister-2022-06-23/.

[4] Anges Helou, 'With Turkish drones in the headlines, what happened to Ukraine's Bayraktar TB2s?', *Breaking Defense*, 6 October 2023. https://breakingdefense.com/2023/10/with-turkish-drones-in-the-headlines-what-happened-to-ukraines-bayraktar-tb2s/.

intelligence and weapons delivery pipeline matured. In June, Western partners also agreed to train Ukrainian forces abroad.

President Joe Biden outlined the American strategy for supporting Ukraine in a May article in the *New York Times*. He set the following end state for Ukraine as: 'a democratic, independent, sovereign and prosperous Ukraine with the means to deter and defend itself against further aggression'.[5] Biden also articulated the boundaries of United States (US) support, noting that the US would not directly engage in the war (no troops engaged in attacking Russian forces), or encourage or enable Ukraine to strike beyond its borders.[6]

By late March 2022, Russia's northern axes of attack met strong Ukrainian resistance and failed. Russian forces were withdrawn from northern Ukraine when it was obvious that the Kyiv, Chernihiv, and Sumy axes of attack could not succeed. Russian units involved on those fronts sustained high casualties and significant equipment loss. Withdrawn units were reconstituted in Belarus and Russia and, by the beginning of June, were being redeployed to a smaller front line which stretched from Kharkiv to Crimea. In the south, the cities of Melitopol and Mariupol fell to Russian occupation in March and May, respectively, and Russian forces were able to connect to the Crimean peninsula by land. The southern front was comparatively more quiet than other areas.

As of June 2022, Russia's priority operations concentrated on Luhansk and Donetsk, where they would attempt (and fail) to execute a deep envelopment of the Donbas, with one of the main prongs attacking south based out of Izium and Lyman. As part of their summer 2022 consolidation, Russian forces began to dig in, and established better electronic warfare (EW) and air defence zones that would provide greater protection for their own forces and introduce new challenges for the AFU's systems. They achieved this by establishing a layered air defence zone and better electromagnetic spectrum management that jammed communications and some precision munitions.[7]

Russia faced a critical manpower shortage that was evident after the opening months of the war. At this time, Russian forces had a strong artillery advantage over AFU forces and used artillery to take and hold their positions and compensate for manpower shortfalls. However, the artillery support arriving from the West would undercut those advantages as the summer progressed. By May 2022, after the shock of the initial operational failures, and accounting of casualties and destroyed equipment, Russia began deploying personnel from its military bases in Central Asia and the Caucasus to fill critical manning gaps.

By June, the Russian military was aware that it needed to replace losses and create a strategic reserve, however, the Russian state had not yet organized to recruit

[5] Joseph R. Biden, 'What America will and will not do in Ukraine', *New York Times*, 31 May 2022. https://www.nytimes.com/2022/05/31/opinion/biden-ukraine-strategy.html.
[6] Ibid.
[7] Koshiw, 'We're almost out of ammunition'.

enough personnel and the Kremlin was likely reticent to mobilize at this time. Russian leaders initiated a volunteer recruiting effort to recover losses and denied they would mobilize Russian citizens (although they did conscript men of all ages in occupied Luhansk and Donetsk).[8] This early recruiting drive, which included combat bonuses and poorly provisioned regional 'volunteer battalions' failed to attract sufficient numbers of recruits—it did not replace casualties, let alone fill out new units to generate a strategic reserve.[9]

During the first six months of the war, the military had not yet closed legal loopholes that allowed contract service personnel to resign on various grounds (such as poor service conditions or moral opposition), and small numbers of officers and those enlisted resigned from occupied Ukraine or refused to deploy from their garrisons in Russia.[10] Local commanders began ignoring such resignation requests by July and August, until the loophole was formally closed in September. Russia's manpower problems would worsen over the summer despite Russia's initial attempts to redress problems and would culminate in a mobilization of 300,000 men in late September.

Case Study: Kharkiv Counter-Offensive

For narrative purposes, this chapter describes the Kharkiv and Kherson counter-offensives as two separate operations, but it should be noted that they were planned as complimentary operations. They overlapped in time: shaping operations would begin in August with the main operational phase beginning in early September 2022. The Kherson offensive was planned first, but once it became clear that the Russian command intended to deploy forces Kharkiv to Kherson, Ukrainian officials decided to create a new surprise supporting counter-offensive in Kharkiv, to exploit the emerging opportunity of a weak spot on the Russian front.[11]

According to senior Ukrainian officials, their intention was to publicly discuss the Kherson counter-offensive while keeping the Kharkiv plans and movements concealed as long as possible. Ukrainian strategists intended to lure Russian forces to Kherson, and once they relocated, use the element of surprise and attack the front line where they were weak, in Kharkiv.

[8] Dara Massicot, 'Russia's post-war military recruiting strategy emerges', *War on the Rocks*, 16 August 2024. https://warontherocks.com/2024/08/russias-post-war-military-recruiting-strategy-emerges/.

[9] Ibid.

[10] Dara Massicot, 'Russian military wartime personnel recruiting and retention 2022–2023', RAND Corporation, RRA2061-4, 16 July 2024. https://www.rand.org/pubs/research_reports/RRA2061-4.html.

[11] Isabelle Khurshudyan and Paul Sonne, 'Inside the Ukrainian counteroffensive that shocked Putin and reshaped the war', *Washington Post*, 29 December 2022. https://www.washingtonpost.com/world/2022/12/29/ukraine-offensive-kharkiv-kherson-donetsk/.

Disposition of Forces and Operational Preparations in Kharkiv

In the spring of 2022, Russian forces tried to advance in the Kharkiv region, but their forces in the area were depleted after reaching Balakliia and after the failed attempt to cross the Severodonetsk River in May. Russian force disposition in Kharkiv was sparse in the summer of 2022, as the sector was a low-priority area of the front, and most AFU units were engaged elsewhere. Russia's main operational goal as of June 2022 was elsewhere: conducting a deep envelopment of the Donbas in Luhansk and Donetsk oblasts where they also ran into resistance and AFU defences.

Russian units around Kharkiv by the summer were a mixture of understrength or low-readiness units that lacked prepared defensive positions. Some units were at strength but were assigned more territory to hold than their unit could support, weakening their overall combat potential. Russian forces in Kharkiv at the beginning of June 2022 were spread out into two main clusters, one arrayed east of Kharkiv City and another to the south near Izium. Russian forces were thin between Vovchansk and Kupiansk, even though this was a critical ground line of communication (GLOC) that supported to their stronghold at Izium.

Russia had around ten tactical battalion groups in the area north and east of Kharkiv in June, according to open-source information. These units were a mixture of ground forces units, mobilized units from Luhansk and Donetsk, and some BARS (reserve, or Boyevoy Armeyskiy Rezerv Strany) units. Units included the 25th Guards Separate Motorized Rifle Brigade (SMRB), 200th SMRB, the 138th Motorized Rifle Regiment (MRR), and 27th Guards SMRB.[12] Deployed east of Kharkiv and near Kupiansk were elements of the 106th Guards Airborne Division from the Russian airborne forces (VDV), 13th Guards Tank Regiment (GTR), 423rd Guards MRR, and 3rd Motorized Rifle Division (MRD).

The manning levels of this force grouping varied significantly, according to Russian logbooks, recovered after the units fled. For example, the 200th SMRB had only 40% of its authorized personnel and only 30% of its equipment. Yet other nearby artillery units had 90% manning.[13] However, other units subordinate to the 11th Army Corps (AC) near Balakliia, which commanded many units in the area, had highly variable manning and equipment levels. The 11th AC's manning was overall 71% of authorized strength, one week prior to the AFU counter-offensive.[14] For example, the 11th AC units, as of 30 August 2022 ranged from 20–100% of their authorized wartime manning, according to recovered Russian documents, with a

[12] 'Deep State map', Deep State blog, 1 June 2022 update. https://deepstatemap.live/en#10/50.2164584/36.7252350 (accessed 31 January 2025).

[13] Massicot, 'Russian military'.

[14] Mari Saito, Maria Tsvetkova, and Anton Zverev, 'Abandoned Russian base holds secrets of retreat in Ukraine', *Reuters*, 26 October 2022. https://www.reuters.com/investigates/special-report/ukraine-crisis-russia-base/ (accessed 31 January 2025).

majority of units above 60%. In one unit with storm (assault) battalions, one storm battalion was at 88% manning, while the second storm battalion in the same parent unit had only 20% of its personnel.[15] A nearby BARS unit was staffed at 23%, but a tank battalion was at 93% strength.[16]

These manpower levels suggest that the 11th AC was not poorly manned or equipped, and only a minority of its units were understaffed. Therefore, other factors like panic and operational surprise are more likely factors that explain why this sector collapsed so rapidly in September.

The largest Russian force grouping in the area was further to the south near Izium, which was intended to be the northern pincer of a deep envelopment of the Donbas.[17] Elsewhere, Wagner forces were brought into the war to assist, with attacks on Popasna and Bakhmut. Russia's offensive tempo south of Izium accelerated in August. Senior Ukrainian officials estimated that Russian force strength near Izium contained around twenty-four battalion tactical group (BTG) equivalents by August 2022.[18] It contained elements of several units like the 1st Guards Tank Army (which had sustained critical damage in the opening weeks of the war, leading to the sacking of its commanding officer), the 76th Guards Air Assault Division (GAAD), 2nd Guards MRD, 5th Guards Tank Brigade (GTB), 37th Guards SMRB, 252nd Guards MRR, 752nd Guards MRR, 38th SMRB, 64th SMRB, 1st Guards MRR, and 1st GTR.[19]

Prior to the offensive, the AFU engaged in a successful strike campaign on Russian command posts, logistical hubs, and weapons storage sites in July, after they received HIMARS from the West. Russia's then-Defence Minister Shoigu ordered Russian forces to prioritize the destruction of HIMARS units. Russian intelligence services quickly prioritized gathering information on HIMARS launchers, yet it would take a month for Russia to relocate many of its command posts outside of strike range.[20] On 21 July, the Russian FSB informed the 11th AC leadership that the AFU was preparing to move HIMARS into its area of responsibility, and intended to strike specific command posts, according to recovered Russian documents.[21] Three days later, the AFU struck several positions, suggesting the FSB's intelligence was accurate. Within two weeks of the warning, Russian forces deployed POL-21 EW systems to the area to jam inbound Ukrainian precision munitions.[22]

Further, recovered Russian documents from staff meetings in late July revealed that the 11th AC either determined or was informed by Russian intelligence that the AFU wanted to 'take control of Balakliia'. These reports indicate that Russian intelligence

[15] Ibid.
[16] Ibid.
[17] @JominiW, X, 2 August 2022. https://x.com/JominiW/status/1554647164692926464/photo/1 (accessed 18 January 2025).
[18] Khurshudyan and Sonne, 'Inside the Ukrainian counteroffensive'.
[19] 'Deep State map', 1 June 2022.
[20] 'Russia targets Ukraine's missiles as Western-supplied weapons bite', *Reuters*, 18 July 2022. www.reuters.com/world/europe/russian-defence-minister-prioritise-destroying-ukrainian-missiles-2022-07-18/.
[21] Saito et al., 'Abandoned Russian base'.
[22] Ibid.

was accurately picking up on crucial information, and that commanders were made aware of AFU plans but may have lacked actionable specifics on timing.[23]

Regardless of the new EW systems in the area brought in to protect them, Russian command posts in this area were struck again by HIMARS in August. According to recovered Russian documents, this force grouping had an insufficient number of reconnaissance drones, which may have limited their ability to detect Ukrainian movements prior to the offensive, and it also lacked enough functional armoured personnel carriers, which may explain why they could not engage attacking AFU units and why so many were abandoned during the counter-offensive.

In July, news began to spread in the Ukrainian information space that a planned counter-offensive in Kherson would occur. Russian forces took notice. In mid-August, Western intelligence shared information that half of Russian forces in Izium would be diverted to Kherson, as the Russian command saw signs of a Ukrainian counter-offensive there, according to Ukrainian officials.[24] Ukrainian officials decided in August that Kharkiv would become a new secondary counter-offensive effort, designed to take advantage of the weakness caused by the Russian deployments to Kherson.

All planning for the Kharkiv operation was closely guarded. Counter-offensive plans for Kharkiv and Kherson were not disseminated to many commanders until late August, according to participating Ukrainian officers, and some field-grade officers were told about it within one to two weeks of the operation itself.[25] As predicted by Western intelligence, mid-August, Russian forces were removed from the Izium area to reinforce Kherson, in advance of a future Ukrainian attack on Kherson.[26]

Operations Begin

The Kharkiv counter-offensive began on 6 September, when the AFU attacked several areas at once, rather than a massed attack. The Kharkiv counter-offensive depended on deception and surprise for success in its initial stages. To support that goal, in the days leading up to the operations, the AFU used deception to lay pontoons at crossing points to distract the Russians, 30 kilometres south of where their true crossing points would be.[27] According to General Oleksandr Syrsky, operational surprise and rapid manoeuvre were critical elements of the Kharkiv operations: 'everything depended on the first day—how far we could break through. The farther we went the less they

[23] Ibid.
[24] Khurshudyan and Sonne, 'Inside the Ukrainian counteroffensive'.
[25] Ibid.
[26] General Staff of the AFU, Facebook post, 31 July 2022. https://www.facebook.com/GeneralStaff.ua/posts/pfbid0QqNVn7xbH6mFvadsx5DE54U5sJrjD9y2eyeYF32pu9j6qGn66Xbpo9v73nZAXvuPl (accessed 9 January 2025); 'Kharkivska Nastupal'na Operatsiya: Yk Tse Bylo', Sluhati Radio, 3 August 2023, https://www.armyfm.com.ua/ua/harkivska-strategichna-nastupalna-operaciya-yak-ce-bulo/.
[27] Ibid.

could do, the more their units would be cut off and isolated under psychological pressure'.[28]

Ukrainian forces attacked first near Balakliia, Izium, Shevchenkove, and Kupiansk on 6 September.[29] They attacked in multiple areas to sow confusion and chaos and to exploit multiple gaps in Russian lines. On the first day of the counter-offensive, the AFU attacked command posts, fuel storage, and other logistics locations with GMLRS strikes. According to the commander of the Ukrainian 25th Airborne Assault Brigade, 'They [Russian forces] were panicking because we attacked all front-line positions at once—the entire front line itself was enormous—and everywhere there was a breakthrough'.[30] The 11th AC headquarters was struck between 6 and 8 September, causing a temporary breakdown in command and control.

By 7 September, the AFU reached Balakliia, where they encountered Russian ground forces units. They also encountered national guard units (OMON and SOBR), as well as Luhansk militia forces, which were not highly combat capable and retreated quickly.[31] On 8 September, AFU units reached the outskirts of Kupiansk. By 9 September, Russian forces began abandoning their equipment, while others were caught in ambushes. AFU commanders were informed during time that many Russian forces had started a retreat from Izium and so AFU orders were modified to accelerate attacks to pursue them there.

By 10 September, Ukrainian forces reported that between ten and twelve Russian BTGs broke apart near Izium and had lost partial or full combat capability in the process. Abandoned weapons and supplies later recovered by Ukrainian forces near Izium suggest that Russian forces there were well supplied, and likely broke due to panic.[32] The retreat was chaotic, according to eye witnesses, who described Russian soldiers abandoning their equipment on the road and running on foot.[33] According to open-source information, around half of the equipment left behind was fully functional.[34] On 12 September, AFU forces kept up their pursuit until Russian forces had crossed over the border north of Kharkiv.[35] In less than a week, the AFU had liberated around 3,800 kilometres of territory.[36]

Russian forces were dug in around Lyman, and it would take a few more weeks for AFU forces to push out Russian forces in Lyman and dug-in units at Borshchova.[37] In contrast to the other forces in the Kharkiv area, those in the Lyman area did not flee in

[28] Khurshudyan and Sonne, 'Inside the Ukrainian counteroffensive'.
[29] 'Харківська наступальна операція'.
[30] Khurshudyan and Sonne, 'Inside the Ukrainian counteroffensive'.
[31] 'Харківська наступальна операція'.
[32] Khurshudyan and Sonne, 'Inside the Ukrainian counteroffensive'.
[33] Saito et al., 'Abandoned Russian base'.
[34] @ragnarbjatur, Post on X, 13 September 2022. https://x.com/ragnarbjartur/status/1569595181371908096.
[35] 'Харківська наступальна операція'.
[36] Ibid.
[37] Jason Beaubien, 'Ukraine's offensive in Kharkiv was hard and bitter, say soldiers who did the fighting', *NPR*, 29 September 2022. https://www.npr.org/2022/09/29/1125278321/ukraine-offensive-russia-borshchova-kharkiv-oblast.

panic and had installed surveillance systems to monitor and track AFU movements. Ultimately, they were driven out and left at night by early October 2022.[38]

Case Study: Kherson Counter-Offensive

The AFU's goal for the Kherson counter-offensive, planned during the summer of 2022, was to liberate territory and destroy or capture Russian forces on the right bank of the Dnipro River.[39] In the telling of senior Ukrainian officials, the Kherson counter-offensive was designed to be the main counter-offensive action for the summer and fall of 2022.[40] However, as discussed above, when it became clear that Russian forces were reinforcing Kherson at the expense of the Kharkiv front, the Ukrainian General Staff decided to add a second counter-offensive.

In contrast to the Kharkiv counter-offensive operation, which relied on surprise, speed, and taking advantage of sparsely defended front lines, the Kherson counter-offensive depended on precision strikes to disrupt ground lines of communications (in this case, river crossings), artillery fires, and when possible direct engagements, to push Russian forces east across unfavourable terrain, across the Dnipro River. It was a risk to conduct two major operations at once due to artillery shortages in the force, and leadership in Kyiv had to be convinced.[41]

Russian forces in the south made rapid gains in the Kherson region in the opening weeks of the war and had several months to create dug-in defensive positions. As of June 2022, Russian forces occupied the city of Kherson and areas around it and were attempting to press westward. They also made good use of the terrain, created lines of defences and trenches in treelines, hardened trenches, placed pillboxes, laid mines, and were well supplied by the southern military district's distributed logistics system in the region.[42] The Russian units in the area were well manned as of June 2022, because they did not experience the type of extensive losses that their counterparts faced in the northern invasion routes in February and March.

Russian forces in Kherson also used airfields and other bases on Crimea to reinforce and resupply units in Kherson from two directions. This sector was commanded by one of Russia's more competent generals, General Sergei Surovikin, who was also commander of the Russian Aerospace Forces (VKS). Russia's main challenge in this area was sustaining forces on the west side of the Dnipro River where the city of Kherson was, also known as the right bank, supplied only by small number of road and rail bridges.

The forward grouping of Russian forces across the Dnipro River consisted of several BTGs from higher readiness units from the airborne forces, special forces, and

[38] Khurshudyan and Sonne, 'Inside the Ukrainian counteroffensive'.
[39] Ibid.
[40] Ibid.
[41] Ibid.
[42] Jeffrey Gettleman, 'In Ukraine's south, fierce fighting and deadly cost', *New York Times*, 24 September 2022. https://www.nytimes.com/2022/09/24/world/europe/ukraine-south-kherson-russia.html.

ground forces. By the summer, the group of forces on the right bank of the Dnipro River included elements of the 11th Separate Guards Assault Brigade, 126th Coastal Defence Brigade, 7th GAAD, 56th Guards Air Assault Regiment, the 34th Separate Mountain Motorized Rifle Brigade, and the 205th SMRB.[43] Other supporting elements included BTGs from the 429th MRR near Vasylivka and 503rd MRR. The 58th Army headquarters was based in Melitopol with the 19th MRD at Tokmak.[44] Russian force posture would remain the same through June and July.

By early July, the AFU regained control of Snake Island, which reduced Russia's ability to threaten shipping, and ensured that the island could not be used as a launch site for ground-based missiles. The AFU conducted shaping operations to the north and northwest of Kherson City in July. For any counter-offensive to be successful in the area, Ukrainian planners knew they had to disable the main road and rail bridges and a dam that Russia used to resupply its forces on the right bank of the Dnipro—the Antonivka and Daryivka bridges and the Nova Kakhovka dam.

The AFU launched precision strikes at these locations in advance of the operation, first to test how many munitions would be needed to damage or destroy the bridge, and also to learn more about Russia's ability to jam or intercept these precision weapons for when the main attack would begin.[45] The AFU would strike these locations three times in July and August. The Russians, sensing the vulnerability, responded in between strikes by setting up pontoon crossings and modifying air defences and EW systems. They also installed metal radar angle reflectors that would appear as pontoon crossings on radar imagery or, perhaps, were designed to confuse the seeker on the missile itself to disrupt accuracy.[46] Despite these counter-measures, the AFU were able to damage the bridges from late August onward.

Ukrainian forces attacked the Black Sea fleet headquarters and other airfields in Crimea in July and August, prompting some relocation of assets.[47] By early August, Ukrainian forces counter-attacked near Velyka Oleksandrivka, and Russian military leaders concluded that Ukraine had begun a counter-offensive to push them across the Dnipro River.

Russian military leaders decided to mount a defence and were not yet considering abandoning the territory, so they bolstered their reconnaissance assets and forces in Kherson. To hold its gains, Russian forces added additional missile, artillery, and infantry units to the Kherson area to support the forward units.

[43] 'Deep State map', 1 June 2022; @JominW, X post, 31 May 2022. https://x.com/JominiW/status/1531802127303168001/photo/1 (accessed 12 January 2025); President of Russia, 'Honorary guards designation conferred on 34th Detached (Mountain) Motorized Rifle Brigade', Kremlin, 5 December 2024. http://en.kremlin.ru/acts/news/75764 (accessed 18 February 2025).

[44] @JominiW, X post, 31 May 2022. https://x.com/JominiW/status/1531802125294088195/photo/1 (accessed 12 January 2025).

[45] Khurshudyan and Sonne, 'Inside the Ukrainian counteroffensive'.

[46] Glavkom, 'Rosiia buduie na Khersonshchyni mosty-pryvydy: shcho tse i jak pratsiuie', 2 August 2022. https://glavcom.ua/publications/rosija-buduje-na-khersonshchini-mosti-prividi-shcho-tse-i-jak-pratsjuje-865125.html.

[47] @Osinttechnical, Post on X, 9 August 2022. https://x.com/Osinttechnical/status/1557077182110998528 (accessed 15 January 2025).

Ukraine forces reportedly struck the command posts of five Russian units during the first week of August.[48] Russian forces knew they needed to target HIMARS launchers as best as they could and devoted additional reconnaissance strike assets to targeting them in Kherson. The AFU used frequent relocation and deception to protect these assets, including a fleet of wooden decoy HIMARS to confuse Russian reconnaissance.[49]

By 29th August, the AFU had liberated villages and advanced on positions known to be manned by low-readiness pro-Russian militia units from the Donbas, who quickly retreated.[50] However, the second line of Russian defences posed several challenges to AFU forces, according to Ukrainian officers who fought there. The head of Ukraine's military intelligence noted that the second line of defences were better prepared and manned by capable Russian units: 'the most trained and most capable Russian units are currently in Kherson. A large share of them are from airborne troops of the Russian Federation, Russian special operation forces and the naval infantry'.[51]

In late August and September, Russia added several BTGs to the forward grouping of forces. The VKS, operating out of Crimea, supported Russian forces by launching glide bombs against advancing AFU units, until Russian artillery and missile units were brought into position. By 10 September, Russia had sent two artillery battalion elements across the Dnipro River to reinforce its positions near Velyka Oleksandrivka, other units were sent to Vasilivka, and VDV elements were moved closer to Kherson City.[52]

For nearly 50 miles north of Kherson, Russian defensive positions had been hardened, and equipment was dug into concealed firing positions, and Russian forces made good use of terrain and mines, according to Ukrainian units that fought there.[53] Ukrainian forces fought against prepared Russian positions approaching Kherson, and noted that the Russian airborne forces fought hard, counter-attacked frequently, used effective counter-battery fire, and used limited numbers of Shahed attack drones, which were just emerging on the battlefield in August.[54]

[48] @JominiW, Post on X, 16 August 2022. https://x.com/JominiW/status/1559710649747488773 (accessed 4 February 2025).

[49] John Hudson, 'Ukraine lures Russian missiles with decoys of U.S. rocket system', *Washington Post*, 30 August 2022. https://www.washingtonpost.com/world/2022/08/30/ukraine-russia-himars-decoy-artillery/.

[50] Jim Sciutto, 'Ukrainian counteroffensive underway in Russian-held south, former president says', *CNN*, 29 August 2022. https://www.cnn.com/europe/live-news/russia-ukraine-war-news-08-29-22.

[51] Howard Altman, 'Ukraine's intel chief on how the war ends, Putin's nuclear threats, Iranian drones, and more', *The War Zone*, 28 October 2022. https://www.twz.com/ukraines-intel-chief-on-how-the-war-ends-putins-nuclear-threats-iranian-drones-and-more.

[52] 'Deep State map', Deep State blog, 10 September 2022, update 4. https://deepstatemap.live/en#10/47.2620760/33.4485350 (accessed 25 January 2025); @JominiW, Post on X, 16 August 2022. https://x.com/JominiW/status/1559710631774781445.

[53] Khurshudyan and Sonne, 'Inside the Ukrainian counteroffensive'.

[54] Yuri Butusov, 'Taktyka oborony rosiiskykh VDV: vazhkyi shliakh do vyzvolennia Khersonu', 4 November 2022. https://censor.net/ua/r3378332https://censor.net/ua/r3378332; Carlotta Gall, 'Where Russian fire pours down, his men sneak forward', *New York Times*, 1 November 2022. https://www.nytimes.com/2022/11/01/world/europe/ukraine-war-kherson.html.

Ukraine's advances were slow against this prepared defence and AFU casualties were high.[55] The senior Ukrainian commander in the area was removed in October.[56] However, the operational environment in Kherson was very different from the circumstances in Kharkiv, and lack of progress could be attributed to multiple factors (better Russian defences, air support, experienced units) and not only command problems.

Because Russia had no strategic reserve at the time, reinforcing Kherson meant pulling units from elsewhere on the front. As discussed above, the AFU attacked in Kharkiv to exploit this opportunity. A week after the Kharkiv front collapsed, President Putin and Defence Minister Shoigu declared a 'partial' mobilization of 300,000 men on 21 September, due to the deteriorating situation on the front and manpower shortages within the units on the front lines.

Next, on 30 September, following sham referendums in the occupied territories, President Putin publicly signed documents and illegally declaring that four Ukrainian oblasts—Luhansk, Donetsk, Zaporizhzhia, and Kherson, would be annexed into the Russian Federation.[57] General Surovikin, the commander of the southern sector, was named overall commander of the Russian Group of Forces in Ukraine on 8 October.

By the first week of October, the Russians slowed the AFUs advance by targeting small bridges and key roads. Russian forces previously laid traps and used treelines and other terrain features wisely. As one local commander described it, 'You can see every 10 meters they had firing positions in the tree line', he said. 'They had such a thick defense, I cannot describe it. Every foxhole was firing, every bush was firing'.[58] Russia's belief that they could defend their positions on the right bank of the Dnipro most likely ended by early October, when AFU precision strikes seriously damaged bridges and the Nova Kakhovka dam. In mid-October, Russian forces were falling back to their next prepared line of defence, and attempted one last ground counter-attack in the region, which was unsuccessful.

The decision to retreat would not be an easy one, politically speaking, within the Russian system. Russian military institutions teach officers at several points in their careers that retreating from their positions is shameful and it is their duty to hold positions unless there is no possibility of survival or success.[59] President Putin just two weeks prior declared Kherson part of Russia. Nevertheless, the decision to withdraw came quickly after the Nova Kakhovka dam crossing was destroyed. On 19 October, the commander of Russian forces in Ukraine, General Surovikin, announced a civilian withdrawal from Kherson, due to the destruction of bridges that made sustaining

[55] Gettleman, 'In Ukraine's south'.
[56] Khurshudyan and Sonne, 'Inside the Ukrainian counteroffensive'.
[57] Paul Kirby, 'What Putin's annexation means for Ukraine's regions', *BBC News*, 30 September 2022. https://www.bbc.com/news/world-europe-63086767.
[58] Gall, 'Where Russian fire pours down'.
[59] Anika Binnendijk, Dara Massicot, et al., 'Russian military personnel policy and proficiency: Reforms and trends, 1991–2021', RAND Corporation, RR1233-6, 2023. https://www.rand.org/pubs/research_reports/RRA1233-6.html; Alexander Vedenin, 'Building resilience [Формируя Стойкость]', *Ural'skie Voennye Vesti*, No. 29, 2021.

the area difficult.[60] The Russian artillery and MLRS brought in initially to hold forward positions would now be used as covering fires for a full retreat from the right bank of the Dnipro.

The next day, on 20 October, the full military withdrawal began. Withdrawal would not be discussed by Russian officials in public until 9 November when it was most likely already complete, to preserve operational security. Russian forces and their installed puppet governments in the region would begin a controlled withdrawal across the Dnipro and south to Crimea. The withdrawing forces laid intricate traps and mines on routes, cut the electricity and powerlines behind them, and continued counter-battery fire, to slow down the AFU advance and to cover the passage of Russian forces across the Dnipro via damaged bridges, pontoon bridges, and riverboats.

The Russian withdrawal from the right bank of the Dnipro lasted from mid-October through early November. Ukrainian forces wanted to destroy Russian forces before they could cross the river, but could not close on Russian forces leaving Kherson due to the counter-attacks and obstacles retreating Russian forces left in their path. Ukrainian officials noted that Russian forces had mined roads and adjacent approaches to Kherson City densely, which meant that AFU units could not reach Russian forces before they retreated across the river. They noted that Russian forces 'didn't hold back. They mined with everything, even new means we had never heard of.'[61]

Russian forces did not experience panic or uncontrolled breaks in lines in Kherson, and command and control continued—in contrast to the Kharkiv front collapse that had occurred in September. In the final days of the military withdrawal, VDV forces crossed the Dnipro River under fire while other units would cover for them while retrograding, also under fire.[62] Further, according to Ukrainian officials, Russia surged a few recently mobilized battalions across the Dnipro to assist that had little training.[63] From mid-October to 11 November, Russian forces moved heavy equipment across the Dnipro River, some of it by barge.[64] Ukrainian officials estimated that at least 25,000 Russian forces withdrew across the river, making use of the Antonivsky bridge and a small pontoon bridge adjacent to it.[65]

On 9 November, Ukrainian forces had the city of Kherson in artillery fire range, and General Surovikin and then-Minister of Defence Shoigu held an unusual (and likely highly choreographed) televised briefing. Surovikin claimed that 115,000 residents of the Kherson region had been evacuated since October, that the AFU had

[60] Laurence Peter, 'Ukraine war: Russia admits Kherson "tense" under shelling', *BBC News*, 19 October 2022. https://www.bbc.com/news/world-europe-63309095.

[61] Khurshudyan and Sonne, 'Inside the Ukrainian counteroffensive'.

[62] 'Deep State map', Deep State blog, 1 October 2022 and 11 November 2022. https://deepstatemap.live/en#9/46.4859646/33.2625242 (accessed 8 January 2025).

[63] 'Russia has sent up to 1000 troops to defend occupied Kherson, Ukraine says', *CNN*, 28 October 2022. https://www.cnn.com/europe/live-news/russia-ukraine-war-news-10-28-22#h_3964d32cb7f6a36472d70f1c896e2876.

[64] '"Плывём". Как Россия уходит из Херсонской области', RadioLiberty, 19 October 2022. https://www.svoboda.org/a/plyvyom-kak-rossiya-uhodit-iz-hersonskoy-oblasti/32091813.html.

[65] Khurshudyan and Sonne, 'Inside the Ukrainian counteroffensive'; Altman, 'Ukraine's intel chief'.

sustained heavy losses, and that Russian forces were intercepting a majority of precision strikes in the area.[66] Surovikin stated that the late September AFU strike on the Nova Kakhovka dam risked flooding the area and the 'complete isolation of our group of troops on the right bank of the Dnieper [Dnipro River]'.[67] Surovikin explained the need to withdraw to save the forces on the right bank, and Shoigu 'granted permission' for a withdrawal.[68]

The Russian withdrawal was nearly complete when this conversation was broadcast. Ukrainian forces would enter and liberate the city of Kherson two days later, on 11 November. Once Russian forces were across the Dnipro River by 11 November, they would destroy the remainder of the crossings by detonating explosions on spans of the Antonivsky and Daryivsky bridges, and a portion of the Nova Kakhovka dam.[69] Russian forces set up new defences along the left bank of the Dnipro as a forward line of defence.

The Russian Ministry of Defence claimed that all equipment, weapons, and personnel had been withdrawn.[70] While it is probable that some equipment was left behind, the withdrawal operation of these forces under fire was executed successfully. Russian forces were subsequently reconstituted and then recommitted to other parts of the front. Russian forces would transition to the strategic defensive across the front from November to late January 2023. They would also accelerate the construction of defensive networks (the so-called 'Surovikin Lines'), while mobilized personnel were sent to understrength units and to form the basis of an operational reserve force.

Nuclear Escalation Scare and Fall 2022 Counter-Offensives

Russian military power approached its lowest point of the war by September and October 2022. The Kharkiv front crumbled in days, and Russia's elite airborne forces were crossing the Dnipro River under fire across pontoon bridges and small boats. Mobilization had only just started. It was during this time in September and October 2022 that Russian officials again made public statements about nuclear weapons. In late September, President Putin publicly stated 'in the event of a threat to the

[66] Alexander Tikhonov, 'Для нас жизнь и здоровье российских военнослужащих всегда являются приоритетом', *Krasnaya Zvezda*, No. 126, 11 November 2022. https://ric.mil.ru/Novosti/item/444989/.
[67] Ibid.
[68] Ibid.
[69] Marc Santora and Anna Lukinova, 'An explosion on a crucial Kherson bridge severed the city's last major crossing', *New York Times*, 11 November 2022. https://www.nytimes.com/2022/11/11/world/europe/kherson-bridge-russia-ukraine-war.html.
[70] 'Exclusive: US prepared "rigorously" for potential Russian nuclear strike in Ukraine in late 2022, officials say', *CNN*, 9 March 2024. https://www.cnn.com/2024/03/09/politics/us-prepared-rigorously-potential-russian-nuclear-strike-ukraine/index.html; 'Россия сдала Херсон. Как войска покидают оккупированные территории?' ['Russia has surrendered Kherson. How are troops leaving the occupied territories?'], *BBC News Russian Service*, 11 November 2022. https://www.bbc.com/russian/features-63599675.

territorial integrity of our country and to defend Russia and our people, we will certainly make use of all weapon systems available to us. This is not a bluff'.[71] When the Russian withdrawal from Kherson began on 20 October 2022, senior Russian defence officials also began privately calling their counterparts in Europe and the US, claiming they had information that Ukraine was preparing to detonate a dirty bomb.[72]

The foreign ministers of the US, Britain, and France issued a joint statement condemning 'Russia's transparently false allegations that Ukraine is preparing to use a dirty bomb on its own territory. The world would see through any attempt to use this allegation as a pretext for escalation. We further reject any pretext for escalation by Russia.'[73]

Privately, the Biden administration became concerned about escalation, given the context of two imploding Russian fronts, and sought to pay close attention to any indicators that Russia's calculus on using a tactical nuclear weapon could be changing.[74] In the fall of 2022, they developed contingency plans in the event of nuclear use.[75] Of several scenarios one involved a full collapse of the Russian front line as a 'potential trigger' for tactical nuclear weapons use. One anonymous senior official claimed that

> If significant numbers of Russian forces were overrun […] that was a sort of precursor to a potential threat directly to Russian territory or the Russian state. […] In Kherson at that time there were increasing signs that Russian lines could collapse. Tens of thousands of Russian troops were potentially vulnerable.[76]

Despite picking up chatter that 'at least something that lower levels of the Russian system were discussing',[77] it is not clear that any abnormal readiness changes were observed within Russia's nuclear forces, who were conducting an annual nuclear exercise named Grom at this time.[78] Officials in India, China, the US, and other countries made several public and private statements to Moscow warning against nuclear use. The moment passed without further incident, and without additional details to confirm this period was anything other than a mutual exchange of public and private signalling statements, this period remains an unusual episode to outside observers.

[71] '"This is not a bluff": Putin threatens nuclear response in Ukraine war', *Wall Street Journal*, 21 September 2022. https://www.wsj.com/video/this-is-not-a-bluff-putin-threatens-nuclear-response-in-ukraine-war/AA197C32-F882-4E01-B539-5E5BE6A798F8?mod=itp_trending_now_video_pos5.

[72] New York Times, 'Western Officials Warn Russia Could Use a Dirty Bomb as a Pretext', 25 October 2022, https://www.nytimes.com/2022/10/24/us/politics/russia-dirty-bomb-west-ukraine.html.

[73] 'Joint Statement on Ukraine', US Department of State, Media Note 23 October 2022. https://www.state.gov/joint-statement-on-ukraine-2/, archived version available at https://archive.ph/OcKb8.

[74] 'Exclusive: US prepared'.

[75] Ibid.

[76] Ibid.

[77] Ibid.

[78] Lachlan MacKenzie, 'Six days in October: Russia's dirty bomb signaling and the return of nuclear crises', *CSIS*, 3 September 2024. https://www.csis.org/analysis/six-days-october-russias-dirty-bomb-signaling-and-return-nuclear-crises.

Kharkiv and Kherson Counter-Offensive Conclusions

The Kharkiv counter-offensive was successful for a few factors that would prove to be difficult to replicate in later phases of the war. Ukrainian forces used a combination of deceptions to lure Russian forces to Kherson, while planning strikes on command posts and logistics hubs at the beginning of their operation in Kharkiv. The AFU simultaneously attacked multiple sparsely manned areas in the Kharkiv region, caught Russian forces by surprise, and penetrated their lines in several areas, sowing panic and chaos. At this time in Kharkiv, Russia had not yet constructed a network of defensive positions and minefields, and they did not have enough drones for persistent reconnaissance and rapid targeting that would have assisted situational awareness. The Kharkiv counter-offensive also took advantage of Russia's lack of operational reserves for the first seven months of the war.

The advance warnings from Western intelligence allowed AFU forces the chance to exploit a specific window of time during Russian force redeployment to Kherson. The AFU also made effective use of precision munitions against key command posts and logistic nodes in support of manoeuvres, and Russian forces had not yet adapted to Western precision guided munitions with their EW capabilities or air defences. Russian manpower levels were highly variable in the Kharkiv area. Russian units at Balakliia and Izium had most of their personnel and equipment, according to recovered documents, but panic and chaos set in nonetheless These factors suggest that surprise and lack of situational awareness contributed to this chaos and collapse more than inadequate force strength. These factors in combination allowed the AFU to make, and then consolidate, rapid gains.

The AFU encountered more resistance in Kherson than in Kharkiv, because unlike the conditions in Kharkiv, Russian forces in Kherson were better prepared and it was occupied by more capable forces. Russian military leaders concluded, in part due to Ukraine's deliberate information campaign and visible preparations, that Ukraine's main counter-offensive in 2022 would occur in Kherson, and that it would target vulnerable Russian positions on the right bank of the Dnipro River. To support this group of forces, Russian commanders added additional reconnaissance, infantry, paratroopers, artillery, and close air support. Russian defences and counter-attacks in Kherson stymied the type of rapid advance and surprise that would be a key part of the Kharkiv operation's quick success. Further, rather than the thinly defended lines in Kharkiv, Kherson was occupied by higher readiness Russian forces, and their multiple lines of defences were actively reinforced during the AFU counter-offensive.

Ultimately, Russian forces were hobbled by the disadvantages of geography and advancing AFU units. Once the AFU damaged or destroyed multiple crossing points over the Dnipro River with precision fires, Russian forces had no other option but to withdraw, or risk encirclement and destruction of a large percentage of the VDV and other capable forces. Russian forces executed a controlled withdrawal across the Dnipro River and did not collapse as they did on the Kharkiv front.

Both operations (Kherson in particular) contained warnings for future AFU counter-offensives in 2023. First, Russian forces were hard to dislodge in the hands of a capable commander, pervasive reconnaissance, effective use of terrain, dug-in positions, and mines. Second, after Russia mobilized in late 2022 and improved its volunteer recruiting methods, they were able to rotate units off the front lines as needed, keep front lines manned at sufficient levels (overall), create operational reserve forces, and reinforce hotspots as needed from reserves. Russia's drone and loitering munition advantage over Ukraine increased as their defence industry scaled up production after the September 2022 mobilization decree.

These factors worsened by the summer of 2023, and are critical in explaining why the Ukrainian counter-offensive in 2023 failed to achieve objectives. Due to frequent coverage in the press, the Ukrainian 2023 counter-offensive was not a surprise for Russian forces and outside observers. The nature of the operation itself meant that forces had to concentrate to attack, allowing Russia to fortify and prepare. The AFU's own challenges going on the offense have also been documented, not least of all because some participating units lacked the right experiences: one of the lead AFU brigades had 70% of personnel with no combat experience.[79]

During the 2023 counter-offensive, the AFU struggled to cut through extremely difficult conditions without close air support. Russian landmine density (sometimes mines were stacked several deep, remotely detonated or reseeded), and defence positions built into high ground and other advantageous terrain meant that advancing AFU units were targeted relentlessly during the day and at night with Russian reconnaissance strikes, enabled by drones and FPV munitions.[80] Within four days, 60% of de-mining equipment was destroyed and the AFU had to abandon mechanized assault for small teams of ten infantry to clear the fields of mines.[81] The counter-offensive would culminate a few months later.

In summary, the operational factors that contributed to the Kharkiv counter-offensive's success in Kharkiv and Kherson in 2022 were no longer present by June 2023, and the factors that challenged the AFU during the Kherson counter-offensive were amplified. The end of the Kharkiv and Kherson counter-offensives, and Russian mobilization would mark a new phase in the war in Ukraine—a shift to positional and attritional warfare, and a sustainment competition between Russia and its backers, and Ukraine and its backers.

[79] 'In Ukraine, a war of incremental gains as counteroffensive stalls', *Washington Post*, 4 December 2023. https://www.washingtonpost.com/world/2023/12/04/ukraine-counteroffensive-stalled-russia-war-defenses/.

[80] 'Miscalculations, divisions marked offensive planning by U.S., Ukraine', *Washington Post*, 4 December 2023. https://www.washingtonpost.com/world/2023/12/04/ukraine-counteroffensive-us-planning-russia-war/; 'In Ukraine, a war'.

[81] 'In Ukraine, a war'.

17
The 'Spring Offensive' 2023

Viktoriya Fedorchak

Introduction

The Russian aggression against Ukraine has continued for twelve years, with more than four years of full-scale invasion. Much has been written on causes, various stages, and the conduct of the warfighting during the full-scale phase. Many more things will be discovered about numerous battles and the names of many lives taken from the Ukrainian nation, and many lessons learned will be identified and incorporated into military practice from different stages of the warfighting during this war. The purpose of this chapter is to discuss one episode of this war—that is, the case of the Ukrainian counter-offensive of 2023. The chapter explains the preliminary preparatory stage on both sides prior to the commencement of the counter-offensive in June 2023. It is then followed by the main developments, tactics used, enemy counter-actions, factors contributing to the state of affairs, and further points for consideration. The main factors influencing the character of the campaign and its outcomes are discussed.

Accordingly, some of the factors include the multi-domain nature and the need to improve cross-domain integration in the battlespace; the importance of terrain characteristics and complexity of breaching the mine barriers; trends in counter-battery fires; increased use of drones by the Russian forces; the important role of electronic warfare (EW); the significance of people as a crucial element of resilient fighting power; and finally, the need for building critical mass especially in a war of attrition. Some points for further consideration are also identified.

Preparations for the Counter-Offensive on Both Sides

The 2022 year of warfighting in Ukraine brought many surprises. Ukrainians survived. The combination of Western and Soviet equipment and the Ukrainian people created resilient fighting power that not only withstood Russian invasion but also pushed back and regained territories during the Kharkiv and Kherson counter-offensives in the summer and autumn of 2022.[1] Thoughtful planning and

[1] Viktoriya Fedorchak, *The Russia-Ukraine War: Towards resilient fighting power* (London: Routledge, 2024), 81–99.

Viktoriya Fedorchak, *The 'Spring Offensive' 2023*. In: *Russia's War in Ukraine and Modern Warfare*. Edited by: Mikael Weissmann and Niklas Nilsson, Oxford University Press.
DOI: 10.1093/9780198970903.003.0017

the effect of surprise allowed the Ukrainian side to achieve a lot during these counter-offensives. On the negative side, the primary limitation of the principle of surprise is that the same trick does not give the same effect if used twice. While the Ukrainian counter-offensive of 2022 was a surprise to a great extent, the counter-offensive of 2023 was more of an anticipated event for everyone involved—from the fighting sides and ordinary people to Ukraine's allies. In essence, it was an expected next stage of the war, but the timing depended on the ability of the Ukrainian armed forces (UAF) to accumulate sufficient capabilities for the mass offensive.

Just as with any more kinetic and intense episode of warfighting, the Ukrainian counter-offensives of 2022 significantly degraded Russian personnel, resulting in the announcement of Russian mobilization on 21 September 2022 with many newly mobilized troops immediately being dropped on the front line. Instead of futile efforts to regain recently lost territories, Russia focused on strengthening its presence in already well-established occupied territories of Donbas and holding lines across the front. In order to deny the Ukrainians the opportunity to regroup and strengthen their capabilities:

> Russia compensated for its ground losses with aerial and long-range bombing of the entire territory of Ukraine, with a specific focus on targeting civilian infrastructure. Besides its demoralising effect and the exhaustion of the Ukrainian ammunition stockpile, the offensive was also aimed at keeping Ukrainian leadership and rear support busy, distracting from the next stage of the war at hand—regrouping and consolidation for further actions.[2]

Furthermore, as of 3 December 2022, in an interview with *The Economist*, General Valerii Zaluzhnyi confirmed that both sides were gathering capabilities in preparation for further action in 2023. He suggested that Russians were regrouping for the next attack, potentially trying to capture Kyiv by taking advantage of a strike from Belarusian territory. Russian attempts to propagate negotiations while continuously bombing Ukrainian cities were only means of buying time to mobilize more troops and equipment for further attacks.[3] This became evident in the forthcoming months of January 2023, with the intensification of Russian assaults across the entire front line.[4]

In a December interview, General Zaluzhnyi emphasized the key strategic objectives that the UAF aimed to achieve. The priority was to hold the lines and preserve already liberated territories from being retaken by the Russians. The rationale was quite simple: It is significantly more difficult to liberate an area than to hold it. Second, any further counter-offensive and regaining of Russian-occupied territories would require the accumulation and readiness of significant assets. Thus, the objective was to at least return to the borders as of 23 February 2022. Finally, it was essential for

[2] Ibid., 90.
[3] 'An interview with General Valery Zaluzhny, head of Ukraine's armed forces', *The Economist*, 15 December 2022. www.economist.com/zaluzhny-transcript.
[4] Fedorchak, *The Russia-Ukraine War*, 90.

Ukraine to strengthen air defence with more advanced systems.[5] This was conditioned on an estimated effectiveness of the Ukrainian air defences during autumn 2022 at the 0.76 coefficient of efficacy along with increased use of more advanced long-range ballistic missiles by the Russians. Regarding the liberation of Crimea, the way to reclaim it was by seizing 84 kilometres towards Melitopol to undermine Russian ground access to Crimea. This objective would require at least 300 tanks, 600–700 armoured personnel carriers (APCs), and 500 Howitzers.[6]

During most of 2022, the UAF employed the principle of getting the most with what it had or what was given across domains and capabilities, but the establishment of the relatively fixed front lines and the consolidation of Russian presence in those territories was making liberation of territories more difficult with each passing day. In turn, a breakthrough in any part of the front lines would require more and heavier capabilities in terms of both equipment and people. For the Ukrainian counter-offensive, time was not on the Ukrainian side. The more time was spent on preparation and accumulation of fighting power, the more time the Russians had to strengthen their fortifications and lay more mines as their first lines of defence, making liberation efforts more strenuous. Thus, from the perspective of linear progression of the two trends, the counter-offensive was to take place when at least some critical mass of the capability was established in order to be able to conduct the counter-offensive with greater effectiveness without giving the enemy time for strengthening its defences even more. Critical mass in modern warfare can be defined as:

> the ability to rapidly produce and/or have sufficient numbers of military capabilities to deploy, modify, sustain and integrate into the force structure of a given operating environment according to the operational requirements. Thus, the scale would largely depend on the adversary and the type of conflict or war at stake.[7]

For the Ukrainian side, the winter and spring 2023 months were used to establish sufficient critical mass suitable for the liberation campaign in a highly contested environment.

In the meantime, the Russians used those months to their advantage in several ways. First, they increased their mobilization efforts by combining various methods, including conscription, prison recruitment, and financial contracts.[8] Second, the Russians continued strengthening their lines of defence across the entire front line with various means, such as 'a multi-layered Russian network of anti-tank ditches, mazes of trenches, concrete "dragon's teeth" barricades, steel "hedgehog" obstacles, spools of razor wire and minefields'.[9] Third, they managed to resupply their troops

[5] 'An interview with General Valery Zaluzhny'.

[6] Fedorchak, *The Russia-Ukraine War*, 90.

[7] Ibid., 18.

[8] Anatolii Chernysh, 'Glimpse into mobilization in Russia: Overview, methods and prospects', *Foreign Policy Council: Ukrainian Prism*, 22 March 2023. https://prismua.org/en/english-glimpse-into-mobilization-in-russia-overview-methods-and-prospects/.

[9] 'Ukraine's counteroffensive against Russia in maps: Latest updates', *Financial Times*, 8 July 2024. https://www.ft.com/content/4351d5b0-0888-4b47-9368-6bc4dfbccbf5.

after the experiences of Ukrainian counter-offensives in 2022. However, despite Russian efforts to strengthen their fighting power during those few months, a reverse phenomenon became evident during spring 2023—sabotage activities against the Russian mobilization campaign and mass surrender to the UAF of the newly deployed Russian troops.[10] Accordingly, the pace of Russian surrenders was 3,000 per month in March and April 2023.[11]

Main Developments in the Ukrainian Counter-Offensive 2023

The counter-offensive started on 4 June 2023, with the Ukrainian column attacking Russian positions near Novodarivka in the Zaporizhia district. The attack illustrated the enduring challenges that would be faced during the entire counter-offensive—Russian preparation for the mechanized assaults included minefields, anti-tank guided missiles (ATGMs) and drones for intelligence, surveillance, and reconnaissance (ISR) purposes. The attack also showed immediate losses in equipment due to a lack of sufficient numbers of critical mass and air support to reinforce the mechanized assault.

On 5 and 6 June, Ukrainian attacks on Russian positions intensified, with columns consisting of Leopard tanks, Bradley fighting vehicles, and minesweepers deployed in three directions across the front line. The primary direction was Tokmak in the Zaporizhia district. The second direction was Velyka Novosilka in the Donetsk region. The third direction was Bakhmut. The primary objective of the attacks was to breach Russian defence lines and gain a more favourable strategic standing. The direction of Tokmak was essential since successful advancement would allow passage all the way to Melitopol, cutting off the Russian land route to Crimea. However, the Russians heavily defended this strategically significant area. There were three layers of Russian defences that included the immediate front line with minefields, reducing any significant movement of mechanized units; many kilometres of trenches with thousands of soldiers defending them with ATGMs; and, unlike the first year and a half of the full-scale invasion, a greater number of uncrewed aerial systems (UAS), improving Russia's situational awareness regarding the movement of troops in the battlespace, fire coordination, and integration of efforts. Another layer of defence was the enforcement of units with reserves and air assault troops. Hence, the decision to divide the counter-offensive into three directions was conditioned by the need to spread Russian troops across the front line and distract the significant numbers that had accumulated in the Tokmak area.

On 6 June, the Russians blew up the Kakhovka dam on the Dnipro River, resulting in the flooding of 620 square kilometres of territory across four districts: Kherson,

[10] Fedorchak, *The Russia-Ukraine War*, 93.

[11] Taras Levchenko, '"There were appeals from the entire units": How do the Russian soldiers surrender?' *Radio Freedom*, 10 May 2023. www.radio svoboda.org/a/yal-viyskvirf-zdayutsya-u-polon/32405035.html.

Mykolaiv, Dnipropetrovsk, and Zaporizhia, which had a direct impact on 100,000 residents. The overall damage to people, infrastructure, and the ecosystem was estimated at $13.79 billion.[12] Besides the immediate and long-term impact on human lives and the ecosystem, this event had significant implications on the development of the Ukrainian counter-offensive and what could have become a Ukrainian amphibious assault from the Kherson region by crossing the Dnipro River. It is essential to first consider the utility of this action in the overall counter-offensive; then it will be relevant to discuss why the breach of the dam made the action unfeasible if not impossible.

The amphibious assault from the Kherson region had a number of functions. To begin with, that part of the Zaporizhia region was less well-defended by the Russians with fewer accumulated reserves, primarily because of the natural watershed of the Dnipro River and the focus on strengthening the defences of the northern part of the region with the three lines of defence. While amphibious attacks in various parts of the Zaporizhia would not have gained significant success on their own, in combination with the mechanized attacks of the counter-offensive from the other side of the Zaporizhia region, they would serve as an additional distraction to dilute Russian defences, engaging the reserves, which, in turn, could not be used to reinforce Russian defences against Ukrainian mechanized assaults. Another consideration referred to the strategic significance:

> An offensive in Kherson Oblast would provide Ukrainian forces with the shortest possible land route to the Crimean Peninsula, the liberation of which is one of Ukraine's stated strategic goals for the war. Kherson Oblast is also fortified relatively sparsely compared to the neighboring Zaporizhzhia Oblast, meaning that maneuver forces would face fewer layers of entrenched defenders in the case of a breakthrough.[13]

The destruction of the dam made the amphibious assaults unfeasible in their immediate execution for several reasons. First, in the actual Kherson City area, the increased water level of 5 metres required the amphibious forces to retreat from their initial locations; for any further action, the waters had to recede. Second, the very characteristics of the river had been altered with faster currents and changes in the banks, making previously suitable landing locations inapt. The same can be said about the emptied reservoir itself, which turned into a swamp. Third, any crossing of the river was complicated by the debris from different elements of infrastructure, trees, and even mines that moved with the flooded waters. Therefore, any crossing of the river would require significant reconnaissance, time, and resources. Finally, due to the absence of functional bridges, any crossing and establishment on the other

[12] 'Kakhovka dam destruction inflicted US$14 billion damage and loss on Ukraine: Government of Ukraine-UN report', *United Nations Ukraine*, 17 October 2023. https://ukraine.un.org/en/249742-kakhovka-dam-destruction-inflicted-us14-billion-damage-and-loss-ukraine-government-ukraine.

[13] Riley McCabe, Alexander Palmer, and Joseph S. Bermudez Jr, 'The impact of the Kakhovka dam breach on the new Ukrainian counteroffensive', Center for Strategic and International Studies, 16 June 2023. https://www.csis.org/analysis/impact-kakhovkadam-breach-new-ukrainian-counteroffensive.

bank of the river would require continuous support for the Ukrainian fighting power under the conditions of flooding and the distraction of resources on a newly shaped geography, but the provision of effective logistics across the river would be a greater challenge than before.[14] Accordingly, it was only in October 2023 that various reports were made of Ukrainian amphibious assaults across the river.[15]

On 7–8 June, intensified Ukrainian attacks in the Tokmak direction took place, which employed the pre-planned mechanized assaults in columns facing severe fire from the Russian side. Although mine clearing vehicles were used, Bradleys and Leopard tanks would face losses. The situation was further aggravated by Russian artillery and ATGMs. This was greatly facilitated by the increased use of Russian Orlan-10 surveillance drones to more quickly spot approaching Ukrainian columns, allowing timely fire coordination for the Russian artillery, helicopters, and drones.[16] While the Ukrainian artillery provided fire cover, it was not immune to attacks by the Russian K-52 helicopters and Lancet loitering munitions. While the Tokmak direction was the most contested, more progress was achieved in the area of Velyka Novosilka, where the Ukrainian troops managed to breach Russian defences in various areas, and, in further battles, liberating the Neskuchne, Blahodatne, and Makarivka villages. Overall, seven villages were liberated in this direction.

The complexity of breaching Russian defence lines and the heavy toll on the equipment in the Tokmak direction required a change in tactics to reduce losses of equipment and personnel. As a result, the focus was placed on smaller infantry units consisting of ten people to gradually degrade the Russian lines while engineers cleared the area. Artillery and heavy vehicles were to serve as fire cover for the troops and strike further into Russian defences. This change reduced the number of losses in terms of equipment and personnel, but it also reduced the intended and expected speed of advancement: 'the rate of advance during this period [June 2023] was one tactical advance for three days of fighting, with each advance moving the line of control approximately 700–1200 metres forwards'.[17] Nevertheless, progress was achieved across various directions of the counter-offensive, with the liberation of Pyatihatki on 18 June and Rivnopil on 26 June.

The months that followed were characterized by ongoing fights across all three directions, with Ukrainian troops achieving some territorial gains. For most of August, there were fights for Robotyne, which resulted in its liberation on 28 August 2023. Autumn 2023 showed further fights with more modest territorial gains. The counter-offensive ended on 1 December 2023, having achieved various territorial

[14] Ibid.

[15] 'Russian offensive campaign assessment, October 18, 2023', Institute for the Study of War, 18 October 2023. https://www.understandingwar.org/backgrounder/russian-offensive-campaign-assessment-october-18-2023.

[16] Sebastien Roblin, 'Drones on the front line for Ukraine's counteroffensive', *Inside Unmanned Systems*, 19 October 2023. https://insideunmannedsystems.com/drones-on-the-front-line-for-ukraines-counteroffensive/.

[17] Jack Watling and Nick Reynolds, 'Stormbreak: Fighting through Russian defences in Ukraine's 2023 offensive', *RUSI*, September 2023, 14. https://static.rusi.org/Stormbreak-Special-Report-web-final_0.pdf.

gains but not to the extent of fulfilling its posed objectives of a significant breakthrough and change of parity in the war.

The campaign illustrated numerous developments on the technological side and from the perspective of adaptation. First, while the Russians slowed down Ukrainian progress and had achieved the destruction of 20% of Ukrainian heavy vehicles in the first two weeks of the counter-offensive,[18] they suffered significant losses in artillery and tanks themselves, with the loss of artillery of particular concern. Second, the Russians deviated from their mining doctrine and instead of a traditional 120 metre depth for mines opted for a 500 metre depth to significantly reduce any kind of rapid breaches of their defence lines. Also, the practice of laying two anti-tank mines, one on top of the other, was used in order to boost the damaging effect and compensate for the lower density of the minefields. A third area of Russian adaptation was EW and the improvement of communications. A significant change took place from the perspective of the traditional Russian doctrine of mass fire—the previous numeric fires proved to be unfeasible during the first year and a half of the full-scale invasion. The lack of munitions and a significant logistical footprint required a shift towards effects instead of numeric indiscriminate attacks. As a result, the focus was placed on manufacturing Krasnopol laser guided 152 millimetre shells and a wider use of unmanned aerial vehicles (UAVs) for reconnaissance and fire coordination.[19]

As was previously mentioned, the use of drones on both sides had significantly increased. While Russia lacked drone production in the previous stages of the war, it managed to increase its drone fleet with an increase in national production capacity as well as supplies from its allies. From the obvious roles of ISR and artillery and tank spotting and the use of kamikaze drones and loitering munitions to destroy tanks and armoured vehicles, improvements were made by Russia in integrating various types of fire in the battlespace, including UAS, artillery, and rotary-wing aviation, with further employment of fighter aircraft in dropping gliding bombs. Hence, compared to previous Ukrainian counter-offensives, this time Russian aviation was more involved in the battlespace, causing significant challenges to the Ukrainian troops and Ukrainian aviation. On the Ukrainian side, the availability of the Army Tactical Missile System (ATACMS) allowed the targeting of Russian airfields in Berdyansk and Luhansk on 7 October 2023, destroying nine helicopters—seven Ka-52 and two Mi-8—and damaging eight Ka-52 and seven Mi-8.[20]

Ukrainian aviation was used with great care during the counter-offensive due to the scarcity of both rotary and fixed-wing aircraft. While some assets were used for close air support, a few significant long-range attack sorties employing Storm Shadow missiles were conducted behind Russian defence lines. On 10 and 12 June, Ukrainian aviation conducted missile strikes on Zaliznyj port, Schaslyvceve, and

[18] 'After suffering heavy losses, Ukrainians paused to rethink strategy', *The New York Times*, 15 July 2023. https://www.nytimes.com/2023/07/15/us/politics/ukraine-leopards-bradleys-counteroffensive.html.

[19] Watling and Reynolds, 'Stormbreak', 15–19.

[20] Anna Holodnova, 'The OSINT group Oryx calculated the losses of the Russians after the strikes on the airfields in Berdyansk and Luhansk', *Babel UA*, 23 October 2023. https://babel.ua/news/99821-osint-grupa-oryx-pidrahuvala-vtrati-rosiyan-pislya-udariv-po-aerodromah-u-berdyansku-i-lugansku.

Prymorsk, thereby targeting Russian command and control (C2) locations. On 22 June, Ukrainian Su-24 struck the Chonhar road bridge—an important logistics connection between Crimea and Russian-controlled parts of the Kherson region. On 29 July, another strike was aimed at the Chonhar railway bridge. One 6 August, the Chonhar and Tonka protoka bridges were hit. The increased presence of Russian aviation and the lack of sufficient Ukrainian counter-measures in the aerial domain became the most evident during this counter-offensive since the Russians could take full advantage of the vertical depth being way beyond the reach of the ground-based air defences (GBADs).

In one of his last interviews as commander-in-chief of the armed forces of Ukraine, given to *The Economist* in November 2023, General Valerii Zaluzhnyi concluded that the war was moving from manoeuvre to positional warfare, which was already evident during the summer counter-offensive in 2023.[21] He argued that the initial calculations that bleeding Russian troops would make the war untenable for Putin were false: 'That was my mistake. Russia has lost at least 150,000 dead. In any other country such casualties would have stopped the war ... Let's be honest, it's a feudal state where the cheapest resource is human life. And for us ... the most expensive thing we have is our people'.[22]

Although the speed of troop advancement was expected to be much higher, the relative technological parity achieved by the two sides contributed to significant delays in the advancement of the counter-offensive across various directions, not to forget well-established Russian defences and extensive minefields. In this regard, for a breakthrough to take place, a technological marvel was needed. Another important consideration was the necessity for both the Ukrainian industry and allied supplies to keep up with the tempo of the warfighting taking place in Ukraine and the speed of Russian modernization of their technologies. General Valerii Zaluzhnyi also addressed the increasing role of sensors on both sides, making battlespace even more transparent: 'The simple fact is that we see everything the enemy is doing and they see everything we are doing. In order for us to break this deadlock we need something new, like the gunpowder which the Chinese invented and which we are still using to kill each other'.[23]

Overall, the much-anticipated counter-offensive of 2023 did not achieve its posed objectives, and its tempo was far from the planned or desired one. It could not have the same surprise effect as the counter-offensives of 2022 because it was in the middle of the public and media hype; furthermore, the battlespace had become even more filled with sensors and ISR capabilities. The planned breach of 80 kilometres into Russian defences in the Robotyne direction, which could allow further access to Crimea, was only around 15–17 kilometres. The Russians prevented the planned Ukrainian amphibious assaults from the Kherson region by blowing up the Kakhovka dam. The targeting of Russian logistical routes required a greater availability of

[21] 'Ukraine's top general on the breakthrough he needs to beat Russia', *The Economist*, 1 November 2023. https://web.archive.org/web/20231101191731/https://www.economist.com/europe/2023/11/01/ukraines-commander-in-chief-on-the-breakthrough-he-needs-to-beat-russia.

[22] Ibid.

[23] Ibid.

long-range missiles and suitable aircraft to counter Russian presence in the air. This campaign illustrated the move from manoeuvre to positional war and the actual prolongation of the war under the given circumstances.

Factors Contributing to the State of Affairs and Points for Consideration

This section addresses various factors contributing to the establishment of the described situation. Some of these are more situational characteristics of the given operating environment, while others are more enduring factors present throughout the entire war, which are becoming more evident with each stage of the war. Some of the mentioned factors also illustrate the significant shift in warfighting and the changing character of war; still others confirm the enduring traits in warfighting established since the addition of the aerial domain into the battlespace.

Control of the Air in Multi-Domain Battlespace

Allied doctrines on joint operations and national equivalents always operate under the assumption and/or expectation that control of the air is established either by national or allied assets. It might be partial (air superiority) or overwhelming (air supremacy) control. Swedish doctrine also identifies the third option of situational, localized control of the air space at a given time. Control of the air is the inherent role of air forces, and it is aimed at providing freedom of manoeuvre to friendly forces across other domains and denying that freedom to the enemy. This is basic knowledge about air power taught in military colleges around the world. Nevertheless, in the last three years of the full-scale invasion, the land-centric nature of the war in Ukraine, and the establishment of mutual air denial due to the increase in drones and GBADS has led to debates on the present and future of warfare that are focused on the reduced role of aviation. However, what some analysts are missing is that if capabilities are insufficient in one domain, more pressure is put on others. In other words, the cost of not having sufficient numbers of rotary and fixed-wing aviation is the blood of ground troops, especially the infantry. The modern battlespace is multi-domain:

> The battlespace is integrated by its very essence, and the side that can better integrate its assets, reduce duplication and achieve effective deconfliction can obtain more productive layered effects across different domains and levels of warfare and, hence, attain a military advantage. This aspect is of even greater significance under conditions in which there are significant gaps across military equipment in various domains. The layered approach of combining arms in one domain or across domains can compensate for immediate gaps in other domains.[24]

[24] Fedorchak, *The Russia-Ukraine War*, 20.

According to General Zaluzhnyi, at the beginning of the war, the UAF had 120 tactical aircraft, with only forty technically suitable for employment, and thirty-three medium- and short-range anti-aircraft missile battalions, but only eighteen had sufficiently serviceable equipment.[25] Thanks to the assistance of partner nations, Ukrainian aerial and anti-air defences were reinforced and strengthened, allowing them to degrade corresponding Russian assets equivalent to thirteen regiments (brigades) of army aviation and more than 550 pieces of GBADS.[26] The survivability of Ukrainian aviation and GBADs was also conditioned by the continued employment of the principles of dispersal, mobility of ground defences, and the training of pilots in a close-to-war environment during the interim eight years of war.[27] In addition, in the last few years of the full-scale invasion, the lack of firepower in the aerial domain has been compensated by the more sporadic and precise long-range attack sorties of the Ukrainian air force (UkrAF), GBADs, and drones that provided the cheap and fast mass required to fill the aerial ISR and the firepower gap at the tactical level.[28] However, as with anything in the war, any innovation is followed by the enemy's adaptation and innovations of their own, and the cycle continues.

While in the first two years of the full-scale invasion, Russians lacked integration of their various assets and services and had fewer drones and relevant skills to fully integrate them into the warfighting, 2023 illustrated that they had learned their lessons and had begun to adapt and innovate to gain advantage from their capabilities. From the aerial perspective, a greater presence of their aviation in dropping gliding bombs on various Ukrainian positions during the counter-offensive, the wider availability of ISR, attack and kamikaze drones, and close air support for their troops with helicopters illustrated their ability to make the warfighting environment even more contested for the UAF, especially with further wearing out of the aerial assets of the UkrAF. Since Russia continues to produce and modernize its aircraft, the disparity between Ukrainian fixed-wing aviation and Russian aviation continues to increase. As for F-16s, they would have made a significant difference during the counter-offensive of 2023 and the war itself, just as would all those aircraft Ukraine had after the collapse of the Soviet Union (aircraft of which Ukraine was disarmed),[29] but wishful thinking does not intercept Russian bombers and does not stop missiles.

As some of the attack sorties conducted with Storm Shadow during the counter-offensive illustrated, a more significant presence of aviation would have allowed for further degrading of Russian reserves and logistics in the rear, which could have

[25] Valerii Zaluzhnyi, 'Modern positional warfare and how to win in it', 1 November 2023, 2. https://web.archive.org/web/20231101184848/https://infographics.economist.com/2023/ExternalContent/ZALUZHNYI_FULL_VERSION.pdf.

[26] Ibid.

[27] Viktoriya Fedorchak, 'The mass approach in the air war over Ukraine: Towards identifying a critical mass', *Handlingar och Tidskrift*, 1 (2023): 110–26. https://kkrva.se/hot/2023:1/fedorchak-the-mass-approach-in-the-air-war-over-ukraine.pdf.

[28] Fedorchak, *The Russia-Ukraine War*, 100–19.

[29] Viktoriya Fedorchak, 'The development of the Ukrainian air force 1991 on the road to 2022: Phoenix rising from the ashes', *The Norwegian Air Power Journal—Luftled* 2 (2022): 18–21. https://issuu.com/konsis/docs/luftled_nr_2_2022?fr=sN2I0YjQ4MzI1NjQ&mode=embed&viewMode=magazine&pageNumber=1.

weakened Russia's ability to sustain the offensive. The same goes for the close air support, which remains vital in the ongoing war. Regarding GBADS, as significant as they are in this war, their availability is always scarce since there is always the need to choose between the protection of civilian infrastructure, strategic and military objects, and the front line, especially with the changes in Russian tactics with long-range attacks focusing on waves, intensity, and gliding bombs.[30] Nevertheless, the system of Ukrainian air defences and the increased number of engaged targets still keep the enemy uncomfortable in the Ukrainian skies, as 'his aviation tries not to enter the kill zone of our anti-aircraft defences and utilizes air assets mainly from long ranges, which significantly reduces their effectiveness'.[31]

Breaching the Mine Barriers and the Terrain

Another obvious factor undermining the counter-offensive was the significantly strengthened Russian mine barriers and defences, with the depth in the most crucial areas reaching 15–20 kilometres. From the Ukrainian side, although some Soviet era de-mining equipment was still available, the bulk of capabilities like M58 MICLIC, Wisent 1, and NM 189 were sent by allies. Considering the scale of the de-mining activities, the available numbers were still insufficient. The situation was further complicated by the persistent presence of Russian reconnaissance drones, which were constantly monitoring any de-mining activities of the clearing teams. Even in the areas with successful breaches, the Russians could quickly restore the mine barriers with the use of the remote mine-laying system 'Zemledeliye'.[32] As General Zaluzhnyi said in 2022, it is ten to fifteen times more difficult to regain territory than to hold it.[33]

It is also essential to address the distinctive specifics of the terrain. The areas of the counter-offensive, especially in the Tokmak direction, are primarily flat fields with a few short strips of tree plantings and narrow farm roads. The Russians had dug into the area for a few years, and they had mined the key routes for advancement (especially roads) and established ambushes in the tree plantings; it should also not be forgotten that they had built three layers of defences specifically aimed at opposing a mechanized assault. Although under ideal circumstances, hollow terrain would be ideal for the mechanized assault, the problem was also the lack of sufficient natural cover to hide the advancement of the mechanized assaults. Combined with the widely spread availability of drones on the Russian side, the movement was taking place in relatively plain sight, with fast responses from Russian artillery, aviation, or ATGMs. Accordingly, large advancements were easily spotted, which suggested the change of tactics towards smaller infantry units with the support of firepower from other assets. Another consideration was the watersheds. The blowing up of the Kakhovka

[30] Viktoriya Fedorchak, 'IAMD lessons learned and identified in the Russia-Ukraine war', Presentation at NATO IAMD CoE Annual Conference, Chania, Greece, 19–20 July 2024.
[31] Zaluzhnyi, 'Modern positional warfare', 2.
[32] Ibid., 2–3.
[33] 'An interview with General Valery Zaluzhny'.

dam gave the Russians natural barriers and freed more troops for reinforcement of the defences in the Tokmak direction. Similarly, Russians retreated into the Balhodatne and Makarivka area, moving behind the Mokri Yaly River and blowing up the dam there, too. According to Valerii Shershen, the spokesperson for the Tavriia Front defence forces, with those actions, the Russians were once again trying to slow the progress of the counter-offensive.[34]

Counter-Battery Fires

At the beginning of the full-scale invasion, Russia had significant superiority in numbers of artillery, taking full advantage of amassed fires. With the introduction of Western missiles and artillery in the Ukrainian arsenal, precision strikes counter-battery had a significant effect on degrading Russian artillery. In this regard, Ukraine had an advantage since it applied shoot-and-scoot tactics before the Russians could strike back fast enough, and the Russians had lower artillery density. Hence, smaller areas could be covered. In terms of targeting, Ukraine followed the Western approach of precision strikes with significant care in the selection of targets to increase the effect of the strikes. In contrast, the Russians used the traditional, indiscriminate mass attacks.[35]

That was the case in the first year and a half of the war, but the Russians learned from their mistakes and adapted accordingly. First, the Russians began applying dispersal by guns, taking advantage of firing from the maximum range; means of EW and global positioning system (GPS) jamming were used to undermine the precision advantage of Western advanced technologies. Second, various drones were used in a counter-battery, which were challenging to intercept. Third, the density of Russian artillery had been increased by employing more outdated Soviet World War Two artillery systems like D-1 and Cold War gun-Howitzer D-20. Fourth, Russia began producing and employing more 122 millimetre Krasnopol precision-guided munitions (PGMs).[36] Finally, Russians learned from Ukrainian innovations and improved their ability 'to use rapid artillery fires from decentralised fire positions'.[37] The initial Ukrainian innovation was the app GISA-Arta to coordinate fire on a designated target from dispersed artillery. Russian adaptation was the introduction of 'the "Strelets" fire control and communication system, which reduced the reaction time of Russian artillery to 2–3 minutes from the previous 20–30 minutes. Twenty minutes might allow mechanised forces to complete a surprise attack. Three minutes does not'.[38]

[34] Kateryna Tyshchenko, 'Occupiers blow up dam in Donetsk Oblast, causing flooding', *Ukrainska Pravda*, 11 June 2023. https://www.pravda.com.ua/eng/news/2023/06/11/7406385/.

[35] Fedorchak, *The Russia-Ukraine War*, 81–100.

[36] Zaluzhnyi, 'Modern positional warfare', 3–4.

[37] Gustav Gressel, 'Beyond the counteroffensive: Attrition, stalemate, and the future of the war in Ukraine', European Council on Foreign Relations, Policy Brief, 18 January 2024, 15. https://ecfr.eu/publication/beyond-the-counter-offensive-attrition-stalemate-and-the-future-of-the-war-in-ukraine/.

[38] Ibid.

Proliferation of Drones among the Russian Armed Forces

Although the war in Ukraine is not the first one to be characterized by the use of drones by both sides and the significant role of drones in warfighting,[39] it is most certainly an illustration of the greater spread of drones with varied functionality, operation, and effects across domains. The first year of the full-scale invasion illustrated the disparity between Ukrainian UAS and Russian equivalents. Initially, the Russian military production of drones was focused more on the ISR function rather than in support of more kinetic tactical requirements of the battlespace. During the second year of the full-scale invasion, the Russians refocused on strengthening their mass with the production of cheap drones across various functions.[40] In the recent strategy, they announced significant production of drones, with the aim of one million by 2026 and 1.5 million by 2035.[41]

As was already mentioned, in the battlespace of the counter-offensive 2023, Russian drones posed additional challenges moving towards technological parity on both sides. For instance, the combination of Orlan and Lancet drones was used against Ukrainian artillery and mechanized assault teams. Accordingly, Orlan ISR drones would be used to spot the targets and then coordinate fire for Lancet kamikaze drones or any other source of firepower to target it. Russia is working on extending the range, payload, and functionality of its drones.[42] In essence, the increased presence of ISR drones on both sides meant that attacking forces could be spotted on their way to battle, even before the actual engagement, allowing preparation of a strike from the available source of fire. This was prominently illustrated with the Ukrainian mechanized assault and their failure to breach the Russian defences in the first days of the counter-offensive. Furthermore, both sides used first-person view units (FPVs) to carry anti-tank rockets to attack tanks and armoured vehicles. The primary advantage of FPVs compared to traditional anti-tank missiles is their superior range. They allow a focus of fire against moving mechanized assault units over much larger distances and wider areas.[43]

Electronic Warfare and the Battle of Signatures

From the very beginning of the Russian modernization of its military in 2008, one of the areas of technological improvement was EW, primarily because it was viewed as a

[39] Ulrike Franke, 'Drones in Ukraine and beyond: Everything you need to know', European Council on Foreign Relations, 11 August 2024. https://ecfr.eu/article/drones-in-ukraine-and-beyond-everything-you-need-to-know/.

[40] Viktoriya Fedorchak, 'AI and unmanned platforms in Ukraine', *Luftled*, 1 (2024): 32–25. https://issuu.com/konsis/docs/luftled_nr_1_2024?fr=sN2Y5NTcyMDE2NTE&mode=embed&viewMode=magazine&pageNumber=1.

[41] Pavel Luzin, 'Russian military drones', Foreign Policy Research Institute, November 2023, 23. https://www.fpri.org/wp-content/uploads/2023/11/russian-military-drones-.pdf.

[42] Roblin, 'Drones on the front line'.

[43] Gressel, 'Beyond the counteroffensive', 6.

sufficient means of undermining Western technological superiority across domains. In 2009, Russia created a separate branch of its armed forces—the EW troops. According to General Zaluzhnyi, during the war, the Russians used at least sixty types of modern EW equipment, constantly introducing new solutions, automation functions, and software updates.[44] They even established mass production of the means of 'trench EW'. For instance, the 'Silok' jamming station was used against UAVs. In some directions, such as Kupyansk and Bakhmut, they established layered EW systems, taking full advantage of mobility.[45]

On the Ukrainian side, most of the EW assets were from the Soviet legacy, with new assets accounting for only 25%. The anti-UAS systems, such as 'Bukovel-AD' (Ukraine), 'Enclave', 'Nota' (Ukraine), 'Kharma', 'Mandat', and R- 330KV1M (Ukraine), 'Polonaise' (Ukraine), and Antidron KVSG- 6 (Ukraine), illustrated effectiveness in the battlespace. This arsenal was strengthened by partner capabilities like VAMPIRE (United States (US)), Blighter Surveillance Systems A422 (United Kingdom (UK)) and portable anti-drone systems (anti-drone guns), including EDM4S Sky wipers (Lithuania), Nightfighter (UK), and Drone Defender (US). Also, Ukrainian engineers designed various smaller 'trench' EW devices for the front line.[46] The modernization of this segment of the UAF remains paramount since Russia has significant assets and places emphasis on EW. This was once again evident during the counter-offensive and continued Russian jamming of various equipment.

With the proliferation of drones, more sophisticated PGMs are developing hand in hand with counter-measures in the EW tool kit, along with the more kinetic means of their interception and elimination. The effectiveness of detection and countering significantly depends on the level of technological sophistication of the drones and PGMs. Thus, different defence technologies have sensors adjusted for the different signatures of incoming threats.[47] In the Ukraine war, the phenomenon of the battle of signatures is already underway. Mick Ryan coined the term as one of the trends for the future of warfare, arguing that each type of military equipment and each military organization would have their own signatures across visual, aural, and electromagnetic spectra. Accordingly, future warfighting will be about concealing one's own signature and accurately identifying distinctive signatures of the enemy in different domains of warfare.[48]

Placing this into the context of the Russia-Ukraine war, this battle is already taking place. For instance, the newly developed Russian version of the Shahed drone was aimed at being faster than its predecessor, which was procured from Iran. It was to reach the speed of 800 kilometres per hour in order to mimic ballistic missiles and, hence, engage and drain more sophisticated Ukrainian air defences. However, they could not reach the speed of 800 kilometres per hour. The significance of this aspect of warfighting is evidenced in Ukrainian modernizations in capabilities to counter

44 Zaluzhnyi, 'Modern positional warfare', 4.
45 Ibid., 5.
46 Fedorchak, *The Russia-Ukraine War*, 112.
47 Ibid., 128–9.
48 Mick Ryan, *War transformed* (Annapolis, MD: Naval Institute Press, 2022), 82–4.

high-precision weapons. The national EW system 'Pokrova' was established across the country in order to get Russian Shahed drones off course.[49] To improve C2, situational awareness systems like 'Graphite' and 'Quartz' were incorporated. The first system sends and displays data on small UAVs, and the second processes data on radio-electronic assets.[50]

On People

Another essential aspect to consider is the importance of human resources in warfighting. From the Russian perspective, people are seldom prioritized and are often treated as cannon fodder. The social and ethnic stratification and unequal treatment of its personnel and newly mobilized assets were evident across each stage of the war. Furthermore, training and investment in skills development were never the primary priority of the Russian military, since modernization has always been perceived as the improvement of technologies instead of getting skilled personnel. While full-scale mobilization in Russia (especially in the western parts of the country) was counter-productive for political reasons, recruitment and mobilization among prisoners and disadvantaged social and ethnic groups proliferated. Despite the poor quality of their assets, the Russians still managed to pull significant numbers to strengthen their defences in southern Ukraine, with less sophisticated troops being dropped on the front line, and more elite and trained troops remaining at the rear as reserves.

On the Ukrainian side, the approach was very different—with Ukrainian people being the primary advantage in warfighting, bringing their skills, knowledge, and innovations into the battlespace, which often saved lives. The more the war progressed, the more Ukrainian soldiers gave away their lives in different segments of the front line. The numbers would be smaller than the Russian losses, but the difference is that Ukraine was losing its golden generation—brilliant minds and first-class talents. Western audiences do not see the faces and names of the Ukrainian losses, but we, Ukrainians, do in every city, on every platform, every single day ...

This distinction is important for understanding the human resources situation during the spring counter-offensive. First, simply amassing all Ukrainian assets into an attack within a single axis against the most well-defended Russian areas would have been suicide and could have resulted in more significant losses to the UAF. Ukraine could not afford to lose more people; hence the decision for various axes of attack and the consequent change in tactics when mechanized assaults proved ineffective. Second, besides the challenge of amassing required technological capabilities, the UAF had issues with pulling in significant numbers of ready and trained personnel prepared for this type of assault, not to forget the enduring challenge of cross-unit

[49] 'Pokrova EW system, which successfully neutralizes Shaheds, has been launched in Ukraine', *Militarnyi*, 13 February 2024. https://mil.in.ua/en/news/pokrova-ew-system-which-successfully-neutralizes-shaheds-has-been-launchedin-ukraine/.

[50] Zaluzhnyi, 'Modern positional warfare', 5.

integration and cohesive performance, which requires time for preparation and training. Third, according to the then commander-in-chief of the armed forces of Ukraine, General Valerii Zaluzhnyi, there were issues with mobilization due to the enduring lack of transparency in mobilization legislation,[51] with various privileged categories of the Ukrainian population finding ways to avoid mobilization. Fourth, training of reserves could not be effectively conducted on the territory of Ukraine, due to continuous Russian attacks across the entire territory of the country, with any concentration of military personnel becoming an easy target. This was also greatly conditioned by the pro-Russian sabotage activities deep into Ukrainian territory, which would continue to coordinate fire and give away locations of the Ukrainian military across the country. Fifth, the scarcity of combat-experienced soldiers to join the existing warfighting across units and tasks undermines not only the ability to strengthen the existing positions but even to allow proper rotation of people who were at the front line in warfighting without having had any rotation for almost two years or so. Finally, the preparation of the reserves generally takes time, but having them dispersed across other countries makes it even more complex and time consuming. After all, time proves to be the scarcest asset that simply was not available during the continuous strengthening of the Russian defences on the territories they occupied.

Mass, War of Attrition, and Industry Production Capacities

The last four years of the full-scale invasion illustrated that although Western advanced technologies provided layered effects across domains, as the war progressed and despite losses, the Russians began to adapt to the Ukrainian innovations and the advantages the Western technologies provided. Although the discussion on the supremacy of the artificial mass versus the traditional numeric superiority was in favour of the fewer but more advanced technologies, the situation with the counter-offensive in 2023 and the then situation in autumn 2025 illustrate the need to have sufficient numbers of the more advanced technologies in order to overcome what General Zaluzhnyi called the established technological parity between Ukraine and Russia. The counter-offensive showed that with the established front lines, Russia had time to build defences and reinforce them with the accumulated equipment and to speed up the tempo of its production of various capabilities; it became a question of speed of equipment and ammunition delivery to sustain the effectiveness in warfighting of both sides. In other words, time once again became a crucial factor in sustaining warfighting, especially under the conditions of positional and attrition warfare.

The importance of timeliness of assets production and delivery were emphasized regarding numerous capabilities on both sides. However, the disparity lies in the differences in the origins of weapons supply. Despite being heavily sanctioned, Russia turned on its military industry to work at wartime capacities. The rates of Russian drone production have already been mentioned. According to different estimates,

[51] Zaluzhnyi, 'Modern positional warfare', 4–5.

Russia can produce a hundred cruise and ballistic missiles monthly. According to the Russian official statements, before the war, it could produce around 120 tanks, a hundred infantry fighting vehicles (IFVs), and a hundred APCs, but since the beginning of the full-scale invasion, it is estimated that production has increased by 50%. Despite the planned boost in mass production capacity, the Russians are rapidly losing their equipment, resulting in a need to dig deep into their equipment reserves, not all of which are combat suitable and still require repairs.[52]

In the case of Ukraine, its supply of equipment is diversified. Part of the equipment is produced in Ukraine, which can provide some stability of supply and wartime production tempo. However, due to the constant bombing, blackouts, and fire, coordination, continuity, and the scale of equipment production can be disrupted. Production also depends on the type of equipment that is in question. The same situation exists with repair of the equipment and the challenge of finding some spare parts. Another source consists of the Western partners' provision of more sophisticated Western equipment. While various previous-generation reserves of the Western countries allowed Ukraine to build significant capabilities, often the bilateral arrangements and the time and means of equipment delivery would prolong the time needed for the equipment to reach the front line. Furthermore, the counter-offensive of 2023 illustrated that for a massive breakthrough, a more significant critical mass has to be built utilizing if not more advanced technologies, then a greater number of the existing ones. However, one of the problems is that the delivery of various equipment significantly depleted national reserves. Manufacturing of the more advanced technologies takes time, while military industries during peacetime continue to operate on a contract basis. Numerous pieces of equipment operating in the Ukrainian battlespace required repairs that were often conducted if not in the countries of origin, then in Ukraine's neighbouring countries. It all takes time.

In addition, while the international military industry experiences a significant increase in procurement orders, and some industries have significantly increased their production, the numbers of the gradually increasing supply do not match the combined demands of the warfighting in Ukraine and national demands from the Western countries to strengthen their own defences. For instance, according to 2024 data, the US aims to increase its annual shell production from 240,000 shells per year to 1,000,000 rounds by 2025. European artillery production was around 650,000 rounds in 2023, half of which was prepared for Ukraine. For a comparative perspective of the tempo of warfighting in Ukraine, Ukraine used on average 5,000 shells daily (1.8 million per year). During the counter-offensive, 8,000 rounds were used per day (which equals three million shells needed per year). It is worth mentioning that Ukraine relies on imported ammunition at 85%.[53]

During the first year of the full-scale war, various additions of advanced Western technologies made a significant impact, even against Russian numeric superiority. However, with the Russian preparation for a long war, consolidating its assets and

[52] Gressel, 'Beyond the counteroffensive', 10.
[53] Ibid., 12.

adapting to the Western technologies, a positional and attrition war requires different approaches to everything from production of the military equipment and the logistics of its supply and repair to employment in the battlespace. Everything takes time, but time is running out.

Conclusion

Overall, it can be concluded that after the successes of the Ukrainian counter-offensive 2022, the much anticipated 'spring counter-offensive 2023' was a vastly different phenomenon. Time was a crucial factor. While Ukraine required time to receive a sufficient number of military equipment units from the partner nations in order to sustain a breakthrough, the Russians used that time to strengthen their defences. The initial plans of mechanized assault were disrupted by the wider use of Russian drones for ISR and fire coordination and the significant depth of the mine barriers. The amphibious assault was impossible due to the Russians blowing up the Kakhovka dam. Although the infantry-centred approach allowed Ukraine to regain some territories, it was far from the intended high tempo of advancement and the reaching of the posed operational objectives. Numerous factors contributed to this outcome of the offensive, illustrating the gradual shift of the warfighting in Ukraine from manoeuvre to positional war with the obvious implications on its conduct now and in the future. In May 2024, Russia launched its offensive in Kharkiv. Although its progress was slowed down, the fighting north of Kharkiv could be compared only to Bakhmut and Avdiivka. As of autumn 2026, various phenomena identified in this chapter continue to be manifested in the ongoing war: even with the Western aircraft operating in Ukraine, greater numbers are needed to establish effective control of the air and provide greater freedom of manoeuvre in the battlespace. Higher intensity of the Russian attacks illustrates the vital importance of a sufficient number of munitions across Ukraine's air defence assets. The enemy continued to take advantage of the terrain, both in intensifying front-line attacks and targeting Ukrainian positions from the Russia-controlled territories. Artillery remains one of the crucial elements of firepower on the battlefield, yet the tactics of its employment are adjusting to the tactical requirements. Drones evolve with greater reliance on artificial intelligence (AI). EW has entered every segment of warfighting and defence in Ukraine. And people are still the ones to pay the highest price in blood and lives. The war continues.

18
The Ukraine War in 2024
The Kursk Incursion and Beyond

Markus Balázs Göransson

Introduction

After Ukraine's failed 2023 counter-offensive, a narrative spread in the West that the war in Ukraine had developed into a stalemate, with little prospect that rapid manoeuvring could deliver substantial gains. It was argued that technological advances and the heavy use of fortifications had made offensive action against prepared defences exceedingly costly, requiring a strategy of attrition. To an extent, Ukraine accepted this logic as it shifted to defence in depth in 2024 and sought to impose disproportionate costs on Russia's forces. But it also acted in ways that revealed the limits of the logic. Ranged attacks against Russian territory, successful operations against Russia's naval forces in the Black Sea, and a surprise incursion into Russia's Kursk Province showed that the strength of contemporary land fortifications did not alter the fact that states retain options in war. At the time of writing, the war in Ukraine remains both dynamic and costly. Meanwhile, international developments, particularly the actions of the new United States (US) administration of President Donald J. Trump, have transformed the diplomatic context and raised questions about the durability of Western support for Ukraine.

The present chapter discusses developments in the war between January 2024 and early March 2025 and opens with a brief discussion of the impact of Ukraine's failed 2023 counter-offensive on views about the effectiveness of manoeuvre warfare in the war in Ukraine. This is followed by a description of Ukraine's shift to defence in depth in late 2023 and early 2024, before a third section notes that this defensive attritional approach aligned with ideas that were being proposed in the West that Ukraine would be best served by husbanding its limited forces and attriting Russian forces in defensive operations while building strength in the rear. The section also identifies crucial problems with these ideas. The fourth section discusses the Kursk incursion of August 2024, which brought manoeuvre back into the war and exposed some of the flawed assumptions often associated with the concept of the *transparent battlefield*. The subsequent two sections consider Ukraine's stunningly successful operations in the Black Sea theatre and its ranged attacks against energy and military infrastructure in Russia,

Markus Balázs Göransson, *The Ukraine War in 2024*. In: *Russia's War in Ukraine and Modern Warfare*.
Edited by: Mikael Weissmann and Niklas Nilsson, Oxford University Press.
DOI: 10.1093/9780198970903.003.0018

before the final chapter reflects on the implications of the election of Donald J. Trump in November 2024 for Western support for Ukraine.

The Aftermath of the 2023 Counter-Offensive

For much of 2024, Ukrainian forces were on the back foot as Russian troops pushed forward, seizing territory incrementally at large expense. This was due not only to the uneven correlation of forces, with Russia having a greater access to men and materiel but also reflected Ukraine's strategy of trading land for time, trying to leverage its defensive positions to exact large losses from its adversary.

The failed 2023 Ukrainian counter-offensive cast a shadow over operations in 2024. Ukraine's inability to make headway against fortified Russian positions in the summer and autumn of 2023 prompted new appraisals of the balance of power between Ukraine and Russia. But it also called into doubt manoeuvre itself as viable approach against an entrenched enemy under contemporary conditions of warfare.

Since the beginning of the large-scale war, several changes had taken place in command, control, communications, computers, intelligence, surveillance, and reconnaissance (C4ISR) that had transformed the tactical outlook on the battlefield. The heavy use of sensors (e.g. reconnaissance drones and satellite systems) and the effective integration of reconnaissance-fire complexes appeared to have vastly reduced the possibility of achieving superior speed and tactical surprise. Now the battlefield seemed to have become *transparent*, a place where it had become so easy to remotely observe and target advancing troops that it was idle to talk of cover and concealment. In such a context, effective manoeuvre warfare, it was suggested, could scarcely be conducted at acceptable levels of risk. Military observers in the West and Ukraine as well as in Russia[1] drew the conclusion that manoeuvre, for now, was out as an appropriate approach; the best way to deploy military power was instead to wear down the adversary's capabilities while protecting one's own in a form of static defence adapted to the battlefield of the 2020s.[2]

Ukraine's Defensive Attrition

Ukraine seemed to have accepted this logic to an extent. During the first half of 2024, it adopted an attritional approach in its operations in the land arena. As the Russians pressed on, buoyed by their successes against the 2023 counter-offensive, Ukrainian troops sought to ward off their attacks while trying to impose lopsided casualty ratios from their defensive positions. Ukraine's newly appointed commander-in-chief Colonel-General Oleksandr Syrskyi spelled out the approach in an interview on 13

[1] Cf. Krisztián Jójárt, 'The war against Ukraine through the prism of Russian military thought', *Journal of Strategic Studies* 47/6–7 (2024): 816–17. https://doi.org/10.1080/01402390.2024.2414079.

[2] Cf. Robert Rose, 'Biting off what it can chew: Ukraine understands its attritional context', *War on the Rocks*, 26 September 2023. https://warontherocks.com/2023/09/biting-off-what-it-can-chew-ukraine-understands-its-attritional-context/.

February: 'The purpose of our defense operation is to exhaust the enemy's forces, and inflict maximum losses, by using our fortifications, technical advantages, drones, electronic warfare, and by holding prepared defense lines.'[3]

The approach was exemplified by Ukraine's defence against the Russian push against the town of Avdiivka in the winter of 2023–4. The Russians had sought to take this town north of Donetsk since the start of the large-scale invasion (and indeed since their first invasion in 2014) but began a concerted attempt to encircle it in October 2023. In response, the Ukrainians used their defences, superior tactics, and apparently stronger morale to hold them up until February 2024, when a withdrawal was finally ordered. By then, the Russians had suffered massive casualties. Some estimates based on reports by pro-Russian military bloggers suggest that Russian losses in the Battle of Avdiivka exceeded Soviet ones in the Soviet-Afghan War of 1979–89,[4] when 15,000 Soviet soldiers were killed. Within a month of the fall of Avdiivka, Ukraine claimed to have killed or injured a further 500 Russian soldiers and damaged or destroyed many Russian weapons systems, including as many as twenty armoured vehicles and tanks, near the town.[5]

Similarly, in Robotyne, one of few settlements of note that had been liberated in the 2023 counter-offensive, Ukrainian troops fought counter-attacking Russian forces for several months until their defensive positions along with the village itself were destroyed, forcing a retreat. In Urozhaine, too, Ukrainian defenders held Russian forces at bay until July, by which time, according to a Ukrainian military spokesperson, the Ukrainians had exacted a heavy price from the Russians:

> To achieve success and some victories in the destroyed towns and villages, the enemy suffers considerable losses every day. Several hundred of them have been killed, and the result is a destroyed settlement and a pile of corpses of Russian soldiers around it who will not return home to their families.[6]

The Limits of Defensive Attrition

Ukraine's defensive attritional approach aligned with ideas circulating in the West about how Ukraine should best use its forces. Observers argued that Ukraine's best option given Russia's numerical advantage was to wear down Russian forces, while rebuilding an offensive capacity that could be deployed at a favourable time in the future. Ukraine, it was argued, should conduct defence in depth, while reconstituting

[3] Kateryna Denisova, 'Syrskyi: Ukraine shifts to defense to drain Russian troops', *The Kyiv Independent*, 13 February 2024. https://kyivindependent.com/syrskyi-ukraine-shifts-to-defensive-to-drain-russian-troops/.

[4] Thomas Gibbons-Neff and Anatoly Kurmanaev, 'In Ukraine, Russia is inching forward death by death', *The New York Times*, 27 February 2024. https://www.nytimes.com/2024/02/27/world/europe/russia-deaths-avdiivka-strategy.html.

[5] Valentyna Romanenko, 'Russia lost a regiment's worth of soldiers on Avdiivka front but continues to press on', *Ukrainska Pravda*, 13 March 2024. https://www.pravda.com.ua/eng/news/2024/03/13/7446333/.

[6] Alona Mazurenko, 'Ukrainian defenders withdraw from Urozhaine, Donetsk Oblast', *Ukrainska Pravda*, 18 July 2024. https://www.pravda.com.ua/eng/news/2024/07/18/7466288/index.amp.

its forces through an increase in Western supplies, a ramping up of domestic military production and a stepping up of recruitment.[7] Some likened the situation on the front in the Donbas to that on the Western front during World War One,[8] where powerful defences had similarly made manoeuvre inappropriate and necessitated a strategy of attrition until effective new tactics and technology were developed and deployed. Under the current conditions, the wisest approach, it was suggested, was to preserve one's resources to buy time, while draining the adversary of manpower and materiel.

This was a compelling argument to an extent. Ukraine's offensive capacity in early 2024 seemed largely spent and it had little option but to shift to the defence in the land arena. But there were also problems with treating what was in essence a Hobson's choice as a viable long-term strategy. The idea that Ukraine could trade land for time and then use that time to build military advantage over the Russians in the mid- to long term rested on three bold assumptions, none of which appears to have been borne out with time.

First, there was the belief that sufficient Western supplies could be ensured in the long term. Yet, the pipeline of foreign military aid was unreliable in the short term and at risk of partially drying up in the long. Already in the winter of 2023 and 2024, the US Congress failed to approve a new military aid package, delaying the transfer of US materiel, including advanced weapons, to Kyiv.[9] This exposed the fragility of US support for Ukraine, while it raised the question of what might happen if Donald Trump won the November US presidential elections. Many observers assumed that Trump, who was seen to entertain good relations with Vladimir Putin and had criticized US aid to Ukraine,[10] was likely to take actions that would be damaging to Ukraine.[11]

The situation was compounded by the fact that Europe would struggle to make up for a reduction in US support. Even though there was arguably greater political will to help Kyiv in some European capitals than in Washington, DC, where rivalries between the Democratic and Republican parties had chewed into the Ukraine

[7] Cf. Franz-Stefan Gady and Michael Kofman, 'Making attrition work: A viable theory of victory for Ukraine', International Institute for Strategic Studies, 9 February 2024. https://www.iiss.org/online-analysis/survival-online/2024/01/making-attrition-work-a-viable-theory-of-victory-for-ukraine/.

[8] Michael E. O'Hanlon, Alejandra Rocha, Sophie Roehse, and Mallika Yadwad, 'What next on the war in Ukraine?', The Brookings Institution, 22 October 2024. https://www.brookings.edu/articles/what-next-on-the-war-in-ukraine/; Mykhailo Krygel, 'Remarque in trenches near Kherson: Impressions from the soldier and film producer, Volodymyr Yatsenko, on the nature of fear in warfare, rebuilding a nation, and life after victory', *Ukrainska Pravda*, 13 November 2022. https://www.pravda.com.ua/eng/articles/2022/11/13/7376017/; Olivia Olander, 'War in Ukraine "essentially trench warfare," senator says after Kyiv visit', *Politico*, 1 August 2023. https://www.politico.com/news/2023/01/08/war-ukraine-trench-warfare-angus-king-00076925.

[9] Yang Tian, 'House approves $61bn aid for Ukraine—What we know so far, and what happens next', *The Guardian*, 21 April 2024. https://www.theguardian.com/world/2024/apr/21/house-approves-61bn-aid-for-ukraine-what-we-know-so-far-and-what-happens-next.

[10] Jones Hayden, 'Trump threatens to cut US aid to Ukraine quickly if reelected', *Politico*, 16 June 2024. https://www.politico.eu/article/donald-trump-ukraine-russia-war-threatens-cut-aid-election-2024/.

[11] Rajan Menon, 'A Trump presidency would leave Ukraine to its fate—because he has China in his sights', *The Guardian*, 25 July 2024. https://www.theguardian.com/commentisfree/article/2024/jul/25/ukraine-war-russia-us-trump-election; Joanna Kakissis, 'What will Trump's presidency mean for Russia's war on Ukraine?', *NPR*, 7 November 2024. https://www.npr.org/2024/11/07/nx-s1-5181985/2024-election-trump-russia-ukraine-war.

issue, the Europeans' capacity to do so was much slighter. This was due not only to Europe's historically smaller military sector but also to the fact that European countries had dragged their feet in expanding their military industrial production after the start of the large-scale war. Ammunition production and intelligence, surveillance, and reconnaissance (ISR) capabilities, all of which were in great need in Ukraine, remained in particularly low supply.[12] True, some positive steps had been taken. European defence spending had increased since February 2022 and the EU had unveiled its first ever common defence industrial strategy,[13] while Czechia, for its part, claimed to have identified 800,000 artillery shells on world markets that could be purchased and sent to Ukraine.[14] But Russia's military production still outstripped Europe's[15] and Ukraine continued to depend on US targeting data for its advanced mid-range missile systems as well as US developed air defence systems.

The second assumption was that Ukraine could impose losses on the Russians that were not only greater than its own but also more difficult for Moscow to sustain. While Russia did seem to be suffering significantly larger casualty numbers than Ukraine, along with visible economic pain, it appeared to be replacing its losses more quickly than expected. A policy of enlisting volunteers, foreign troops, and contract soldiers brought in a steady flow of personnel, reportedly between 20,000 and 30,000 each month in 2024,[16] while the Kremlin avoided the kind of social upheaval that it had witnessed after it had ordered a partial mobilization of 300,000 men in September 2022. This is not to say that Russia's recruitment model was sustainable. After all, access to Russian volunteers was likely to diminish with time. But it did suggest that Moscow had managed to push its strategic culmination point (i.e. the point where its military capabilities in the war would begin to decline) into the future. And it certainly gave pause to suggestions that Ukraine, which struggled with reconstituting its armed forces,[17] stood to win from a strategy of attrition. Indeed, it was not only Ukraine that was able to build offensive strength in the rear but also Russia. On 10 May, Russia launched a surprise attack into Ukraine's Kharkiv region

[12] Sophia Besch, 'Understanding the EU's new defense industrial strategy', *Carnegie Endowment*, 8 March 2024. https://carnegieendowment.org/posts/2024/03/understanding-the-eus-new-defense-industrial-strategy?lang=en.

[13] 'First ever defence industrial strategy and a new defence industry programme to enhance Europe's readiness and security', European Commission, 5 March 2024. https://ec.europa.eu/commission/presscorner/detail/en/ip_24_1321.

[14] Elsa Court, 'Ukraine receiving shells from Prague-led initiative on schedule, Czech foreign minister says', *The Kyiv Independent*, 29 August 2024. https://kyivindependent.com/czech-deliveries-on-schedule/.

[15] Alexandr Burilkov and Guntram B. Wolff, 'Defending Europe without the US: First estimates of what is needed', *Bruegel*, 21 February 2025. https://www.bruegel.org/analysis/defending-europe-without-us-first-estimates-what-needed.

[16] Paul Schwartz, Dmitry Gorenburg, and Olga Thomas, 'Russian military mobilization during the Ukraine war: Evolution, methods, and net impact', *CAN*, October 2024; Margarete Klein, 'How Russia is recruiting for the long war: Covertly mobilising volunteers while preparing for a new round of compulsory mobilisation', *Stiftung Wissenschaft und Politik*, 27 June 2024. https://www.swp-berlin.org/10.18449/2024C24/.

[17] Shaun Walker, 'Ukraine faces difficult decisions over acute shortage of frontline Ttroops', *The Guardian*, 21 December 2024. https://www.theguardian.com/world/2024/dec/21/ukraine-faces-difficult-decisions-over-acute-shortage-of-frontline-troops.

and reportedly took more land than it had over the previous eighteen months.[18] As many times before, the Russian forces suffered large casualties and Ukrainian troops eventually retook part of the lost territory, but the offensive indicated that time was not necessarily on Ukraine's side.

The third assumption behind proposals that Ukraine should shift to attritional warfare was that Ukraine first needed to wear down Russian defences before it was able to manoeuvre around them. This was likely true for the Donbas, where Russia had a very strong system of fortifications, but not for the land arena as a whole. The analogy between Ukraine in 2024 and the Western front during World War One is flawed because there are important differences between the two theatres. On the Western front in World War One, German and Allied forces were hemmed in by natural obstacles (the Alps in the south and the English Channel patrolled by the British fleet in the north). The absence of flanks that could be manoeuvred around led armies to slug it out largely through frontal attacks against fortified defences, although there were attempts to open new fronts elsewhere, notably in Gallipoli.

In contrast, the front in Ukraine is interrupted by natural obstacles only in the south, where the Black Sea forms a formidable impediment to large operations against the Russian-occupied southern and southeastern Ukrainian coastline. In the north, the Russian flank is more open. The land border between Ukraine and Russia is roughly 1,900 kilometres, and as of 2024 around 1,500 kilometres remained unoccupied by Russia; all of it runs through open landscapes.

It was not the impenetrability of the border that kept Ukraine from mounting large offensives across it, but political constraints combined with military calculations. The administration of US President Joe Biden, fearing Russian escalation, had repeatedly signalled its unwillingness to countenance Ukrainian attacks against internationally recognized Russian territory. In an essay in the *New York Times* in May 2022 under the heading 'What America Will and Will Not Do in Ukraine', Biden bluntly said: 'We are not encouraging or enabling Ukraine to strike beyond its borders'.[19] One year later, his National Security Council spokesperson John F. Kirby said, 'We don't encourage, we don't enable and we don't support strikes or attacks inside Russia.'[20]

US objections to cross-border actions decreased with time but did not disappear. When militant groups consisting of anti-Kremlin Russian nationals crossed into Russia's western Belgorod Region in May and June 2023, probably with support from Ukraine's military intelligence services,[21] the White House said it would investigate

[18] Arnel Hecimovic, 'The battle for Vovchansk in Ukraine—in pictures', *The Guardian*, 21 May 2024. https://www.theguardian.com/world/gallery/2024/may/21/the-battle-for-vovchansk-in-ukraine-in-pictures.

[19] 'President Biden: What America will and will not do in Ukraine', *The New York Times*, 31 May 2022. https://www.nytimes.com/2022/05/31/opinion/biden-ukraine-strategy.html.

[20] Helene Cooper, Eric Schmitt, and Julian E. Barnes, 'Biden administration shrugs Ooff Ukraine's attacks in Russia', *The New York Times*, 5 June 2023. https://www.nytimes.com/2023/06/05/us/politics/ukraine-attacks-inside-russia.html.

[21] Tom Balmforth and Anna Dabrowska, 'Ukraine war: Belgorod incursion may stretch Russia's defences', *Reuters*, 24 May 2023. https://www.reuters.com/world/europe/cross-border-incursions-ukraine-may-stretch-russian-defences-2023-05-24/.

reports that US equipment had been used in the attack.[22] And when Ukraine stepped up its ranged attacks against oil refineries in Russia (more on this below), the White House raised concerns about their impact on global oil prices.[23] Meanwhile, the US continued to forbid Ukraine to use some of its most advanced weaponry against Russian territory. The prospect that Washington, DC, would consent to a major operation into Russian lands seemed distant.

For its part, Moscow was complacent about the vulnerability of its border. Rather than bolster it, it redeployed air defences from the border area to Ukraine and sent untrained conscripts to man it together with Federal Security Service (FSB) personnel, Rosgvardia units, and territorial defence troops to free up professional soldiers for service in the war.[24] The US-based Institute for the Study of War later assessed that Russia had treated the Ukraine border as a 'dormant front of the theatre'.[25] Now, Russia was in for a rude awakening. Ukraine would show that there were still flanks along the front around which its mechanized forces could manoeuvre.

The Kursk Incursion

During the high summer, Ukraine massed forces in Sumy Province near the Russian border under the guise of preparing for a possible Russian attack. The forces included elements of experienced units such as the 99th Mechanized Battalion of the 61st Separate Steppe Brigade[26] and the 225th Assault Battalion of the 115th Mechanized Brigade,[27] which had been pulled from other parts of Ukraine. The build-up of Ukrainian troops was likely visible to Russian military observers, some of whom appear to have warned their superiors of a possible attack,[28] yet the Russian military leadership took no adequate counter-measures. Possibly, the Russian command had come to believe that Ukraine lacked the capacity and will to mount a major operation at a time when Ukraine's forces were under heavy attack in the Donbas and the

[22] Ostap Yarysh and Hanna Tverdokhlib, 'White House probes claims US-made gear was used in Russia raid', *Voice of America*, 24 May 2023. https://www.voanews.com/a/white-house-probes-claims-that-us-made-gear-was-used-in-russia-raid/7107837.html.

[23] John Hudson, 'Ukraine's attacks on Russian oil refineries deepen tensions with U.S.', *The Washington Post*, 15 April 2024. https://www.washingtonpost.com/national-security/2024/04/15/ukraine-russia-oil-refinery-attacks/.

[24] Angelica Evans, 'Ukraine's Kursk incursion: Six month assessment', Institute for the Study of War, 6 February 2025. https://www.understandingwar.org/backgrounder/ukraine%E2%80%99s-kursk-incursion-six-month-assessment.

[25] Ibid.

[26] Diana Bondarenko, '"Misto Kontrolyuet'sya ZSU." Ukrains'ky Biitsi Zakhopili Sudzhu na Kurshchini' ['"The city is controlled by the armed forces of Ukraine": Ukrainian fighters captured Sudzha in Kurshchyna'], *Fokus*, 9 August 2024. https://focus.ua/uk/voennye-novosti/662413-misto-kontrolyuyetsya-zsu-ukrajinski-biyci-zahopili-sudzhu-na-kurshchini-video.

[27] Thomas d'Istria, 'Incursion ukrainienne en Russie. "Le moral est monté d'un cran"' ['Ukrainian incursion into Russia: "the morale has increased by a notch"'], *Le Monde*, 13 August 2024. https://www.lemonde.fr/international/article/2024/08/13/incursion-ukrainienne-en-russie-le-moral-est-monte-d-un-cran_6279134_3210.html.

[28] Elsa Court, 'Gerasimov ignored warnings of Ukrainian troop buildup near border, Bloomberg reports', *The Kyiv Independent*, 8 August 2024. https://kyivindependent.com/gerasimov-ignored-intelligence-warnings-of-ukraines/.

US had communicated its disapproval of military actions that could trigger Russian escalation.

Whatever the reasons, Russia's troops were caught by surprise when Ukrainian forces rolled over the border into Kursk Province on 6 August. The Ukrainians swiftly seized the crossing point at Sudzha and then pushed further, taking a large bulge of land that extended some 35 kilometres at its deepest.[29]

The Russian forces fell into disarray and failed to check the Ukrainian forces as they proceeded to take control of settlements and communication lines and entrenched themselves in expectation of a coming Russian counter-attack. Scores of Russian soldiers surrendered to the Ukrainians (nearly 600 during the first three weeks of the invasion, according to Ukrainian sources),[30] indicating a lack of morale and preparedness. On 19 August, Ukrainian President Volodymyr Zelenskyi stated that Ukrainian forces had taken more than 1,200 kilometres of Russian territory, including more than ninety settlements.[31]

The success of the Ukrainian incursion owed much to the principles of concentration of forces, manoeuvre, and surprise. Ukraine concentrated some of its best forces at a weak point in Russia's defences, then used manoeuvre to outflank the Russian defenders. Ukraine's ability to achieve surprise was particularly striking, given the ubiquity of satellite, drone-based, and other reconnaissance that enabled countries to quickly spot troop build-up. A likely reason for Russia's surprise was that it had misread Kyiv's intentions, which were audacious to the point of being implausible. Even if Ukraine could not hide its amassment of forces, it could conceal its intentions. As any chess player will agree, under conditions of omniscience, unexpected intentions remain a major vehicle of surprise.

In addition, Ukraine employed new tactics of drone and electronic warfare, using a combination of measures to weaken Russian surveillance and drone defences. Among other things, it disabled Russia's system of aircraft-type reconnaissance drones possibly with new first-person view (FPV) unmanned aerial vehicles (UAVs) and then exploited the Russian loss of reconnaissance to move electronic jammers into place and use them to disrupt Russian communications.[32] Impressively, the Ukrainians did this seemingly without hampering their own ability to operate drones in the battle space, giving them effective drone dominance over the Russians.[33]

[29] Evans, 'Ukraine's Kursk incursion'.

[30] Elena Giordano and Veronika Melkozerova, 'Ukraine says it's seized nearly 600 Russian POWs in Kursk attack', *Politico*, 27 August 2024. https://www.politico.eu/article/ukraine-offensive-kursk-prisoners-of-war-600-russia-invasion-kyiv/.

[31] Kateryna Denisova, 'Ukraine in control of 92 settlements in Russia's Kursk Oblast, Zelensky claims', *The Kyiv Post*, 19 August 2024. https://kyivindependent.com/ukraine-in-control-of-92-settlements-in-russias-kursk-oblast-zelensky-claims/.

[32] David Hambling, 'Ukraine's Kursk offensive blitzed Russia with electronic warfare and drones', *Forbes*, 9 August 2024. https://www.forbes.com/sites/davidhambling/2024/08/09/ukraines-kursk-offensive-blitzed-russia-with-electronic-warfare-and-drones/.

[33] Michael Winberg, 'Unmanned systems in Kursk—Shaping the battlespace and denying the enemy access', KKRVA: Royal Swedish Academy of Military Sciences, 23 August 2024. https://kkrva.se/unmanned-systems-in-kursk-shaping-the-battlespace-and-denying-the-enemy-access/.

With time, the Russians halted the Ukrainian offensive by rushing in conscripts from other parts of Russia and professional soldiers from less prioritized parts of the front in Ukraine, specifically Zaporizhzhiya and Kharkiv.[34] On 10 September, Russia launched a counter-attack and slowly began to claw back some of the territory it had lost. By March 2025, Russian forces had regained most of the territory at large expense, although Ukraine continued to hold small slivers of land along the border.

One striking development during the Kursk offensive was the introduction of North Korean troops. On 18 June 2024, Russia and North Korea had signed a treaty on defence and security cooperation, which stipulated that the countries should provide each other with military and other assistance without delay in the event of armed aggression against either country.[35] This gave a legal opening for the dispatch of North Korean soldiers to the Kursk theatre. In mid-October, some 1,500 North Korean soldiers were deployed to Russia, according to South Korean media.[36] Thousands more were trained in the Democratic People's Republic of Korea (DRK) for later deployment, according to Ukrainian authorities.[37] By January 2025, reportedly around 12,000 North Koreans had been inserted into the fighting around Kursk. It represented a dramatic escalation of North Korea's involvement in the war, which had previously consisted of sending military supplies, particularly artillery shells, to Russia. But its impact on the fighting in Kursk remained unclear. Inexperienced and prone to adopting near-suicidal tactics, the North Koreans suffered enormous casualties,[38] though they reportedly acquired new skills of electronic and drone warfare with time[39] and were more fearless than their Russian comrades in arms.[40]

Much of the discussion about the Kursk offensive has concerned Ukraine's reasons for launching it. There are several possible reasons: to draw Russian forces away from the major lines of operation in the Donbas, to destabilize the Russian leadership, to improve Ukraine's bargaining position ahead of expected peace negotiations, to increase Ukrainian morale, to call their bluff on Russian threats of escalation, and to reintroduce an element of manoeuvre into operations, among others. Moreover,

[34] Olha Hlushchenko, 'Russia relocates several units from Pokrovsk front to Kursk Oblast', *Ukrainska Pravda*, 1 September 2024. https://www.pravda.com.ua/eng/news/2024/09/1/7472900/.

[35] Helen Regan, Gawon Bae, and Yoonjung Seo, 'North Korea says Kim and Putin's defense pact permits all available means to assist each other if either nation attacked', *CNN*, 20 June 2024. https://edition.cnn.com/2024/06/20/asia/russia-putin-kim-deal-military-assistance-intl-hnk/index.html.

[36] Hyung-Jin Kim and Kim Tong-Hyung, 'South Korean intelligence says North has sent troops to aid Russia's war in Ukraine', *AP*, 18 October 2024. https://apnews.com/article/north-korea-russia-ukraine-war-troops-south-spy-9cd563c5570f68e9b314976009682810.

[37] Ibid.

[38] Nick Paton Walsh, Rebecca Wright, Daria Tarasova-Markina, Victoria Butenko, and Brice Laine, 'Suicidal tendencies and '80s battlefield tactics: How North Korean soldiers are operating in Russia's war on Ukraine', *CNN*, 30 January 2025. https://edition.cnn.com/2025/01/28/europe/north-korean-soldiers-fighting-for-russia-intl-cmd/index.html.

[39] Joanna Kakissis and Volodymyr Solohub, 'North Koreans fighting for Russia against Ukraine have grown skilled in drone warfare', *NPR*, 16 June 2025. https://www.npr.org/2025/06/16/g-s1-71531/north-korea-soldiers-russia-ukraine-war-drones.

[40] David Brennan, '"They do not surrender": Ukraine commander details fight with North Koreans in Russia', *ABCNews*, 8 April 2025. https://abcnews.go.com/International/surrender-ukraine-commander-details-fight-north-koreans-russia/story?id=120403500.

Ukraine's calculations may have evolved over time. In the end, the Kursk offensive may have contributed to slowing Russian operations in other parts of the front line and demonstrated the continued relevance of manoeuvre in the war.

The Black Sea

The Kursk offensive was not Ukraine's only important success in 2024. Other feats were achieved in the maritime arena, where Ukraine had undertaken offensive operations using innovative technology since the beginning of the large-scale war. Its most important maritime achievement was arguably its success in 2023 to force open the shipping routes to allow its large grain exports to reach world markets, which it had done by coercing Russia's naval ships into a narrower perimeter of operations, out of range of the shipping lanes. As North Atlantic Treaty Organization (NATO) Secretary-General Jens Stoltenberg said in February 2024:

> The Ukrainians have been able to inflict heavy losses on the Russian Black Sea Fleet. They have actually been able to push the Black Sea Fleet away from the western part of the Black Sea. And this is a great achievement, a great victory for the Ukrainians. The fact that they've been able to push away the Russian Black Sea Fleet and open a corridor, so they are able to export grain and other commodities to the world market.[41]

Throughout the large-scale war, the main technological enablers of Ukrainian maritime operations have been naval drones and enhanced anti-ship missiles. Remotely piloted uncrewed platforms, such as the Magura V5 kamikaze maritime drone, have been particularly important for Ukraine's success in targeting Russian vessels at range. By November 2023, Ukraine claimed to have destroyed fifteen Russian vessels and damaged a further twelve since the beginning of the large-scale invasion, many of them while they were docked in port in Sevastopol on the Crimean Peninsula.[42] In autumn 2023, it was reported that because of this vulnerability Russia had begun relocating part of its Black Sea fleet from the Crimean Peninsula to Novorossiysk on the eastern shore of the Black Sea.[43]

Ukraine continued to notch up successes in the Black Sea in 2024. By early February, it claimed to have disabled around 30% of Russia's Black Sea fleet.[44] Shortly

[41] 'Pre-ministerial press conference', NATO, 14 February 2024. https://www.nato.int/cps/en/natohq/opinions_222594.htm.

[42] 'Ukraine says it has destroyed 15 Russian naval vessels in Black Sea during war', *Reuters*, 17 November 2023. https://www.reuters.com/world/europe/ukraine-says-it-has-destroyed-15-russian-naval-vessels-black-sea-during-war-2023-11-17/.

[43] Chris York, 'Russia continues to redeploy Black Sea fleet from Crimea, Ukraine promises "New steps" to target it', *The Kyiv Independent*, 21 March 2024. https://kyivindependent.com/russia-continues-to-redeploy-black-sea-fleet/.

[44] Kateryna Denisova, 'Last Russian patrol ship left occupied Crimea, Ukraine's navy says', *The Kyiv Independent*, 15 July 2024. https://kyivindependent.com/last-russian-patrol-ship-left-occupied-crimea-ukraines-navy-says/.

afterwards, it reported sinking the Russian landing ship the *Tsezar Kunikov*,[45] and a few weeks later, the *Sergey Kotov* patrol ship. The latter was reportedly sunk by a Magura V5 maritime drone,[46] a platform that Ukraine's authorities at the end of 2024 credited with the disabling of altogether fifteen Russian vessels since it was first used.[47] Meanwhile, Russia's relocation to the eastern shore of the Black Sea continued. In July 2024, Ukrainian Vice Admiral Oleksiy Neizhpapa, said: 'The Russian navy's Black Sea Fleet has been forced to rebase nearly all its combat-ready warships from occupied Crimea to other locations.'[48]

Hence, Russia remains on the defensive in the Black Sea. Even Novorossiysk is within range of Ukrainian attacks. In 2023, the battleship *Olenegorsky Gornyak* was damaged in its port,[49] and in 2024 and 2025 Ukraine has repeatedly targeted the port facilities with maritime and air drones, which Russia has claimed to have repelled.[50]

Russia retains some options. It continues to maintain a capacity to strike Ukrainian territory with cruise missiles from warships in the Black Sea, and as *Bellingcat* reported in July 2024, it has accelerated development of an alternative naval port in Abkhazia, which would give it increased shelter from Ukrainian attacks.[51] But even so, Russia has patently been losing the adaptive struggle in naval warfare and seen its room for manoeuvre in the Black Sea decrease drastically.

Ranged Attacks

On land, Russia has continued to target Ukrainian energy and military infrastructure as well as residential areas with cruise missiles, drones, and other long-range platforms. Ukraine, in turn, has carried out a growing number of attacks on Russian infrastructure and military bases, some at very large distances from Ukraine. Developments in ranged warfare shows Ukraine's capacity for technological adaptation and

[45] Radio Svoboda, 'U Genshtabi ZSU Pidvertdili Znishcheniia Rossis'koho VDK "Tsezar Kunikov" ['The general staff of the AFU confirmed the sinking of the Russian landing ship "the Tsezar Kunikov"]', 14 February 2024. https://www.radiosvoboda.org/a/news-henshtab-vdk-rf/32819008.html.

[46] Irina Balachuk and Valentina Romanenko, 'V Chornomu Mori Pidbili VDK "Tsezar Kunikov", Vin Zatonuv—Dzerela [In the Black Sea, the landing ship "The Tsezar Kunikov", was hit, it sunk— Dzerela]', *Ukrainska Pravda*, 14 February 2024. https://www.pravda.com.ua/news/2024/02/14/7441777/.

[47] 'Russia deploys new tactic to counter Ukrainian naval drones in the Black Sea', *The New Voice of Ukraine*, 2 February 2025. https://english.nv.ua/nation/russia-tests-new-drone-strike-tactic-to-counter-ukrainian-naval-drones-in-black-sea-isw-50486536.html.

[48] Chloé Hoorman and Elisa Vincent, 'Ukrainian naval drone attacks force Russian fleet out of Crimea', *Le Monde*, 22 July 2024. https://www.lemonde.fr/en/international/article/2024/07/22/ukrainian-naval-drone-attacks-force-russian-fleet-out-of-crimea_6694576_4.html.

[49] James Waterhouse and Kathryn Armstrong, 'Russian ship hit in Novorossisysk, Black Sea drone attack, Ukraine sources say', *BBC*, 4 August 2023. https://www.bbc.com/news/world-europe-66402046.

[50] 'Ukrainian maritime attack on Black Sea port Novorossiysk repelled: Russia', *Al Jazeera*, 3 July 2024. https://www.aljazeera.com/news/2024/7/3/ukrainian-maritime-attack-on-black-sea-port-novorossiysk-repelled-russia; David Brennan, 'Russia reports massive Ukraine drone attack on strategic port and logistics hub', *Abcnews*, 26 February 2025. https://abcnews.go.com/International/russia-reports-massive-ukraine-drone-attack-strategic-port/story?id=119198093.

[51] Logan Williams, 'Construction accelerates at planned Russian navy base in disputed Abkhazia', *Bellingcat*, 20 July 2024. https://www.bellingcat.com/news/uk-and-europe/2024/07/30/construction-accelerates-at-planned-russian-navy-base-in-disputed-abkhazia/.

its ability to respond asymmetrically to Russia's quantitative superiority. Recognizing its potential in this area, Ukraine has prioritized its long-range strike ability against Russia, declaring in late 2024 that it intended to produce 30,000 long-range drones and 3,000 missiles over the coming year.[52]

On 2 April 2024, Ukraine conducted its deepest strike into Russian territory by that point when it targeted a drone production facility and an oil refinery in the province of Tatarstan.[53] Three days later, it launched a massive drone attack in the provinces of Rostov, Krasnodar Krai, Belgorod Kursk, and Saratov, reportedly damaging or destroying three Tu-95 strategic bombers at the remote Engels-2 airfield (750 kilometres from the Ukrainian border).[54] Later, Ukraine has repeatedly targeted oil refineries, military bases, and other objects both closer to the Ukrainian border and further afield, including Moscow. Several times, it has launched drones against far-away Murmansk Province some 1,700 kilometres from Ukraine.

If Ukraine's ability to strike deep inside Russia has dealt a psychological blow to a Russian population largely shielded from the war, it has also dealt a material blow. According to an analysis by Reuters in February 2025, Ukrainian drone attacks have destroyed around 10% of Russia's refining capacity. Coupled with the effect of Western sanctions on Russia's energy sector, Reuters suggests, this is likely to force Russia to reduce production,[55] putting further pressure on Russia's strained economy. Since that assessment, Ukraine has increased the number and range of its attacks on Russian refineries.[56]

Donald Trump's Election

Russia appears to regard Western support for Ukraine as the central source of Kyiv's resilience against the Russian invasion. With the election of Donald Trump as the forty-seventh president of the US on 5 November 2024 that support has been reduced significantly.

In mid-February, a major realignment of US policy on Ukraine took place. It began on the 12th, when Trump held a ninety-minute phone call with Vladimir Putin. This was the first direct communication between the US and the Russian president since the beginning of the large-scale war and one step in mending relations between the

[52] 'Ukraine aims to produce 30,000 long-range drones, 3000 missiles, Zelensky says', *The Kyiv Independent*, 19 November 2024. https://kyivindependent.com/zelensky-missiles-drones/.

[53] Laura Gozzi, 'Ukraine war: Deepest Ukraine drone attack into Russian territory injures 12', *BBC*, 2 April 2024. https://www.bbc.com/news/world-europe-68712158.amp.

[54] Chris York and Oleksiy Sorokin, 'Source: Ukraine hits Russia's Engels air base. Can it change how Russia attacks?', *The Kyiv Independent*, 5 April 2024. https://kyivindependent.com/source-russias-engels-air-base-was-hit-by-ukraine-heres-why-its-important/.

[55] Anna Hirtenstein and Florence Tan, 'Insight: Russia braces for oil output cuts as sanctions and drones hit', *Reuters*, 13 February 2025. https://www.reuters.com/world/europe/russia-braces-oil-output-cuts-sanctions-drones-hit-2025-02-12/.

[56] Maria Ilyushina and Francesca Ebel, 'Russian gasoline production buckles under Ukrainian drone strikes', *The Washington Post*, 2 October 2025. https://www.washingtonpost.com/world/2025/10/02/russia-oil-refining-attacks-shortage/.

White House and the Russian leader. On the same day, US Defence Secretary Pete Hegseth, speaking at a meeting with the Ukraine Defence Contact Group, a coalition of fifty-seven states supporting Ukraine, ruled out NATO membership for Ukraine and called the restitution of Ukraine's 2014 borders an 'unrealistic' prospect.[57] Three days later, the White House's Ukraine envoy General Keith Kellogg dealt a blow to European hopes for playing a prominent role in diplomatic efforts on the war by saying European countries would not participate in planned peace talks on Ukraine.[58] Meanwhile, President Trump sent a clear signal that Washington's views on Ukraine had undergone a dramatic shift when he seemed to blame Kyiv for starting the war[59] and later called Zelensky a 'dictator' for not calling elections during war time (which is not permitted under Ukraine's constitution).

Members of the administration have made efforts to backpedal on some of the statements. US Secretary of State Marco Rubio has tried to reassure European states that they would not be shut out of a US-brokered settlement on Ukraine,[60] while Pete Hegseth, at a press conference in Brussels, suggested NATO membership for Ukraine was, in fact, still on the table. Later, Trump himself has sent conflicting signals, at times appearing to strongly support Ukraine. In one social media post, he said Ukraine could take back all land seized by Russia since 2014 and even implied Ukraine might be able to take additional territory beyond this. Yet it is evident that a change of course has taken place. A clear sign was the face-to-face meeting of Rubio, US National Security Advisor Mike Walz, and US Middle East Envoy Steve Witkoff with Russian Foreign Minister Sergey Lavrov and senior Russian presidential aide Yuri Ushakov in Riyadh, Saudi Arabia, on 18 February. Another was the scandal-ridden press conference between Trump and Zelensky attended also by Vice President J. D. Vance and Rubio, where Trump explicitly positioned himself as an arbitrator between Ukraine and Russia, rather than a supporter of Ukraine. Finally, Trump hosted Vladimir Putin for face-to-face talks on US territory at a summit meeting in Alaska in August 2015.

Another momentous event came on 4 March 2025, when Trump declared that the US would cut off all military aid to Ukraine. This applied to aid that had previously been decided but not yet been disbursed, including aid that was in transit.[61] Although

[57] 'Opening remarks by Secretary of Defense Pete Hegseth at Ukraine Defense Contact Group', US Department of Defense, 12 February 2025. https://www.defense.gov/News/Speeches/Speech/Article/4064113/opening-remarks-by-secretary-of-defense-pete-hegseth-at-ukraine-defense-contact/.

[58] 'Europe to be excluded from Russia-Ukraine peace talks, US envoy confirms', *Al Jazeera*, 15 February 2025. https://www.aljazeera.com/news/2025/2/15/ukraine-peace-talks-will-not-include-europe-us-envoy-says.

[59] David Smith, 'Trump blames Ukraine over war with Russia, saying it could have made a deal', *The Guardian*, 19 February 2025. https://www.theguardian.com/world/2025/feb/19/trump-ukraine-war-russia-could-have-made-a-deal.

[60] Michael Schwirtz and Adam Entous, 'In private remarks on Russia, Rubio tries to reassure Europeans', *The New York Times*, 20 February 2025. https://www.nytimes.com/2025/02/20/us/politics/rubio-russia-europe.html.

[61] Andrew Roth, 'US suspends all military aid to Ukraine in wake of Trump-Zelenskyy row', *The Guardian*, 4 March 2025. https://www.theguardian.com/world/2025/mar/04/us-military-aid-ukraine-pause-trump-zelenskyy-updates.

aid was later reinstated, the decision weakened Ukrainian military readiness and likely dealt a blow to Ukrainian morale.

Since Trump took office, in other words, we have seen a warming of relations between Russia and the US and a cooling of relations between the US and Ukraine. Still, several questions remain. To what extent are European states able to compensate for reduced US support? What security guarantees can be provided to Ukraine? What is the state of Russian military readiness after almost four years of war? These questions will be discussed over the coming months and perhaps years. How they will be resolved will determine the onward course of the war in Ukraine.

Conclusion

In 2024, Ukraine accepted a logic of attrition in its defensive operations in the land arena. It used the strength of defensive warfare to try to impose disproportionate costs on Russian forces, while it ceded ground and fell back when continued defence was deemed too costly. But Ukraine also attempted to take advantage of the opportunities that opened for offensive action against an adversary that remained both overstretched and slow to adapt. Using new maritime drones, Ukraine forced Russian naval forces in the Black Sea to relocate to safer port in Novorossiysk. Similarly, it used novel long-range aerial drones to strike Russian military and energy infrastructure as far away as Moscow and Murmansk. Most dramatically, Ukrainian manoeuvre forces took a large part of Russia's Kursk Province and held most of it for more than six months. The defence in depth in land operations in southeastern Ukraine was one facet of Ukrainian warfare in 2024. Ukraine continued to exploit pathways of attack elsewhere.

Yet if Ukraine imposed costs on Russia through effective asymmetrical action, it remained vulnerable both to Russia's numerical superiority and to changing political winds in the West. The election of Donald Trump as the forty-seventh president of the US in November 2024 has augured a new era in US foreign policy and has led to a significant repositioning of the US in the war in Ukraine and Russia. Trump's actions so far call into doubt the durability of US support for Ukraine in the war going forward and pose questions about the ability of European countries to offer robust assistance to Kyiv.

PART IV
CONCLUSIONS

19

Russia's War in Ukraine

Taking Stock

Niklas Nilsson and Mikael Weissmann

Introduction

As the most significant high-intensity conflict in Europe since the Second World War, Russia's war in Ukraine will likely be a defining case and reference point for debates on modern warfare in the coming decades.[1] The conclusions drawn are far beyond the reach of any single publication and will likely be subject to reinterpretation over time. Moreover, the richness of documentation and array of different perspectives applied to this long war, and the different phases it has gone through are already fuelling opposing arguments drawing on the same empirics.

Russia's war in Ukraine has stimulated an enormous volume of military, strategic, and political analysis, based on which a wide range of observers have sought to draw enduring lessons about the future of warfare. However, the extraction of clear and universal insights from the Ukrainian battlefield remains complex and must be treated with caution. As is the case with all conflicts, experiences from the war in Ukraine are contingent on the context in which they were drawn.

Managing the multitude of interpretations that can emerge from the same set of observations presents an illuminating challenge while extrapolating experiences. This pertains to divergent views regarding the future utility of manoeuvre warfare, the relative importance of new technologies versus that of conventional and tested means of fighting, and the significance of information and strategic communication in war. Existing evidence supports competing hypotheses across a range of examples.

The fact that the war continues at the time of writing this volume, and the final outcomes are yet to be seen, is a factor adding to the complication. Thus, the durability of conclusions will be realized over time, since aspects of the war will be

[1] Funding has been received from the Joint Advanced Staff and Command Programme Directorate (PL HOP), the Officers' Programme Directorate (PL OP) of the Swedish Defence University and the Swedish armed forces' research programme. During the preparation of this work, the authors have used a number of generative artificial intelligence (AI) and AI-assisted technologies in the writing process, including ChatGPT, Co-pilot, and Grammarly. These tools were employed to support the writing process and to enhance and refine the text. The authors carefully reviewed, edited, and validated the content to ensure its accuracy and integrity and take full responsibility for the final published work. The authors contributed equally to this chapter.

Niklas Nilsson and Mikael Weissmann, *Russia's War in Ukraine*. In: *Russia's War in Ukraine and Modern Warfare*. Edited by: Mikael Weissmann and Niklas Nilsson, Oxford University Press.
DOI: 10.1093/9780198970903.003.0019

debated, reinterpreted, and re-evaluated in light of new evidence and knowledge. In all likelihood, the long-term strategic implications of clearly established tactical or operational implications of the fighting will in all likelihood remain contested and heavily dependent on adversarial adaptation.

However, with these caveats in mind, this volume has covered a wide range of aspects of Russia's war in Ukraine in order to provide an overview of useful lessons to be drawn for our understanding of modern warfare and the challenges faced by modern militaries. The war has gone through several phases since February 2022, and it is now possible to approach those with some analytical distance and evaluate the events in light of subsequent developments. Indeed, Russia's war in Ukraine has provided an extraordinarily rich trove of lessons about modern warfare, not least due to the exceptionally detailed documentation of the fighting that is available to researchers, whose significance will in all likelihood remain over time.

The chapters in this volume are divided into the three parts 'Dimensions of Modern Warfare'; 'Organization, Technology, and Training'; and 'Case Studies'. The main conclusions drawn from the contributions of these sections are as follows.

Dimensions of Modern Warfare

Experiences from the war offer several perspectives on the idealized modes of fighting in traditional domains versus the realities of battle. Manoeuvre is a centrepiece of contemporary Western tactical and operational thinking in the land domain. In this regard, Tuck highlights an important observation of established doctrinal wisdom: supported by the same empirics, the war has sustained two opposing interpretations of manoeuvre warfare. Indeed, given that fighting has gradually become dominated by positional and attritional warfare, this has reopened a fundamental debate about the nature and utility of manoeuvre warfare. Several observers have argued that this feature of war has underscored the enduring importance of attrition in warfare, whereas the transparency and lethality of the modern battlefield have made manoeuvre warfare increasingly risky and costly. However, these are not the only explanations of the increasingly attritional character of the war. Rather than concluding that manoeuvre warfare is becoming obsolete, Tuck argues that the realities of the modern battlefield demand that military organizations formulate credible alternatives and contingencies.

Although the war in Ukraine has been heavily land-centric, the volume presents important observations from the maritime and air domains. In terms of air power, neither side of the war has achieved uncontested air superiority. Instead, to reiterate Fedorchak's observation, the war witnessed a form of mutual air denial, which did not allow either side to establish a conclusive dominance over the airspace. Despite the lack of control over the air domain, both parties consistently leveraged aerial assets to support their ground operations. However, given the risks involved in these operations, the utilization of manned aviation was limited. Ukraine's layered integration of both Soviet era and Western systems in its air defences not only promotes the army's resilience and adaptability but also remains an important feature of the air domain. Furthermore, the rapid proliferation and diversification of drones, including reconnaissance, loitering munitions, and strike unmanned aerial vehicles (UAVs), have been heralded as a transformative shift in the character of air warfare. The very rapid development in this field during the war has demonstrated the significance of innovation and adaptation in warfare, as well as of deconfliction and integration of assets belonging to different domains.

Although the scale of naval engagement has been limited, Sanders observes that Russia's Black Sea fleet played an important role in the first phases of the invasion; however, its relevance gradually eroded over the course of the war. This reflects Ukraine's successful efforts to innovate and neutralize traditional naval superiority. Ukraine's employment of sea mines, land-based assets, and uncrewed surface vehicles (USVs) demonstrates that maritime power extends beyond naval capabilities and can be effectively generated with limited means. Although maritime power remains highly relevant in the twenty-first century, Ukrainian successes in the maritime domain, especially the sinking of the *Moskva*, underscore the critical role of qualitative factors in maritime power, as insufficient attention to training, leadership, and technological integration appears to have been important contributing factors to Russia's loss of the flagship. Moreover, the war demonstrated the continuous use of amphibious operations. Although challenging, Ukraine has been able to leverage these operations with adaptability and impact.

Alongside these observations from traditional domains, the war in Ukraine also provides important insights into the variable operational environments in which the war has been fought. Some of the most challenging and destructive battles have taken place in dense urban environments, which are expected to gain increased centrality in future conflicts with peer or near-peer adversaries. Weissmann and Winton show that Ukraine's experiences underscore the significant shift in how technology influences urban warfare. Drones, electronic warfare, and surveillance tools have changed tactical dynamics, serving as important complementary assets to traditional methods, without fundamentally changing them. Fighting in densely populated urban environments, where the balance between military necessity and humanitarian concern becomes particularly salient, places extraordinary demands on battlefield intelligence. Urban combat likely cannot be avoided; therefore, addressing challenges attached to urban combat requires a shift in conceptualization, training, and execution. Weissmann and Winton suggest an adaptable, multi-dimensional approach to the problem by integrating technologies, tactics, and operational situational awareness.

The war has also produced critical lessons regarding logistics, particularly in terms of Russian operational shortcomings. Kinsey and Ti point to the structural inability of the Russian military to sustain its operations due to centralized logistics, inflexible supply systems, and a disconnection between higher command decisions and front-line needs. The war in Ukraine is a testament to the increasingly transparent battlefield, with heightened vulnerability for the rear areas. This transformation implies an increased demand for logistics functions that employ camouflage, electronic discipline, and tactical dispersion.

Full-scale wars also place enormous strain on military health service capabilities. Bricknell emphasizes that the war in Ukraine has demonstrated the necessity for civil-military integration by mobilizing the entire health sector as part of the war effort. Although specialized, trauma-focused military care capabilities are central in forward areas, rear area care and rehabilitation can leverage civilian health infrastructure. From the point of injury to long-term recovery, clinical standardization is important across the continuum of care.

The war has also highlighted the value of military medical academic institutions as engines of innovation, training, and system oversight. In addition to robust clinical registries and governance structures, these institutions are essential for driving continuous improvements in clinical outcomes.

Organization, Technology, and Training

The long duration of the war in Ukraine has raised important issues regarding the sustainability and endurance of modern militaries, including challenges encountered in terms of maintaining a steady supply of munitions and replacing equipment lost during intensive and attritional fighting. However, the perhaps most central long-term imperative for both sides in the war has been to sustain manpower. The evolution of structure and training practices demonstrated by Ukraine throughout the war has offered several critical observations, for the armed forces of Ukraine (AFU) and its Western partners. As Watling notes, early mobilization driven by necessity provided a surge in the number of volunteers and conscripts that outpaced the state's ability to equip and train them effectively. Although this enabled an immediate response to the existential threat posed by Russia's full-scale invasion, it also incurred long-term costs, particularly in terms of lowering the quality and consistency of combat units.

The AFU initially relied on in-unit training, with limited results compared to rotating units off the front line for training. Watling underscores the importance of a resilient training system with dedicated training battalions staffed by experienced or recuperating personnel who can deliver realistic and up-to-date instructions. Another organizational imperative is the need for legal and procedural flexibility in times of war, allowing for combat-realistic training and the acceptance of higher levels of controlled risk. The diversity in equipment supplied to Ukraine also highlights the necessity of institutionalizing flexibility through training that can equip personnel across multiple platforms without relying on standardization. The longevity of the war and the heavy attrition it has inflicted on the AFU has also raised questions regarding the balance in Western training philosophies between taught operational tempo and force regeneration capacity, and the inherent trade-offs between them. Emphasizing the importance of long-duration exercises can help reframe tactical incentives.

Inclusion and equality are also important features of Ukraine's societal mobilization. Martsenyuk points out that Ukraine has made substantial progress in promoting gender equality within society and its military structures, driven by its commitment to international standards, including North Atlantic Treaty Organization (NATO) norms and the Women, Peace, and Security (WPS) agenda. The adoption of the Second National Action Plan for United Nations (UN) Security Council Resolution 1325 reflects the strong political will of the government to support women's roles in security and defence. Notable successes include legislative support for gender equality, increasing the representation of women in combat roles and leadership positions, and the removal of formal restrictions on women's military education and participation. However, the AFU continues to encounter serious issues, such as gender-based violence, inadequate institutional mechanisms for handling harassment, limited visibility of women in top military and expert roles, and the enduring 'glass ceiling'.

The war has also led to several important observations regarding the predictive reliability, utility, and utilization of intelligence. As Nilsson demonstrates, early deficiencies in intelligence assessments, especially overestimation of Russian capability and underestimation of Ukrainian resilience, underscore the limits of prediction and the difficulty of accounting for intangible variables such as morale, leadership, and adaptability. A much-discussed development has been the United States (US) and United Kingdom (UK) decision to declassify and disclose strategic warning intelligence in the run-up to the invasion to pre-empt Russian disinformation and align Western responses, to a large extent enabled by the growing open-source intelligence (OSINT) ecosystem and real-time commercial satellite imagery.

Intelligence sharing between the US and NATO allies, and Ukraine grew progressively after the full-scale invasion, in close connection with decisions to provide Ukraine with long-range weaponry and political rebalancing of the objectives of helping sustain Ukraine's defence and avoiding escalating the conflict with Russia beyond Ukraine. Intelligence has also seen a transformative technological development during the war. Ukraine has become a live testing ground for commercial artificial intelligence (AI) solutions and cloud-based intelligence processing. These tools offer the potential to process the vast volumes of data generated by modern warfare. However, the outsourcing of technological integration to commercial actors also raises strategic questions about dependency, data sovereignty, and long-term control over battlefield decision-making architecture.

Finally, the analysis and dissemination of experiences from recent wars are key to forecasting the future demands of conflict and the organization of militaries world-wide. In this regard, the war in Ukraine provides a rich trove of information that has had a major impact on military thinking and planning, and will likely continue to do so for decades. As an illustration of a structured analysis and learning process, Hecht and Berelovich describe how the Israel Defense Forces (IDF) instituted comprehensive research into the war in Ukraine, which challenged many of the prevailing assumptions about future wars and reignited debates about the appropriate structure and capabilities of the Israeli military.

Case Studies

The trajectory of the war can, in simplified terms, be viewed as a sequence of phases in which different characteristics of the fighting have been predominant. While such temporal divisions can of course be subject to debate, this volume has elected to highlight four distinct phases of the war, intended to underline important changes in the mode and pre-conditions for the fighting. These include the preparations and points of departure for both sides at the initiation of the full-scale invasion and the initial Russian operational failure in early 2022; Ukraine's successful counter-offensives in late 2022; the failure of Ukraine's major counter-offensive in 2023; and the war's stagnation into positional and attritional warfare, and attempts to break this deadlock through asymmetrical means.

Arnold presents a detailed analysis of the failure of Russia's initial 'Special Military Operation' (SMO), which was predicated on the assumption of rapid Ukrainian capitulation. Russia's failures were systemic. At the political level, Putin's personal miscalculations generated a misalignment between strategic objectives and operational planning. While Ukraine unified its strategic and tactical levels with the singular aim of national resistance, Russia's invasion plan was riddled with inconsistencies and overconfidence.

These three principal factors underpin the early setbacks of Russia. First, the disjunction between the political leadership and military planners was a fundamental vulnerability to the initial strategy. Second, the moral component of Ukraine's fighting power, manifested as societal cohesion and national resolution, was far superior to that of the aggressor. Third, Ukraine applied traditional principles of war, such as the selection and maintenance of the aim, flexibility, surprise, and economy of effort, far more effectively than Russia, which proved decisive in shaping the course of the war in its early stages.

Ukraine's defence proved resilient and adaptive, providing important insights into defence strategies for smaller or asymmetrically disadvantaged states. Ashour and Shelest argue that Ukraine employed a 'hybrid' defence strategy, which enabled the country to compensate for disadvantages in size. Five key lessons can be drawn from Ukraine's ability to resist the Russian invasion. In particular, reservists constitute as cost-effective force multipliers, enhance national resilience, and allow for rapid mobilization and scalability. Preparing and enabling civilian infrastructure for military use, as applied in the defence of Mariupol, served to strengthen the overall defence system and enabled prolonged resistance against superior forces. Ukraine's ability to integrate capabilities across domains, synchronizing land, air, cyber, and maritime capabilities, allowed it to offset Russian mass through agility and innovation. Remote warfare, enabled in Ukraine through the use of UAVs, USVs, unmanned underwater vehicles (UUVs), and unmanned ground vehicles (UGVs), was employed at a scale and sophistication that was previously unseen in state-on-state conflicts. These platforms enhanced the lethality and extended the operational reach of the AFU. Moreover, Ukraine's strategic support for non-state forces, in Ukraine's occupied territories as well as in Russia, and as far afield as in Syria and Mali, enabled horizontal escalation and forced Russian forces to divert resources across a wider battlespace.

The 2022 Ukrainian counter-offensive in Kharkiv achieved remarkable success, conditioned on surprise, rapid manoeuvre, and Russian overextension. While also successful, the Kherson operation by contrast encountered stiff resistance and underscored the difficulty of achieving rapid breakthroughs under more contested conditions. However, as Massicot notes, these offensive accomplishments could not be replicated in subsequent operations. By mid-2023, Russian forces improved their defensive posture through unit rotations, wide and deep mining fields, and increased drone coverage. This transition marked a broader shift from manoeuvre to positional warfare. Russian forces' increasing reliance on drones and loitering munitions further inhibited Ukrainian advances and redefined the tempo and nature of engagements. Massicot characterizes this new phase as sustainable competition, with both sides locked in a prolonged contest for endurance, resource mobilization, and external support.

Fedorchak's analysis of the 2023 spring counter-offensive further underscores Ukraine's structural and tactical constraints. The ambitious plans for mechanized assaults were derailed by dense Russian minefields, pervasive intelligence, surveillance, and reconnaissance (ISR) from drones, and flooding of key terrain following the destruction of the Kakhovka dam. After initial and failed attempts at mechanized breakthroughs, the offensive relied on infantry-centred tactics. However, despite yielding limited territorial gains, the Ukrainian armed forces (UAF) fell short of achieving their operational objectives. This period can be characterized as a contest for time and preparation, as Ukraine's priorities for personnel, training, and deployment and delayed access to essential Western equipment created a window for Russia to entrench its defences.

As the conflict entered 2024 and 2025, and either side failed to accomplish the breakthroughs and openings required to reintroduce manoeuvre to the fighting, Göransson highlights how Ukraine adopted a strategy of elastic defence, leveraging terrain and defence-in-depth to impose disproportionate costs on Russian attackers. In parallel, Ukraine increasingly sought to impose asymmetry on the war. The AFU utilized long-range aerial and maritime drones to disrupt Russian operations far beyond the front lines, including carrying out attacks on infrastructure deep within Russian territory. Ukraine's incursion into the Kursk region also represented a successful surprise attack in an area where the Russian side was unprepared, temporarily leaving Ukraine in control of a sizeable piece of Russian territory.

Despite these tactical successes, Ukraine has remained vulnerable to Russia's numerical advantages and shifts in Western political dynamics. Donald Trump's election as US president in late 2024 introduced new uncertainties about the durability of US support for Ukraine. This change in US politics raised questions about the long-term viability of Ukraine's military strategy in the absence of sustained external assistance.

Expanding the Scope

The multitude of perspectives covered in this volume cannot purport to provide a full picture either of the war or of modern warfare. There are of course notable additional angles of importance, which also speak to the conceptual approaches introduced in Chapter 1. Of these, both sides in the conflict have put significant effort into the informational dimension of the war, including the systematic use of strategic communication, information influence, and psychological operations. Russia's extensive spread of disinformation, both domestically and abroad, aimed to undermine Ukrainian morale and resistance and fracture Western unity and support. *Vis-à-vis* a Western audience (although not globally), however, Ukraine was capable of establishing a powerful narrative of the war's causes and trajectory, which dominated Western understandings of the war especially during the first two years.

Ukraine's agile strategic communication, making innovative use of social media and memes, with a direct appeal to Western publics, played a significant role in sustaining international engagement and framing the conflict. The battle for perception and shaping of the cognitive domain has been a central feature of the war.

Cyber warfare has featured prominently in the war and was an important coercive instrument that Russia employed against Ukraine between 2014 and the beginning of the full-scale invasion. In the cyber domain, Russia sought to target Ukrainian infrastructure (e.g. attacking energy grids, conducting espionage, and disabling command and control networks). Ukraine, with extensive assistance from Western tech firms and cyber defence agencies, mounted significant cyber defence efforts, including migrating government data to secure Western cloud servers. The integration of cyber operations with conventional warfare is undoubtedly an important feature of modern warfare.

The issue of force sustainment is naturally also connected to economic measures as part of the overall war effort. Since 2022, Western states have imposed sanctions targeting Russian banking, energy, and defence sectors; imposed export controls on high-tech components critical to Russia's military-industrial complex; and frozen the assets of Russian oligarchs and officials. Russia responded with energy blackmail and attempted to reroute trade through non-Western partners.

Although the volume focuses on Ukraine, Russia, and the West, the war naturally has wider global ramifications. These include China's ambivalent but strategically calculated position, transfers of weapons to Russia by Iran and North Korea, and Turkey seeking a balancing and mediating role.

Finally, the war has had significant environmental impacts, from the immediate destruction of infrastructure such as the Kakhovka dam to the significant pollution and environmental degradation caused by the fighting. Indeed, climate and environmental security are both a consequence of, and a potential tool in, war, including the potential use of ecological damage and food supply disruption.

In an attempt to provide some synthesized conclusions based on the individual contributions, it is useful to revisit the two conceptual frameworks presented in the introductory Chapter 1 of this volume: the continuum of military operations in relation to the intensity of conflict and the versatility of military forces.

Indeed, the Ukrainian experience over time elucidates how military organizations must have the capacity to operate across a continuum of scenarios, responding to fluid tactical conditions and shifting strategic demands with both adaptability and flexibility. The war between Russia and Ukraine has fluctuated in conflict intensity since 2014, and the significance of military force, relative to other coercive means that Russia has employed against Ukraine, has varied over time. Ukraine's defence against Russia's 2022 full-scale invasion represents the far right end of the military operations continuum, with full-spectrum engagement across all domains: land, air, maritime, cyber, and space. The transition from manoeuvre to entrenched positional warfare underscores the need for scalable doctrinal and operational alternatives. This emphasizes that military organizations must plan not only for optimal conditions but also for attritional and ambiguous environments.

An important insight from the outbreak of the war is the role of the 'pivot zone', the transitional area on the operations continuum where conflicts shift from limited to high intensity, and the inherent difficulty of pinpointing this shift in a war that is not of one's own choosing. Indeed, Ukraine's political and military leadership, while seeing Russia's build-up and receiving warning intelligence from partners, appears to have been surprised by the timing and scale of the attack. Ukraine's ability to navigate this shift, particularly through rapid mobilization and solicitation of international support, was decisive in the country's ability to resist, as was societal cohesion, strategic communication, and intelligence sharing, which supported political and military decision making.

The case studies have highlighted how intensity affects the balance between conventional and unconventional warfare. The war in Ukraine has truly been a full-spectrum conflict in which conventional force employment has worked in tandem with the integration of hybrid and asymmetric strategies, such as non-state actors, cyber operations, and covert action. This demonstrates the necessity for an armed force structure that is adaptable and deployable across intensity levels.

Versatility, which emerges from the interplay between adaptability and flexibility, is considered an ideal model for assessing operational effectiveness. For Ukraine, adaptability through doctrinal reform, rapid integration of foreign equipment, and learning loops from battlefield realities, has enabled adjustments to evolving threats and changes in the operational environment. Flexibility, manifested in tactical innovations and agile responses to Russian actions, has allowed these adaptations to be implemented swiftly and effectively.

The Ukrainian experience illustrates how versatility must be cultivated through seven interdependent factors: doctrine and TTPs, technology, interoperability, cross-domain integration, military identity and culture, education, and training. Deficiencies in any of these factors constrains versatility. For example, problems with training and equipment integration impede offensive capabilities. Conversely, robust civil-military integration in war efforts mitigates systemic strain and improves long-term sustainability.

In summary, the war in Ukraine underscores the centrality of the ability to adapt, learn, and operate across multiple domains and intensities. Versatility, as a function of adaptability and flexibility, when institutionalized across the seven factors of military capability, is pivotal in modern militaries.

Modern Warfare and Military Preparedness in an Unstable Environment

At the time of writing this volume, the war in Ukraine has stretched into its fourth year. Governments and intelligence agencies across Europe are developing increasingly detailed assessments of the threats to Europe and NATO that Russia might pose after a ceasefire or peace settlement. These predictions reflect the common recognition that Russia's geopolitical ambitions and willingness to use force will not be extinguished by the cessation of active hostilities in Ukraine. Instead, many expect Moscow to rearm, adapt, and seek new opportunities to challenge Europe's security. Throughout the war in Ukraine, Europe has witnessed a surge in hybrid activities traceable to Russia, including sabotage, espionage, and targeted assassinations. The cessation of large-scale fighting in Ukraine is unlikely to imply a relaxation of these actions, while potentially opening opportunities for Russia to project the conventional military power of its dramatically boosted defence production.

Indeed, there is no shortage of geopolitical flashpoints underscoring the risk that the entire spectrum of conflict and war scenarios, including full-scale warfare between technologically advanced adversaries, could become a distinct feature of international politics in the coming decades. China's potential to launch a military invasion of Taiwan remains the most significant security concern in east Asia. Recently, Chinese military activity around Taiwan has markedly increased, including frequent airspace incursions and large-scale exercises in the Taiwan Strait. The situation on the Korean peninsula has a continued potential for renewed fighting.

The escalation between India and Pakistan in 2025 underscores the prevalent risk of war between the two nuclear-armed neighbours, whereas Pakistan in February 2026 declared 'open war' against Afghanistan with operations ongoing at the finalization of this manuscript. The first months of 2026 have also seen the US/Israeli war against Iran and a renewed Israeli incursion into southern Lebanon.

The increased risk of international volatility and conflict amid ongoing geopolitical shifts underscores that the continued development of versatile military forces employable across the entire range of conflict scenarios is a critical task for NATO allies and partners.

The evolving features of the modern battlefield, including transparency, data abundance, and information overload, along with the continued pertinence of well-established conventional warfighting, highlight how new innovations are added to existing ones, rather than replacing them. In the war between Russia and Ukraine, fighting has not been predominately characterized by the application of any particular set of tactics, materials, or innovations; it involves everything at once.

The very rapid development of AI in the service of decision making and targeting and the rise of automation are distinctly new technologies that some observers argue will revolutionize warfighting, although scepticism is also rightly raised as to whether this will be the case. Nevertheless, optimism regarding the capacity to integrate the full range of effects that a military can exert, epitomized in the evolution of multi-domain operations, is fuelled by the potential to fuse enormous amounts of information granted by new technologies.

At a time of grave uncertainty regarding international politics and the future of the international system, we are witnessing significant investments in military capability, drawing in large part on experiences from the devastating war in Ukraine and attempts to apply present knowledge to the anticipation of future opportunities granted by technology and doctrinal innovation. This volume has attempted to pull together a set of important observations drawn from the war in Ukraine, in order to hopefully provide a piece of the puzzle, and contribute to established understandings of military power and modern warfare.

Index

For the benefit of digital users, indexed terms that span two pages (e.g., 52–53) may, on occasion, appear on only one of those pages.

Tables and Figures are indicated by *t* or *f* following the page number.

B